KT-426-704

City and Islington College
SFC11956

MARS

A Cosmic Stepping Stone

MARS

A Cosmic Stepping Stone

Uncovering Humanity's Cosmic Context

KEVIN NOLAN

C

Copernicus Books

An Imprint of Springer Science+Business Media

PRAXIS

in Association with

Praxis Publishing, Ltd.

ISBN 978-0-387-34164-4 e-ISBN 978-0-387-49981-9
DOI 10.1007/978-0-387-49981-9

Published in the United States by Copernicus Books,
an imprint of Springer Science+Business Media.

Copernicus Books
Springer Science+Business Media
233 Spring Street
New York, NY 10013
www.springer.com

Library of Congress Control Number:
2008929492

The photocollage on the dust jacket, an artist's conception, includes a NASA image of the Martian surface combined with an image of a night sky with a small blue dot added, representing Earth.
Design by James Sarfati

Manufactured in the United States of America.
Printed on acid-free paper.

9 8 7 6 5 4 3 2 1

To my parents,
Jim and Annie

Contents

Many of the illustrations which appear within the body of the text are also reproduced in color in two color sections in this book.

Preface

Being aware of our origins and the nature of our existence is of paramount importance and occupies the thoughts of every generation and every individual. Through science, religion, and philosophy—among many other means—we have relentlessly pursued those fundamentals. For many, the extent of our existence transcends the physical universe and bounds of detection, and it is also clear that virtually all of the great questions about the physical universe remain unanswered. What was the origin of the Universe? What is the underlying nature of matter, energy, space, and time? What was the origin of life on Earth, what is the universal context for all life, and is there a connection between the two? Questions like these, while probing the bounds of the physical universe, cut very close to our deepest spiritual ponderings.

Yet these questions need not remain unanswered. Through science, we have at our disposal some of the most powerful tools with which to move ever closer toward answers. And although modern science is at times taxing, it is of essential value in our pursuit of serious, fundamental answers. The *process of science*, although a very human endeavor, has stringent in-built fail-safes that impede stagnation and falsehood and allow the truths in nature, however elusive, to be revealed in exquisite detail.

The process of science begins with a theory or hypothesis about the working of some aspect of nature. With robust and persistent testing followed by scrutiny by anonymous peers in a given field, weak theories are discarded, possible theories are revised, and slowly but steadily sound theories emerge that provide valid descriptions of how nature actually functions. Even then science does not rest. It is only when theories stand the test of time that we gain confidence in their use-predicting the behavior and future outcomes of natural systems and working well alongside other loosely related theories or *auxiliary hypotheses.* And if, for any reason, or at any time, our most acclaimed theory is shown to be even slightly flawed, it must be satisfactorily amended or, however reluctantly, rejected. No "established fact" is sacred and no "expert" is impervious to the rigors of science. This is science's greatest strength, and is the reason it has been so immensely effective.

The process of science is thus relentless in peeling back layer upon layer of universal mysteries, revealing their underlying structures and behaviors. But even here it encounters challenges. Many of the great bastions of religion, philosophy, and metaphysics—traditionally central to our purpose—have been challenged if they appear to lack a scientific foundation. It often seems that science is in conflict with that which, based in mystery, is so precious to us.

Despite such contention, we must however ask ourselves whether we wish to probe more deeply into the origin and fate of the Universe, and into our own origins and the cosmological context for all life, by using every means at our disposal. Do we want to know more about such things and are we prepared for the implications to our findings, no matter how contentious and challenging? Do we have a role in the unfolding human quest to know who we are and from whence we came—and do we want to acknowledge it; or are we content to say that the extent of our existence is already well enough known? Do we have a role in shaping our future, or do we just allow Nature to take charge of the future direction of humanity?

These are challenging issues—not just for scientists, philosophers, and priests, but for all of us. And the often exasperating nature of the questions themselves does not help. No matter how much we have wanted to know the origin of the Universe or the cosmic nature of life, finding answers has been incredibly difficult. With virtually all trace of our origin long since faded and with the Universe so prohibitively vast, answers to questions about the nature of life and the origin of the Universe have traditionally been elusive.

This, however, is no longer the case. Over the past century we have crossed a threshold. We have developed scientific hypotheses and some measure of insight into those most mysterious aspects of Nature—from the fundamentals of matter, energy, space, and time to the dawn of the Universe billions of years ago; and from the origin of life on Earth to the seemingly universal nature of the make-up of its building blocks. And while we are not much closer to ultimate answers, we are now more confident of eventually attaining some scientific conclusions. Our powerful process of science is underpinned by hundreds of years of significant groundwork that is now supported by spectacular technology capable of peering into ever smaller domains, to ever greater cosmological vistas and back through billions of years of natural history on Earth, and elsewhere.

Indeed, the question of life elsewhere in the Universe—traditionally relegated to the realm of speculation due to the lack of plausible answers—is now firmly rooted in analytical science and we are perhaps on the verge of a revolution in both thinking and results, the outcome of which may be profound. As we probe the origins of life on Earth and the grand nature of the Universe, we are increasingly finding suggestions that life here is not some isolated incident, but instead seems to be intimately connected to the Cosmos within which we find ourselves. Increasingly, questions about our origin and nature are being answered, but many of those answers are resident in space. The Universe is where we now think we will find many far-reaching answers about our origins and existence.

In view of recent progress and developments, the pursuit of our great questions continues with ever-increasing pace, not only because real and tangible answers may be forthcoming, but also because they will be of supreme relevance to all of us. Thus, our role in life is surely clear: we may not yet have answers to many of the great questions of our fundamental nature, but in being capable of achieving at least some answers, our mandate is to build upon the work of those who came before us and progress toward a more enlightened future.

This book is therefore an attempt to convey a significant development that is currently unfolding in our understanding of, and connection with, the planet Mars, and what it may reveal regarding our origins and the cosmic context for all forms of life.

Acknowledgments

I would like to thank the following people for their support in enabling me to write this book.

First and foremost I would like to express my deep gratitude to my elder sister Sheila for her unbounded love and support during the writing of this book. I also thank my younger sister Enda for her love and pragmatic advice. Although my parents, Jim and Annie, are now deceased, I wish to acknowledge their love and dedication to my sisters and me during their lives. A very special thanks must go to my longstanding friend Nora Hearty who, on August 28, 2003 (the evening after the closest opposition of Mars for nearly 60,000 years), suggested over a meal with friends that I should write book about Mars. Although I had spent the previous few years developing a book about life in the Universe, Nora's suggestion brought my efforts into sharp focus.

I should also like to thank my friend and PhD supervisor Dr Niall Smith, Head of Research and Development at Cork Institute of Technology, Ireland. From his guidance while I was an undergraduate to our current research collaborations and advice, I owe much to Niall. My other lifelong friend, Paul Bracken (former Coordinator to Ireland for The Planetary Society), has also given me significant assistance. Paul's unique and sharp insight, and our many conversations into the early hours of the morning, have been invaluable.

I also thank David Moore FRAS, President of Astronomy Ireland and editor of *Astronomy & Space* (A&S) magazine, for providing me with the opportunity to develop my ideas and skills in an A&S column titled "Cosmic Perspectives." In a similar manner, I thank Dick Ahlstrom, Science Editor for *The Irish Times*, for offering me the opportunity to write occasional articles, including a full-page feature about Mars in May 2003. I also thank Dr Ian Elliott of Dunsink Observatory for inviting me to be secretary to Ireland for the 2001 ESO/ESA/CERN European "Life in the Universe" youth competition, from which much of the impetus to write this book originated. I also thank my friend and colleague Dr Michael Maloney of the US National

Academy of Sciences for his astute observations and pointers in science policy matters within the USA.

I most especially thank Clive Horwood of Praxis Publishing for making this opportunity available, and for his extensive guidance, advice, and endless patience. I also wish very much to express my deep gratitude to Dr Stephen Webb (Praxis Editorial Board member) and particularly Dr John Mason (Subject Advisory Editor for Praxis Publishing) for their tireless reviewing, important insights, recommendations, and overall guidance in the preparation of this book. I also extend my sincere thanks to Alex Whyte for his careful editing of the text, and his advice and guidance.

I sincerely thank Dr Mary Bourke of the School of Geography, Oxford University (and Mars research scientist with The Planetary Science Institute) for her valuable insight into current Mars planetary science. I also thank Mary's colleague, Dr Frank Chuang of The Planetary Science Institute, for his insightful and robust reviewing of the Mars science chapters within this book. I extend my gratitude to Lonny Baker, Alice Wakelin, and Vilia Zmuidzinas of The Planetary Society (TPS) for their help, guidance, and advice in the preparation of the many Mars lectures I have presented on behalf of TPS in Ireland and from which so much of the groundwork for this book was developed. I thank Professor Bruce Murray, Professor of Planetary Science and Geology, Emeritus, California Institute of Technology, former Director of NASA Jet Propulsion Laboratory and co-founder of TPS, for his comments to me that, in part, influenced the theme of this book.

I should also like to thank Stephanie O'Neill, Enterprise Ireland, and television producer Paul Tully, for their support and production of my 2003 Mars lectures. My thanks also go to the following people for sourcing images and in granting permission to use them in this book: Dr Remington P.S. Stone, former Director of Operations at Mt Hamilton Lick Observatory; Robin Witmore, Publications and Web Specialist at UCO/Lick Observatory; Tony Misch of the Mt Hamilton Lick Observatory Laboratories; Dr R.J. McKim, Director of the Mars Section of the British Astronomical Association; Antoinette Beiser, Librarian for The Lowell Observatory; Kevin Righter of the Johnson Space Center; and Corby Waste of The Jet Propulsion Laboratory. A very special thanks to Piotr A. Masek for providing the spectacular images of Mariners 6 and 9, enhanced with his own image-processing software.

I also thank Dr Mary Costello, Dr Brian Murray, and Dr T.J. Ennis of the Institute of Technology, Tallaght, Dublin (ITTD), for discussions on matters of chemistry, biochemistry, and chirality. I thank the Head of the School of Science at ITTD, Dr Ken Carroll, and my colleagues Dr Aine Allen, Dr Eugene Hickey, and Tadhg O'Briain for their continuing support and many informative and useful discussions. I pay homage to the three great teachers who inspired me in science—Mr Harrington, Ms Margaret Mullet, and Br Rory Geoghegan.

Finally I would like to acknowledge eight people in the public domain who have been hugely influential on me throughout my life: Galileo, Einstein, Beethoven, and Debussy as well as Patrick Moore, Carl Sagan, Jean Michel Jarre, and Vangelis.

List of Acronyms

ALH	Allan Hills
APXS	Alpha-Proton-X-Ray Spectrometer
ASPERA	Analyzer of Space Plasma and Energetic Atoms
ATP	Adenine Triphosphate
CCD	Charged Couple Device
CEV	Crew Exploration Vehicle
ChemCam	Chemistry and Micro-imaging
CheMin	Chemistry & Mineralogy X-Ray Diffraction
CLV	Cargo Launch Vehicle
COSPAR	Committee on Space Research
cPROTO	Compensated Pitch and Roll Targeted Observation
CRISM	Compact Reconnaissance Imaging Spectrometer for Mars
DAN	Dynamic Albedo of Neutrons
DNA	Deoxyribonucleic acid
DRM	Design Reference Mission
EDL	Entry, Descent, and Landing
ELT	Extremely Large Telescope
EPAC	Exploration Program Advisory Committee
ERV	Earth Return Vehicle
ESA	European Space Agency
ESAS	Exploration Systems Architecture Study
ESO	European Southern Observatory
EU	European Union
GCMS	Gas Chromatography Mass Spectrometer
GEX	Gas Exchange Experiment
GRB	Gamma Ray Burst
GRS	Gamma Ray Spectrometer
HiRISE	High Resolution Imaging Science Experiment
HRSC	High Resolution Stereo Camera
HST	Hubble Space Telescope
ISRU	In Situ Resource Utilization
ISS	International Space Station
JPL	Jet Propulsion Laboratory

JSC	Johnson Space Center
JWST	James Webb Space Telescope
LEO	Low Earth Orbit
MARSIS	Mars Advanced Radar for Subsurface and Ionosphere Sounding
MAV	Mars Ascent Vehicle
MCO	Mars Climate Observer
MECA	Microscopy, Electrochemistry, and Conductivity Analyzer
MER	Mars Exploration Rover
MGS	Mars Global Surveyor
MOC	Mars Orbital Camera
MOLA	Mars Orbital Laser Altimeter
MPL	Mars Polar Lander
MRO	Mars Reconnaissance Orbiter
MSL	Mars Science Laboratory
MSM	Mars Society Mission
MSR	Mars Sample Return
MTV	Mars Transfer Vehicle
NASA	National Aeronautics and Space Administration
NEO	Near Earth object
NTR	Nuclear Thermal Rocket
OMEGA	Observaoire pour la Mineralogie, l'Eau les Glaces et l'Áctivité
PFS	Planetary Fouricr Spectrometer
RAD	Radiation Assessment Detector
RAM	Random Access Memory
RAT	Rock Abrasion Tool
REMS	Rover Environmental Monitoring Station
RNA	Ribonucleic acid
SAM	Sample Analysis at Mars, Instrument Suite
SDV	Shuttle Derived Vehicle
SNC	Shergottite, Nakhlite, and Chassigny
SSSP	Space Science Strategic Plan
TES	Thermal Emission Spectrometer
THEMIS	Thermal Emission Imaging System
THM	Transfer Habitation Module
TPF	Terrestrial Planet Finder
TPS	The Planetary Society
ULDB	Ultra Duration Balloon
USSR	Union of Soviet Socialist Republics
VLT	Very Large Telescope
VSE	Vision for Space Exploration
WEB	Warm Electronic Body

Prologue

Since the dawn of humanity, we have pursued the truth of our origin and nature. That is one of our defining qualities. It is not enough that we can exist and survive, we must also have a purpose, and such a purpose seems clearer the more we know of our origin, nature, and surrounding context.

Although far from complete, we are now bestowed with the clearest picture yet. The relentless efforts of our forefathers have uncovered the nature of some of the fabrics of the Universe—space, time, matter, and energy. They deduced that the countless stars in the sky, whose nature had perplexed many previous generations, are simply distant suns like our own, and that everywhere in the Cosmos the picture is essentially the same: it is nothing more than a vast collection of ordinary places. The efforts of our forefathers have brought us to the profound realization that there was a beginning—an origin—to the entire Universe and to all life within it, and that both are deeply interconnected. Today, we sense that we can discover that origin and the natural extent of life everywhere.

However, we are still not really aware of who we are. Precisely when, where, and how we came into existence is still a mystery. We are orphans of a cosmic ancestry that has long since faded into antiquity. Countless generations have come and gone since life first began on Earth and no life-form has known its true origins.

But human ancestry has not disappeared in the darkness of time—it has only faded. It is still waiting to be uncovered and today we are equipped like no previous generation to find solutions to our problems. And we are prepared for the challenge. We are already aware of many of the key questions, and slowly but surely, as in a jigsaw, we are putting together the answers.

Of the myriad of questions, two stand out: How did life start on Earth? How abundant is life throughout the Cosmos? These questions in particular occupy our thoughts and efforts; and what we have so far uncovered points to new intricacies regarding the origin and nature of life. We now recognize that the beginning of life on Earth is somehow connected to the universal question of life, and that life on Earth and the cosmic abundance of life are perhaps the same question approached from different perspectives. We are following specific and fruitful lines of exploration, with the broad consensus that we are on the right track. For the first time we are getting a handle on life in a general context.

We have recently taken significant steps forward. We have uncovered evidence of some of the first organisms to have evolved on our planet. We have determined which building blocks make up all life on Earth, and we even know essentially how these building blocks work. We know that the essential items for life were strewn across the early Solar System and that very soon after the birth of our planet, life began here. Planetary science has also revealed many details of Earth's four-billion-plus-year evolution and its association with its planetary neighbors. Giant new telescopes reveal our galaxy in bewildering detail, showing, for example, that much of the galaxy's stock of carbon—the main element of life as we know it—is used up in organic molecules; while vast expanses of water are seen to be perpetually manufactured within the cloud-fields spanning its spiral arms. And with the discovery of new planetary systems almost by the month, the potential for life seems to be everywhere.

The path ahead is nevertheless paved with immeasurable challenges; and answers regarding our origins or the broadest context for life will not be easily found. The Cosmos is unimaginably vast, and we have given ourselves the task of discovering how such complex, yet minute, activity can originate life-forms. How can we accomplish this when our own origin is buried beneath four billion years of rock and even the nearest star systems are at present hopelessly beyond our reach, to say nothing of the remainder of the galaxy? The cosmos to which we seem to be intimately connected will not surrender its secrets easily.

There is, however, a way forward. For all the vastness of the Universe, it turns out that literally on our doorstep there resides the perfect platform—a cosmic stepping stone from which we can take our next critical steps. That stepping stone is the planet Mars.

Mars has always been Earth's companion. Both were born together and shared the same early history. They have continuously accompanied each other around the Sun in a cosmic courtship that brings them close together every two years.

Both planets are similar. Earth is larger, and many of the same planetary processes that led to life on Earth also took place on Mars. After a billion years of high activity, however, Mars literally froze. All that had been active and vibrant came to an end, and for the past three and a half billion years Mars has lay dormant, silently orbiting its parent star along with its sister

Earth, waiting to tell the story of its early history—and perhaps of Earth's too.

As the first humans walked the Earth, Mars and the other planets were indeed true companions, as they were all possible stepping stone to discovering the reality that lies beyond. Throughout the ages, Mars in particular has challenged us to continually re-examine our place in the Cosmos. From realizing that we are not at the center of the Universe to first considering life on other planets, Mars has provided Earth with the impetus to question, and has also supplied some of the answers. Mars is one of our closest companions, but always one step beyond our perceptive capability, and that is the challenge a central position in our future development.

We have awoken. We now know about Mars, its past, and our deep and ancient connection. We realize that it will once again play a central role in our broadening as a species—in knowing of our origin and the most general context for life. As if by cosmic fate, Mars appears to have frozen at roughly the same time that evidence of Earth's origins all but disappeared. If we want to know what happened on Earth four billion years ago, Mars offers a substantial and pragmatic opportunity to find the answers. We can go to Mars; indeed we are already there. Mars is so close that we can reach it in months—by comparison, traveling to the nearest star would take at least thousands of years. For now and for some time to come Mars is therefore our best option for exploring our cosmic connection.

But as we have tried to investigate the Red Planet, it has redefined itself time and again. Just as we think we have reach a conclusion, it reveals yet another perspective, hitherto unconsidered. Although mostly dormant, Mars is far from dead. Recent discoveries reveal a planet rich with activity that might even be supporting life.

Mars, therefore, offers a unique opportunity to pursue the question of minute yet complex activity taking place elsewhere in this gargantuan Cosmos. It also offers the opportunity—which we have already taken—to move substantially closer to answering questions of our own origin. From the first space probes to pass the planet in the 1960s to the planetary orbiters and rovers of today, we have developed the most appropriate long-term program to fulfill our aspirations. We are organized on a world scale, and have developed a phased program that commenced in 1995 and will span 30 years or more. We are conducting three distinct searches: the search for early chemistry leading to life; the search for ancient life; and the search for current life. The program will culminate with a sample return mission and possibly human expeditions, hopefully leading to the clearest answers yet regarding the origin and the cosmic abundance of life.

We are already collectively there, and our connection is set to strengthen and become permanent. We regard Mars as a close neighbor, and in so doing we are already beginning to look at ourselves in a broader context—as only one planet among many. The journey in itself is contributing to a deeper understanding of ourselves in the broadest way possible. The greatest discovery, however, remains to be made. If even one microbe is found on Mars, it will confirm that there is life elsewhere and that we are not alone. The cosmic context for life on Earth will be set as being one planet among many that harbor life. It will confirm our cosmic origins and connection and lay the foundation for an enigmatic future for humanity.

Life in the Universe

Looking into a clear dark sky at night, we are filled with a sense that there is much more beyond. Each of us feels it, even when we are unsure of what we are baring witness to. But the great dark expanse of the sky above represents a reality as tangible as the ground beneath our feet, that perhaps we are only now beginning to fully recognize.

No one can fathom the expanse of space and time that make up the Universe. Even the other planets of our Solar System are millions of kilometers away—too distant for anyone to truly appreciate—yet they are by far our closest cosmic companions. The 3,000 or so stars seen by the unaided eye in the night sky are millions of time more remote; in fact they are so far away that a measurement scale far grander than the humble kilometer is needed to represent their distances. The scale chosen is called a light year; it is the distance that a ray of light, traveling at 300,000 kilometers every second, traverses in one year—10 million million kilometers. The closest star, Proxima Centauri, is over 4 light years or 40 million million kilometers away. When you look at it, you are seeing it as it was just over 4 years ago and as you glance from star to star across the sky you are looking at a different time for each, depending on its distance from the observer. The night sky is a window to many pasts as well as to far-off places.

To the casual viewer, the night sky can seem to be remote and quiescent. The planets, having settled into a well-behaved system over the aeons, move predictably across the sky, while the stars are so remote that no real change in their position can be witnessed by the unaided eye, even over a human lifetime (although they are all moving at great speeds). Every so often an unexpected celestial event draws our attention—such as a passing comet—but most are of fleeting interest. It is perhaps not surprising, then, that we often perceive the sky above as a vast yet placid domain.

There is, however, a reality behind such a perception. For all its vastness, the night sky as observed by the unaided eye represents but the tiniest region of space and time—the here and now—within the grand Universe of which we are a part (see Figure 1). A hint of that grandeur can actually be

K. Nolan, *Mars, A Cosmic Stepping Stone*,
DOI: 10.1007/978-0-387-49981-9_1, © Praxis Publishing, Ltd. 2008

Figure 1: The Hubble Ultra Deep Field. Virtually every object in this image is a distant galaxy comprising billions of Suns, from a time when the Universe was only a few billion years old. [Credit: NASA, ESA, S. Beckwith (STScI) and the HUDF Team]

witnessed on a clear night away from city glare, whereupon a faint, broad band of light is seen to stretch across the sky from horizon to horizon. That band is the combined starlight from billions of remote stars that make up the great Milky Way system of which the Earth, our Sun, our planetary neighbors, and all of the nearby stars of the night sky are but a minute part. Each of the estimated 200,000 million stars of the Milky Way is a sun in its own right, each is separated from the next by light-year distances, and all are gravitationally bound in a gigantic spinning disk called a galaxy that spans 100,000 light years (see Figure 2).

At its center is a spherical bulge characterized by billions of ancient, red giant stars. Surrounding this are colossal spiral arms, each containing further billions of younger stars and vast clouds of gas and dust within which new stars are born. We reside in the Orion spiral arm, some 26,000 light years from the galactic center, aligned with a multitude of other stars along an immense path curving in toward the nucleus. Our region of the

Figure 2: The Milky Way galaxy, if seen from beyond, would look similar to this spectacular spiral galaxy, NGC 1309. [Credit: NASA, ESA, STScI/AURA and A. Riess (STScI)]

galaxy is relatively stable, remaining free from cosmic calamities for millions of years at a time. This is not the case elsewhere in the galaxy, however. Within the nucleus, for example, the stars are so compacted that the explosive death of one can directly affect its closest neighbors, while a supermassive black hole at the center of the galaxy wreaks havoc with its immediate vicinity.

Despite the relative calm of our galactic neighborhood, we have recently identified natural cycles on a galactic scale that may have radically affected our planet in the past and may do so again. As a spinning system, the stars, gas, and dust in the outer regions of the galaxy orbit about the center over enormous timescales, with our Solar System doing so about every 225 million years. Caused in part by the combined motion and gravity of the stars, colossal density waves are also set in motion around the galaxy, but as they move more slowly than the stars themselves, they cause dense accumulations of the stars, gas, and dust that form the beautiful spiral arms characterizing our Milky Way and other spiral galaxies. Such galactic-scale

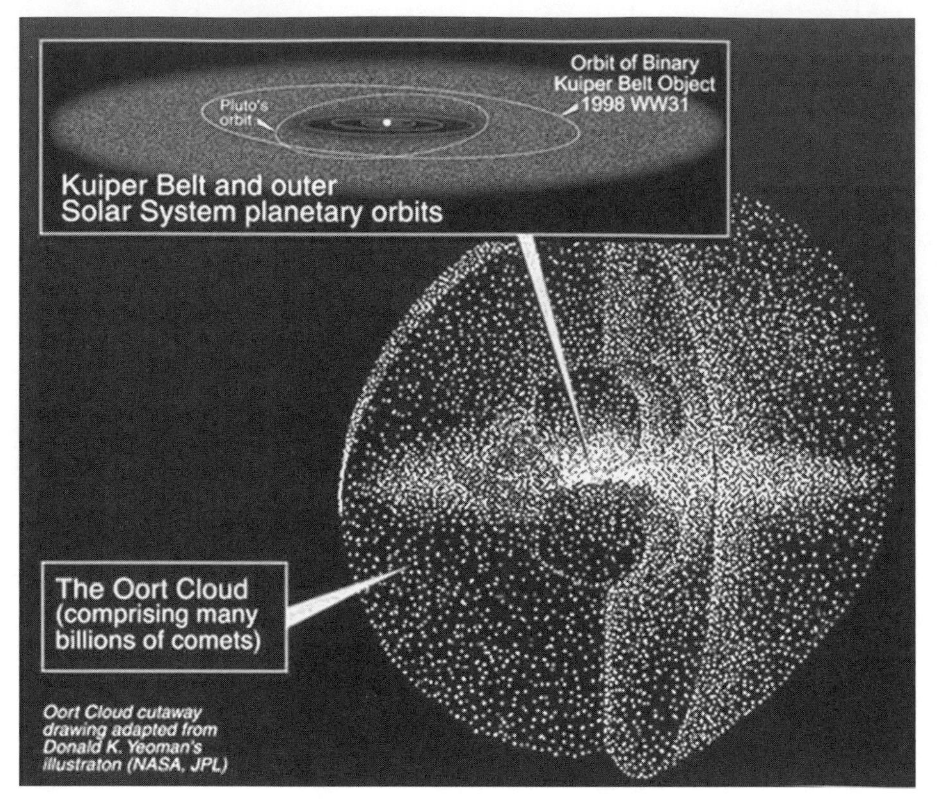

Figure 3: The Kuiper Belt and Opik–Oort Cloud. If many stars retain similar comet clouds, many trillions of comets may occupy and traverse the space among the stars of our galaxy. [Image from Wikipedia: http://en.wikipedia.org/wiki/Oort_cloud. Credit: NASA/JPL]

dynamics means that individual star systems cannot remain in their given galactic environment permanently. Rather, each is tugged and jostled within its spiral arm over timescales of tens of millions of years, with potentially devastating consequences, brought about by the changing gravity upon the billions of comets that we suspect accompany many stars. For example, shrouding our Solar System to a distance of up to a light year we suspect that there is a shell of billions, if not trillions of small icy comets. Called the Öpik–Oort Cloud after the astronomers Ernst Öpik and Jan Oort who proposed it, its constituent comets are so loosely bound to our Sun that, on encountering a galactic density wave, many are pulled from their orbits (Figure 3). Some drift away from the Solar System altogether, while others fall inward, and with the Sun and planets now governing their trajectories, some find their way to Earth where they cause devastation on impact. When we look at the history of major impacts and mass extinctions on our world, there seems to be a correlation to its motion around the galaxy; and if, as we suspect, similar clouds of comets shroud many other stars, it may be that the fate of planets across the galaxy, as with our own, is in part governed by the grand motion and dynamics of the galaxy itself.

Pillars of Creation

While far from understood, our ability to identify such elusive behavior associated with our galaxy is an indication of how well we are beginning to understand the great dark expanse. Traditionally, our telescopes have been severely hampered by Earth's turbulent atmosphere in what they could reveal. Of late, however, revolutionary technologies onboard both ground-based and space-borne telescopes have radically improved our view of the sky and our understanding of the Universe.

Consider just one location within our galaxy called the Eagle nebula—a giant cloud of gas and dust lying at about 7,000 light years toward the center of the galaxy. From ground-based telescopes it appears as a vast field some 20 light years across and energized by nearby stars. From this vantage point the nebula appears lurid and mysterious, with central columns of gas and dust that look superficially like an eagle about to take flight. But when we look more closely, as photographed by the Hubble Space Telescope (HST) in Earth orbit, the Eagle is seen anew, with the three central pillars, known as "The Pillars of Creation," revealed in magnificent detail and characterized by a previously unseen intricate structure (see Figure 4). Of the three pillars, the largest stretches no less than a light year from end to end and is the birthplace of dozens of stars. At its apex we see a finger-like outcrop of gas and dust (see Figure 5) and although appearing small by comparison, more than 20 systems the size of our Solar System could lie side-by-side comfortably across its width. If superimposed upon that outcrop, our Solar System would appear as a tiny dark region (see Figure 5), while to glimpse planet Earth we would need to divide that region into more than 1,000 pieces, whereupon it would appear as a "Pale Blue Dot" (see Plate 2b), as coined by the astronomer Carl Sagan on seeing the now famous Voyager 2 image of Earth taken from the edge of our Solar System, and which he inspired.

Only with such images does the elaborate detail of our Universe, previously hidden from view, begin to appear. Only then can we begin to appreciate at least some of the true character of the sky above. And yet, even such views are far from complete. As mentioned above, so remote yet expansive is the Eagle nebula that even with the Hubble Space Telescope, the identification of objects on a planetary scale within its compass is still

Figure 4: The Pillars of Creation at the heart of the Eagle nebula, lying at approximately 7,000 light years toward the center of our Galaxy. The largest pillar is approximately one light year in length. See Plate 1a in the color section. [Credit: NASA, ESA, STScI, J. Hester and P. Scowen (Arizona State University)]

Figure 5: Seen more closely, the largest pillar reveals an outcrop or ''finger'' of dust and gas pointing directly upward. See Plate 1b in the color section. [Credit: NASA, ESA, STScI, J. Hester and P. Scowen (Arizona State University)]

Figure 6: A black rectangle, superimposed on that outcrop, represents the approximate size of our Solar System. See Plate 2a in the color section. [Credit: NASA, ESA, STScI, J. Hester and P. Scowen (Arizona State University)]

Figure 7: Dividing that tiny rectangle in Figure 6 into a thousand pieces, Earth would appear as a tiny dot (center right), providing some sense of the enormity of the Pillars by comparison. This actual image was acquired by the Voyager 1 spacecraft from a distance of more than 6.4 billion kilometers from Earth. See Plate 2b in the color section. [Credit:NASA/JPL]

hopelessly beyond our view. The Eagle is in reality composed of even finer detail to which we are not yet privy and will not be revealed until we build even better telescopes. Only then can we cross further thresholds of perception and begin to see the Eagle as it really is; only then will it come into focus as a bewildering number of separate and individual locales; and only then will new and forming worlds that are currently invisible to us be observed for the first time.

Similarly, the many thousands of other nebulae and billions of stars that comprise the Milky Way galaxy will also be revealed in an entirely new light. With better telescopes, those other nebulae will also be seen as complex and dynamic environments, while vast numbers of stars across the galaxy will most likely be shown to possess systems of planets, moons, asteroids, and comets. We are now confident that we will eventually find our galaxy to be nothing more than a vast collection of real and tangible places; and that each and every one of the 50,000 million other galaxies comprising the known Universe is similarly an overarching representation of what is in fact a vast system of individual worlds.

While we work rapidly toward obtaining such far-reaching views, current observations of the Universe already show that wherever or however far we look into the Universe, we see that it is in many significant ways similar to our own cosmic neighborhood. We can see that the laws of nature are essentially the same everywhere; and we can see that galaxies and stars at one end of the Universe are basically the same as those at the other end. Gravity has the same effect within a galaxy a billion light years away as it has within our own Milky Way, while the chemical elements making up our world are the same elements that can be seen strewn across the entire Universe. Given such perspectives, we find ourselves already considering a broad and universal context for life. In a Universe perhaps filled by countless other worlds and within which the same natural laws apply everywhere, the possibility of life arising elsewhere cannot be ignored. And so we ask: If the laws of physics and chemistry are ubiquitous, might there also be a universal context for biology?

We must also reconsider our own position in such a grand scheme. A cosmological context for the natural history of our planet points to an origin of life here with a deep-rooted cosmic connection. That the Universe had an origin, that it was different in the past, and that it continues to evolve, all prompt consideration on how life emerged on Earth. And herein lies an opportunity. The further we can see into space, the further we are looking into the past. Our ever-improving instruments will therefore provide ever-improving views of the Universe throughout its various ages, from virtually the moment of its birth to the present day and in exquisite detail, shedding new light upon how it was born, how it has evolved over time, and perhaps how it has paved the way toward life, offering an unprecedented opportunity to probe our own natural history and origin.

Toward Life in the Universe

For reasons still unknown to us, the Universe as we know it originated about 14 billion years ago. At the moment of its birth, in an event called The Big Bang, our initially minute Universe expanded incredibly rapidly, in the process bringing about the fundamentals of nature such as matter, energy, space, time, and so on that would govern its future path for eternity.

In its earliest epoch, the Universe was so physically small that any given event could rapidly take effect across its entirety. As it expanded, however, even significant events had increasingly diminished and delayed impacts elsewhere. As activity across the Universe became increasingly localized, its overall nature began to settle. Today, however, the Universe is so large that it takes billions of years for events to become known across its entirety. For example, many of the powerful Gamma Ray Bursts (GRBs) that are detected by our telescopes today actually occurred at remote locations in the Universe before our planet was even formed, and their energy took billions of years to traverse the Universe to reach us.

Within just thousands of years after its birth, virtually all of the ordinary matter in the Universe comprised the two simplest elements, hydrogen and helium. None of the celestial objects known to us today—galaxies, stars, planets, etc.—had yet emerged and for hundreds of thousands of years the young Universe endured a dark age. Eventually, variations in the distribution of hydrogen and helium (coupled to the underlying structure of elusive dark matter that actually comprises the majority of all matter in the Universe) created dense clumping, with many of the most densely packed clumps forming enormous gaseous globes with masses up to a hundred times that of our Sun and capable of nuclear fusion, becoming the very first generation of stars. Gargantuan lanes of matter were by then emerging across the Universe, and along those lanes stars in their billions formed groupings that would eventually become galaxies. There were still virtually no heavy elements, however, and certainly no planets or smaller celestial objects. The Universe was very different to the one we know today, and was without the possibility of life.

Although the first stars were composed mostly of hydrogen and helium, nuclear fusion within their cores synthesized new and heavier elements. Stars are where the minute realm of atoms is transformed on a universal scale, where, as each star grows older, it becomes increasingly dominated by newly synthesized elements such as beryllium, carbon, oxygen, and silicon—all the way up to iron. By way of emissions from their energetic atmospheres (called stellar wind) and through their explosive deaths, the first generation of stars released new and heavier elements into interstellar space. Subsequently, a plethora of new molecular substances such as water and carbon dioxide were produced, which mixed with the hydrogen and helium that was left over from the formation of the Universe, and gave rise to vast nebulae throughout the forming galaxies, becoming the stellar nurseries for the next generations of stars.

Each new star-forming region could now draw into its vicinity dense clouds composed of the full range of available materials, triggering the manufacture of complex substances including organic materials such as amino acids and nucleic acids—the functional building blocks of life as we know it. Such tremendous quantities of heavy material were now available that, in association with the formation of many new stars, there were dense swirling disks of matter within which the first planets could form, and with them came a significant new regime in nature. Across the Universe, solar systems containing all manner of worlds could now arise containing gas and ice giant planets, large and small rocky planets, rocky and icy moons, asteroids, and comets, among many other types of objects which, in themselves, would give rise to an even more diverse range of environments. From scorching planets close to their parent star, to distant frozen worlds and everything in between, virtually every conceivable environment within the material universe became possible.

And so the Universe evolved into the rich and complex system we see today. Eventually, after billions of years and at least once, the sum total of available resources gave rise to an entirely new natural process. The Universe had become a significantly different place to its original form; and with the birth of our Sun and planet Earth, the emergence of life was now possible. Through natural processes still unknown to us, the natural conditions in the vicinity of our Solar System, and in particular upon the surface of our planet, were conducive to and gave rise to life. After 9 billion years of cosmic evolution, life emerged.

A Living Universe

As far as we can determine, there is nothing special about the nature of our part of the Universe. Broadly speaking, the same laws of nature found here are found everywhere else. There is increasing recognition, therefore, that the question of the origin of life on Earth may be connected to the question of the emergence and abundance of life elsewhere—that when we comprehensively answer one of these questions, we may also answer the other. We are far from having a complete picture, but every new advance reveals a deeper cosmic and planetary context for the origin of life on Earth, and, while only by implication, also increases the plausibility of life having arisen elsewhere.

During the twentieth century, major advances were made in the traditional sciences of physics, chemistry, biology, geology, and astronomy, paving the way by the 1950s and 1960s for visionary scientist such as Frank Drake, Carl Sagan, and others to propose a genuine possibility of life elsewhere in the Universe. Despite conducting many pioneering experiments, those scientists were severely limited by the technology of their time, and so the question of life in the Universe remained largely speculative. Building upon their work, however, more recent advances such as the cross-pollinations of the various sciences and technological advances in computers, optics, and space exploration technology have now made it possible to conduct far-reaching investigations into the origin of life and its cosmic abundance. The search for a cosmic context for life is no longer consigned to speculation and hypothesis—it is now big science, with discoveries that already match, if not exceed, our aspirations and are consistent with a cosmic-scale potential for life.

For example, until the 1990s, astronomers were unable to determine whether any other solar systems actually existed. Today, we are discovering planets around other stars on a regular basis. So capable are our new technology telescopes that in some cases we can detect the tiny wobble in the motion and rotation of a star caused by its accompanying planets, while in others we can monitor the minute dimming of a star as its accompanying planets pass in front of it as seen from our vantage point. Since the discovery of the first such planetary system in 1995 by Michael Mayor and Dider Queloz of the Observatory of Geneva, hundreds of other solar systems have been discovered. One of the more significant discoveries was made in 2007 by Stephane Udry and his colleagues (also of Geneva Observatory) of an Earth-sized planet with some possible Earth-like characteristics circling the star Gliese 581, located approximately 20 light years away in the constellation of Libra. The planet, called Gliese 581 c, is considered to be a rocky planet approximately one and a half times the size of the Earth. Since Gliese 581 is a cool red dwarf star, the newly discovered planet, although orbiting very close to its parent star, lies within its habitable zone (the temperate spherical-shell or "zone" surrounding a star where planetary-surface water and hence life as we know it can be sustained). Hence Gliese 581 c could possess surface

Figure 8: The Orion nebula, where stars with proto-planetary disks, as well as massive quantities of water and various organic materials are being created. See Plate 3 in the color section. [Credit: NASA, ESA, M. Robberto (Space Telescope Science Institute/ESA) and the Hubble Space Telescope Orion Treasury Project Team]

oceans of liquid water and constitutes an important target for future searches for evidence of life elsewhere in the Universe. Such discoveries already provide exciting and tantalizing pointers to the possibility of life elsewhere and in the coming years we hope to discover that many, if not most, stars are accompanied by planets.

Of equal significance have been the results from ground-based observations of the gas and dust within nebulae and the material that occupies the space between the stars. Because each chemical substance has its own light-emitting signature, the various chemical constituents within the gas and dust spread throughout our galaxy can be determined by analyzing their light emission and absorption using an instrument called a spectroscope. For example, when energized, the element sodium emits a

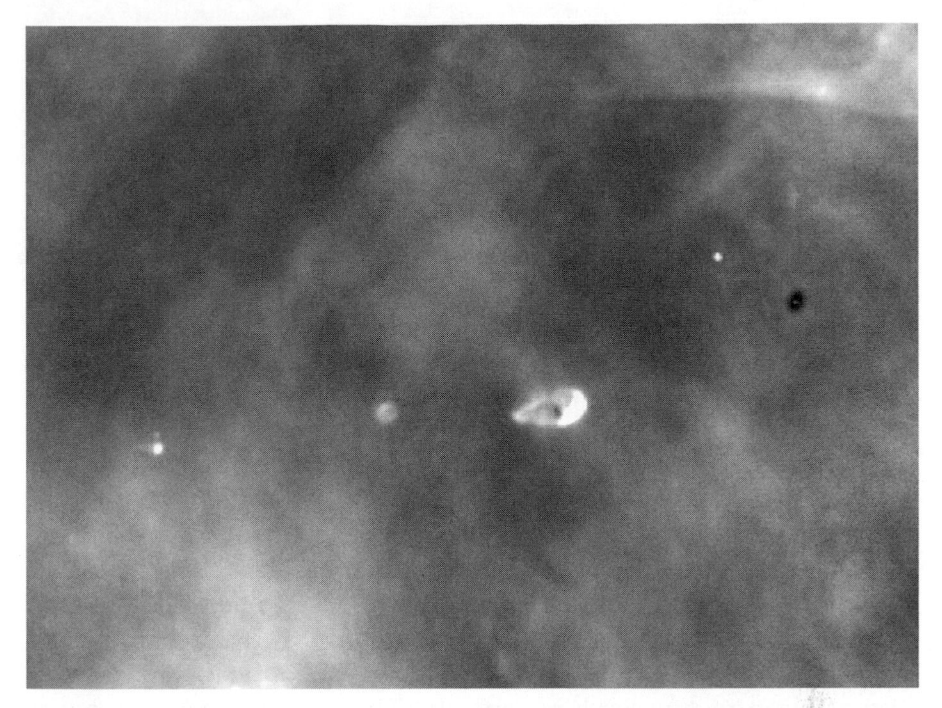

Figure 9: A torus of dust surrounds a new born star within the Orion nebula. Perhaps we are witnessing the origin of a new solar system. [Credit: NASA, ESA, M. Robberto (Space Telescope Science Institute/ESA) and the Hubble Space Telescope Orion Treasury Project Team]

unique shade of orange. Hence, if the light from a distant nebula contains that particular shade of orange, we can infer the presence of sodium. In this manner, dozens of complex materials have been identified within the nebulae of the Milky Way and other galaxies, including organic molecules to the complexity of amino acids and nucleic acids. Indeed, it appears that the majority of carbon in interstellar space is contained within organic molecules.

Equally significant have been results from telescopes launched into space. Since its launch in 1990, the Hubble Space Telescope has worked tirelessly outside the influence of the Earth's atmosphere, trailblazing a new era of astronomical discovery. Among its many spectacular finds has been the identification of almost 200 flattened disks of gas and dust surrounding new born stars within the Orion nebula alone (see Figure 8), many of which may eventually become solar systems (Figure 9). A separate region of that nebula also reveals that water is being synthesized at a rate of 90 times all of Earth's oceans every day. In another exquisite image of the flattened disk surrounding a star labeled HD 107146, Hubble reveals what appear to be gaps in the disk that have possibly been created by currently unseen planets orbiting that star. In a similar manner, the Spitzer Infrared Telescope launched into space in 2003 has provided significant evidence of planets around other stars and of organic synthesis in space. Indeed, not only has Spitzer detected disks around a broad range of newly forming stars (and even including brown-dwarf "failed stars" as small as 15 times the mass of the planet Jupiter), but it can also reveal their various stages of evolution, providing new and significant insights into planetary formation. Of particular importance has been its identification of disks around some stars that are already known to possess planets, showing for the first time a link between flattened gas and dust disks and planet formation. Spitzer has also identified many complex organic molecules being synthesized within such disks, most intriguingly hydrogen cyanide and acetylene—both vital for the synthesis of proteins and genetic polymers—within the habitable zone around the star IRC 46 in the constellation of Ophiuchus.

Of course we do not have to look quite so far for evidence. Since the dawn of the space age our robotic space probes have visited all the planets of our own Solar System as well as dozens of moons, asteroids and comets, revolutionizing our understanding of the origin and evolution of our Solar System in general and Earth in particular. US and Soviet space probes to Venus and Mars—our two closest planetary neighbors—have revealed them to be sufficiently similar to the Earth in make-up, origin, and early history to consider whether life might have arisen on all three in their earliest stages and if life could even exist on Mars today. Findings from the Voyager, Galileo, and Cassini missions to Jupiter and Saturn have been equally intriguing, with Jupiter's moons Europa, Ganymede, and Callisto possibly possessing global subsurface oceans of liquid water that is perhaps suitable for life, while Saturn's moon, Enceladus, displays geysers of icy particles spurting from beneath the surface, perhaps suggestive of subsurface liquid water activity.

Evidence is also available of extensive organic synthesis across the early Solar System. For example, meteorites formed during the birth of the Solar System and which, on occasion, fall to Earth show a range of complex organic materials, while the Stardust space probe, which visited Comet Wild 2 and returned samples of that comet to Earth in 2006, has revealed the presence of organic materials such as chained aliphatic hydrocarbons and nitrogen-rich methylamine and ethylamine, all of which are important to the building blocks of life and could have been delivered to Earth during its formation and early history.

A Time for Exploration

The evidence amassing suggests that we live in a universe characterized by a diversity of worlds and environments capable of complex activity, including organic synthesis. As far as we have been able to search for such activity we have found it, and far from abating, virtually every new search provides new evidence. Although we have not detected life elsewhere and cannot explain how life emerged on Earth, we are still only beginning in this quest and have merely scratched the surface of the various worlds within our Solar System—let alone the other bodies in the remainder of the galaxy—to discover what they have to reveal. The current absence of defining answers does not mean that they do not exist or cannot be found; nor does it provide licence to draw premature conclusions. Rather, our discoveries to date point to a universe with widespread potential for life and to new and hitherto unconsidered questions and lines of investigation. It is clear to all concerned that now is the time for exploration.

We find ourselves in the unprecedented position of being able to ask new and highly focused questions about the origin and cosmic abundance of life, and to seek out answers through revolutionary observations, exploration, and analytical programs. And our questions are neither arbitrary nor naive; they follow from the efforts of those pioneering scientists of the latter half of the twentieth century and from the plethora of current evidence. On the grandest scales, we must conduct surveys of our galaxy to resolutions equivalent to hundreds of times that of the Hubble Space Telescope, through which the character of our galaxy on a planetary scale can finally be ascertained. We must examine hundreds of thousands of stars in our region of the galaxy alone, in order to determine those that possess planetary systems, and for each system determine the number, distribution, types, masses, orbital characteristics, and make-up of its constituent planets. In effect, we must conduct the first census of planetary systems within the Milky Way from which the question of life elsewhere may begin to become clear. Subsequently, we must target candidate planets that suggest the greatest potential for life-related activity, and determine their material make-up, atmospheric constituents, and surface environment. We will also look for evidence of potential biomarkers such as the liquid water oceans and atmospheric gases such as oxygen, ozone, and methane. With the discovery of Earth-sized planets such as Gliese 581 c, we may—around nearby stars and within the lifetime of the youngest members of our society—construct telescopes that are capable of imaging their globes in sufficiently high resolution to reveal surface features that may directly identify the possible existence of life.

Figure 10: The Terrestrial Planet Finder will look for and characterize Earth-like planets orbiting other stars. [Credit: NASA/JPL]

Another priority is to investigate an astronomical context to the origin of life. Here, we must comprehensively quantify and characterize the synthesis of volatile materials such as water and carbon dioxide, as well as organic materials from simple molecules to complex polymers within the interstellar medium, star-forming nebulae, and the protoplanetary disks of new stars. This effort will particularly require new generations of infrared space telescopes that are capable of peering deeply into dense gas and dust nebulae and protoplanetary disks.

We must also investigate the origin of our own Solar System far more comprehensively. Important clues reside among the various worlds of our Solar System, waiting to be discovered. If we wish to know how life began on Earth, we must visit most, if not all, of these worlds, leaving no stone unturned. We must rendezvous with asteroids and comets, travel out to the far-off icy worlds such as Pluto, and conduct long-term investigations on our Moon and the planet Mars in particular—all of which retain a detailed record of activity dating back to the early history of the Solar System.

These and many other lines of inquiry now form the basis for future exploration programs, and long-term and multinational commitments will be required if worthwhile answers are to be obtained. Significantly, both the technological and organizational means are now at our disposal to engage in such bold exploration. Leading universities, research institutes, observatories, space agencies, and related industries are currently gearing up for the exciting challenges and opportunities associated with this endeavor, with a number of global-scale research and exploration programs already in operation. A new field known as "Astrobiology" has emerged, bringing together the many required disciplines of astronomy, physics, chemistry, biology, geology, computing, and mathematics among many others, with the specific objective of tackling the two-pronged question of the origin and cosmic abundance of life. This endeavor is now considered to be among the great ambitions of world science, along with such endeavors such as the determination of the origin and underlying nature of the Universe itself.

The Search for Life in the Universe

The search for our origins and for life elsewhere in the Universe has come of age. The time for debating the validity of this endeavor has passed and it is now time to explore. Over the past decade or so, and for several decades to come, a multitude of programs have emerged that are set on achieving far-reaching answers. NASA's Origins Program—and its associated Space Science Strategic Plans (SSSPs) of 1997 and 2001 for example—has set astrobiology at the forefront of its ambitions, aiming to tackle fundamental questions regarding the origin of the Cosmos, the Sun, the Earth, and life itself. Through consultation with leading scientists, NASA has defined a set of fundamental questions, from which enterprise goals, science goals, roadmaps, programs, and missions have all been derived. Commencing in the 1990s and recently integrated into the US Vision for Space Exploration (VSE), many Origins initiatives are already underway, with many others in preparation. For example, following on from the Hubble and Spitzer space telescopes, NASA's Discovery Program will contain the 2009 launch into space of the Kepler space telescope. Kepler will conduct the first census of planetary systems in our region of the Milky Way by simultaneously monitoring approximately 100,000 stars within the constellation of Cygnus. Using optical fiber bundles to direct the light from individual stars to separate pixels on its CCD (Charged Coupled Device) detector, Kepler will be capable of determining which stars possess planetary systems, revealing for the first time (from a statistically significant sample) the percentage of stars that are accompanied by planets and, perhaps, the percentage of planets that are Earth-sized.

Following on from Kepler, other programs, such as NASA's Navigator Program, include no less than 80 separate initiatives over the coming decade to search for the origin of life and for evidence of life elsewhere. New space telescopes, such as the James Webb Space Telescope (JWST) and SIM Planet-Quest Space Telescope, will provide even better statistics than Kepler, both on the number and character of planetary systems, as well as on the number of Earth-like planets within our galaxy. The Terrestrial Planet Finder (TPF) space telescope array, a project that will most probably be pursued jointly with the European Space Agency (ESA), will target Earth-like planets around the nearest stars for detailed analysis, identify their atmospheric compositions, and possibly detect potential biomarkers (Figure 10). And although not yet budgeted for, the logical outcome of these programs may involve the construction of the proposed Life-Finder and Planet-Imager space telescopes capable of producing quality images of the surfaces of planets around other stars, yielding significant answers regarding the nature of planets and a universal perspective on life. European efforts are equally ambitious. Arrays of large ground-based telescopes, such as the European Southern Observatory (ESO) Very Large Telescope (VLT) array, use adaptive-optics to cancel the turbulent effect of the Earth's atmosphere to provide images that are even sharper than those from Hubble. Also, revolutionary telescopes, such as the planned European Southern Observatory 42-meter Extremely Large Telescope (ELT), will provide unprecedented views of the heavens that will be relevant to the search for planets around other stars and for life elsewhere in the Universe.

Equally far-reaching planetary missions are also underway or planned. The first major reconnaissance of the Solar System from the 1960s to the 1980s visited all the planets, as well as dozens of moons, revolutionizing our thinking about our origins and the nature of life in general. But now we must return to many of those worlds with the ambitious aim of acquiring definitive answers about the origin and nature of our Solar System and the astronomical pretext for life on Earth. In this regard, NASA's Origins and VSE as well as ESA's Cosmic Vision and Aurora programs all have their sights firmly set on various destinations across our Solar System.

For example, as part of its Discovery Program, NASA intends to send three missions to various asteroids and comets: OSIRIS will collect samples from an asteroid and return them to Earth; DIXI will send the existing Deep Impact space probe (which has already visited comet Tempel 1) to a second comet; while Stardust Next will revisit comet Tempel 1 to conduct a second detailed survey. These and other missions, such as the New Horizons planetary probe already *en route* to Pluto, will provide significant insight into the most distant objects in our Solar System and hence produce evidence for the origin and early history of the Solar System itself.

Whether probing deep into our galaxy or surveying our own Solar

System, this new era of space exploration cannot but provide new answers on the universal question of life. But of all the destinations under consideration, one stands out from the rest—the planet Mars. From all we have so far learned about our next-door neighbor, it beckons like no other place. Uniquely, in the two-pronged quest to know of our own origin and of a cosmic context for life, Mars may provide important answers to both. So tantalizingly similar to Earth was Mars' birth, early history, and planetary environment that it is possible that life may have arisen there in its distant past. Its planetary environment is also so tantalizingly close to conditions necessary for life as we know it, that it may even harbor life today.

Thus, in recognition of the great potential that Mars offers, both NASA and ESA are currently setting in place some of the most ambitious programs of scientific exploration ever undertaken, set on uncovering what Mars has to tell about life there. The key to unraveling our most profound questions on the nature of life may be awaiting discovery on our close planetary neighbor.

The Origin of Life on Earth 2

Earth is a spectacular world. It is surely a pinnacle of creation within the Universe, and if our planet is not the Universe's crowning achievement, then an even more breath-taking Universe awaits us.

Our planet has thrived and developed persistently throughout its entire history. Despite events that have on occasion virtually obliterated all life and radically altered its surface, our world has emerged as one of exquisite balance between the gargantuan forces of the surrounding Universe and its own inner make-up, the result of which is the beautiful blue-green globe we now inhabit (Figure 11).

We have already encountered one of the grandest forces of all to affect our world—its quarter-of-a-billion-year journey around the galaxy. But there are many others, including a number of celestial cycles that alter the orientation and movement of our planet in relation to the Sun over timescales of thousands of years. These are known collectively as the Milankovitch Cycles in honor of the Serbian astronomer Milutin Milankovitch who determined many of their characteristics.

The best known of the Milankovitch Cycles is a precession of Earth's axial tilt where, over approximately 26,000 years, a gravitational pull by the Sun and Moon upon Earth's slight equatorial bulge will cause the orientation (but not the size) of the axial tilt with respect to the Sun to vary, just as a spinning-top slowly wobbles when rapidly spinning. For example, where currently the Earth's northern hemisphere tilts most *toward* the Sun at some particular point along its orbit, in approximately 13,000 years (or half of a precession cycle) the northern hemisphere will tilt *away* from the Sun at that same orbital position. Hence, as regular and small adjustments will be required to maintain our calendar over an entire precession cycle, the precession affects the position along Earth's orbit where particular calendar events occur. Another Milankovitch Cycle is called orbital precession where, over a period of approximately 70,000 years, the distant gravitational pull from the other planets causes Earth's orbit about the Sun to precess. This, coupled to axial precession, induces a 41,000-year cyclical change in the

K. Nolan, *Mars, A Cosmic Stepping Stone*,
DOI: 10.1007/978-0-387-49981-9_2, © Praxis Publishing, Ltd. 2008

Figure 11: Earth—a spectacular and vibrant world whose legacy has been defined by internal and celestial forces. See Plate 4 in the color section. [Credit: NASA Earth Observatory]

size of Earth's axial tilt, called its obliquity, from a minimum of about 22 degrees to a maximum of approximately 24 degrees. Perturbations in the gravitational pull of the other planets also affect the roundness or eccentricity of Earth's orbit over a timescale of about 100,000 years, changing the shape of the orbit from a near perfect circle to an ellipse of about 5% eccentricity.

When combined, the various axial and orbital cycles can affect Earth's surface conditions quite significantly. Changes in obliquity, for example, affect the distribution of solar radiation upon the northern and southern hemispheres over time. Changes in eccentricity, on the other hand, affect the amount of solar energy each hemisphere receives along its orbit. At maximum eccentricity, a given hemisphere can receive up to 20% more energy at the Earth's closest point the Sun (called perihelion) compared to when furthest from the Sun six months later (called aphelion), while at minimum eccentricity the amount of energy received throughout the year is more even. Although not conclusive, there is increasing evidence to show that past changes to the world's climate, including significant events such as the ice ages, have been in part caused by the Milankovitch Cycles.

However severely such celestial cycles impact upon our world, they are far from the most significant of external influences. That honor belongs resolutely with our Sun. From the outset, the nature and evolution of our planet has been governed by our Sun's extreme stability over billions of years, coupled to our near-perfect distance from it, residing in the center of the Sun's habitable zone. Despite such idyllic conditions so conducive to life, the Sun also emits lethal doses of electromagnetic and particle radiation that could have severely curtailed life here were it not for Earth's own natural defences—a planetary magnetic field to deflect particle radiation, deep oceans to protect the first water-dwelling life, and a dense atmosphere that evolved to protect subsequent land-based organisms. Thus, while owing our very existence to the Sun, we are also at its mercy. Should the Sun change in any significant way, there can only be cataclysmic consequences for our planet. Although no significant change is expected for millennia, as the Sun grows older it also becomes more luminous, and its habitable zone extends outward accordingly. Consequently, in about a billion years the Sun's habitable zone will have expanded so far that Earth will reside in its inner region and become so hot as to cause the oceans to evaporate, turning our once beautiful blue-green world into a desolate place. Intriguingly, the planet Mars, currently orbiting on the far outer edge of the habitable zone, will then sit favorably within it, perhaps bringing a new vibrancy to that world. The Sun has provided for all that has flourished on Earth, but at some point in the distant future it will cease supporting any form of life. In solar terms, Earth is an aging planet with most of its vibrant life behind it. Our world has circumnavigated the Milky Way galaxy more than 20 times throughout its long history, but it will do so perhaps another four times before it expires.

Our Moon also casts a firm hand over our planet's nature and fate. The Moon is a planet-sized object and the Earth and Moon could almost be regarded as a double planet system. With no evidence of life, or indeed any indigenous surface activity, the Moon is a reminder that even when ideally placed from the Sun, many other planetary factors are required for life as we know it to exist. There is evidence to suggest that our Moon originally formed from a glancing-blow collision between the Earth and another Mars-sized planet in its earliest history, with the Moon taking shape from the remaining debris in only a matter of a few years and settling into an orbit some 20 times closer that it is today. However created, the Earth and Moon have been linked since their earliest history in a gravitational courtship which, through the influence of tidal forces, has radically altered both. Tidal forces occur because the side of our planet facing the Moon experiences a slight increase in gravitational pull, causing a minute flexing of the Earth's shape that produces an upward bulge of several centimeters on land and several meters within the oceans, both on the side facing the Moon and on the side pointing away from it. Because the Earth rotates more rapidly than the Moon orbits, the Earth's tidal bulge exerts a small extra gravitational pull as it moves ahead of the Moon, dragging it ever so slightly forward in its orbit. Conversely, the Moon pulls back on the bulge, slightly slowing Earth's rotation. The result is that energy is taken from Earth's rotation, slowing it by about 2 milliseconds per century (the Earth's day was probably about 6 hours long when the Moon first formed) and is transferred to the Moon's orbital motion, allowing it to drift outward at a rate of several centimeters per year.

Acting over billions of years, lunar tidal forces have been a major factor in shaping the surface of our planet, inducing daily mass movement of the world's oceans that brings about powerful and relentless planet-wide weathering, erosion, deposition, and their associated geochemistry. Critically, tidal forces between the Earth and Moon were in the order of 8,000 times greater in Earth's early history, no doubt affecting tectonic and volcanic activity, stirring the ocean violently and bringing about all manner of energetic surface activity, perhaps relevant to the emergence of life. There can be little doubt that in the absence of the Moon, Earth would have evolved with significantly different oceanic and climatic patterns and perhaps even a different regime for the biosphere and for life itself.

Another significant contributing factor to the character of our world is, of course, its own internal make-up and dynamics. At almost 13,000 kilometers across, our planet has been capable of sustained internal heat production throughout its entire history, generated both by gravitational contraction and from natural radioactive decay of heavy elements at the core. Slowly, heat emanates from the core to the mantle above, heating it to over 3,000°C. Despite such searing temperatures, sustained pressure from above keeps the mantle solid, but at just several hundred kilometers below the surface the upper mantle (called the asthenosphere) becomes plastic and ductile and is driven into gigantic convectively flowing cycles. The outermost layer, upon which Earth's crust sits (called the lithosphere), acts as a lid and responds to the convective currents below by cracking into a dozen or so tectonic plates. At plate boundaries, one plate slides under another in a process called subduction, while on its far side it is replenished from material rising from beneath, usually at mid-ocean ridges which then give rise to sub-aqueous hydrothermal vents, now seen as potentially important to the emergence of life. Over millions of years each plate undergoes drastic change and rejuvenation, creating and moulding continents and releasing huge quantities of water, carbon dioxide, and many other materials to the surface.

Biosphere

Despite the enormity of the forces bearing down from the vacuum of space and upward from the crushing interior, all have conspired to create an interconnecting zone of striking balance upon our planet. Starting at several kilometers below the surface and reaching to perhaps 30 kilometers above, this zone is called the biosphere because it is flourishing with life.

Among the major influences upon the biosphere is its volatile activity. Volatile materials include elements such as hydrogen, carbon, nitrogen, and oxygen as well as compounds such as water, carbon dioxide, methane, and ammonia. They are called volatile materials because they more readily change from solid to liquid to gas than other substances—for example, the changes that take place with water on the Earth's surface. By cycling though the various natural planetary reservoirs in response to changes in heat from the Sun (and from within the planet itself), volatile materials play a vital role in regulating and taming the biosphere while, in the process, bringing about the extraordinary dynamism that is so important to life.

One such cycle is called the hydrological cycle, which regulates Earth's supply of water across its rivers, lakes, seas, and oceans as well as within the crust, in the atmosphere, and polar ice caps. Significant transfers of water, especially between the oceans and the polar ice caps, are driven by changes to the global climate brought about by variations in solar insolation (that is, the amount of solar radiation incident on the Earth's surface). When the planet warms, water from the ice caps melts, with sea levels rising and precipitation increasing. Conversely, when the planet cools, water that finds its way to the polar regions freezes and remains there for thousands of years or more. Earth's hydrological cycle, as dynamic as it may be, is vital to stabilizing the climate, tempering the otherwise devastating effects that would arise from even small changes in the Sun's energy output.

Another key cycle is the carbon cycle, which regulates where and in what chemical form carbon resides—whether as organic carbon in living and dead organisms, as carbon dioxide in the atmosphere or dissolved in the oceans, or as carbonate sediments within the crust. In Earth's early history, carbon dioxide was a major constituent in the atmosphere, but over time virtually all of it has dissolved in the oceans and precipitated out as carbonate sediments. Hence, the carbonate sedimentary rocks within the landscape today are a testimony to an era long since past during which carbon dioxide was a dominant feature of the atmosphere. Despite the loss of all but trace levels, an important carbon dioxide cycle is still maintained today, driven by fresh supplies released from volcanic and tectonic activity, the weathering of carbonate rocks, and from human activity. Although comprising less than four hundredths of 1% of the atmosphere, carbon dioxide is vital to the stability of the planet as we know it, by acting as a heat shield, trapping solar radiation that would otherwise be reflected by the surface back into space. If it were not for such a natural greenhouse effect, Earth's surface would be over 30°C cooler at an average temperature of 18°C below the freezing point of water—too cold for pure liquid water to be sustained on the surface. It is therefore not just Earth's favorable distance from the Sun, but also the presence of trace levels of volatiles, such as carbon dioxide, that sustains our clement environment.

Most of the volatile materials found within our biosphere are also classed as biogenic materials because they are crucial to life. Hydrogen, carbon, nitrogen, and oxygen as well as water, carbon dioxide, and methane, among others, have been intimately linked with the operation of life throughout Earth's history. Indeed, many features on the surface of our world, including its volatile activity, reveal that it is dominated by life. The virtually planet-wide proliferation of vegetation, planetary oceans of liquid water, an oxygen-rich atmosphere, and even our particular weather and climate are all clear indicators of life being a significant feature of our world.

While the proliferation of plant and animal life across the globe is truly a crowning achievement within nature, we have recently become aware of other, significant, regimes of life hitherto unknown to us. For example, where we previously thought there were thousands, we now suspect that there are millions of species of microbial life comprising perhaps half of the total biomass of our planet. And while it was historically regarded as a simpler form of life, we now realize that microbial life is, in general, highly adaptable. Rapid reproduction rates and a high capacity for spontaneous genetic mutation allow microorganisms to adapt more rapidly to changing environments, making them ubiquitous across the biosphere.

Indeed, it seems that life on Earth has adapted to virtually every available niche. It is only in recent decades that entirely different types of ecosystems have been discovered in some of the most extreme environments on Earth, inhabited by types of organisms called extremophiles. For example, extremophile microorganisms have been found within the rocks of the coldest and driest deserts of Antarctica. Other types of extremophiles, called thermophile and hyperthermophile organisms, thrive in total darkness at temperatures well in excess of 100°C and at normally crippling pressures many kilometers down on the ocean floor, feeding from chemical nutrients from hydrothermal vents located at tectonic rifts. Ecosystems have also been found within the most acidic, alkaline, and saline environments known, and one microorganism, *Deinococcus radiodurans*, can even survive radiation dosages typical of nuclear reactors. As varied and extreme as Earth's environments can be, virtually all can support life. It seems that, historically, we have underestimated both the diversity and robustness of life and may continue to do so.

Life on Earth

The discovery of such extreme organisms and ecosystems has provided new insight into the nature of life in general and reveals an intimate connection between life and the planet itself that has perhaps not previously been fully appreciated. Of course we have known of such a connection since Darwin put forward the idea of evolution, but the discovery of extremophile life in particular demonstrates that life does not simply inhabit the world, but is intimately connected to the natural history of the planet as shaped both by its innate activity and its celestial environment. Evolution explains our natural history, by unifying the historical course of past life with the nature of the living world today, while also revealing its deep-rooted planetary connection.

Evolution proposes that an original ancestor—a single species—emerged in Earth's distant past and survived to reproduce within the prevailing planetary conditions. Minor alterations in traits (caused by random changes in genes involved with reproduction) led to successive generations, that had small genetic differences from their ancestors, eventually becoming different species.

Over millions of years, and spreading across the globe, life became evermore diverse, giving rise to a multitude of species. While, for some period of time, a species could cope with the prevailing conditions, significant changes to the environment would put that to the test, and only those that were fortunate enough to have a natural capacity to exist managed to survive. Through countless changes in the environment over Earth's long history, equally countless species have lived and then suffered extinction. On occasion, catastrophic tectonic, volcanic, and celestial impact events have altered Earth's surface so severely as to eradicate large portions of all life, changing its future course. Through relentless alterations to the biosphere, organisms with adaptations to every available niche have emerged, and if life now resides in the most extreme of environments, it is only because those environments were intrinsically involved in its diversification and adaptation from the outset.

The effect of evolution has not been all one way. Along with celestial and internal planetary forces, so too has the proliferation of life played a prominent role in shaping the surface of our world. For example, while our atmosphere now comprises over 20% oxygen, originally there was none. Only with the emergence of water-based photosynthesizing plant life was oxygen gas produced in significant quantities. Even then, oxygen is so chemically reactive that the total produced over the first two billion years reacted with metals dissolved in the oceans and within the crust, and only when Earth's surface was completely oxidized could surplus oxygen begin to accumulate within, and transform, our atmosphere.

Over the course of history, living ecosystems have radically altered the Earth's surface. Chemical alteration of the crust, atmosphere, and oceans, including the regulation of drainage, weather, and climate—and even in part sustaining the world's oceans, which themselves help to lubricate plate tectonic activity—are all outcomes of life. Earth's legacy is an ancient symbiotic relationship between life and the planet itself, both of which have undergone radical change and development over time, transforming our world into the finely tuned and complex one we now inhabit.

The Nature of Life

For hundreds of years we have attempted to classify the multitude of types of organisms now inhabiting our world in such a way as to reveal their underlying nature. An important recent classification was developed by the microbiologist Carl Woese in 1977, both in recognition of an increased understanding of genetics and of the discovery of extremophile life. Woese's classification encompassed three distinct and overarching branches of life called domains—archea, which were previously regarded as an ancient form of bacteria called archeabacteria, are now widely accepted as one of the three domains of all life, along with bacteria (otherwise known as prokaryotes) and eukaryotes.

Eukaryotes, which comprise the multicellular organisms such as plants and animals, are characterized by living cells with an internal membrane-enclosed cell nucleus containing most of the cell's genetic material. Bacteria, on the other hand, are single-celled microorganisms whose genetic material is not contained within a nucleus but instead within the main body of the cell itself. Archea are similar in this respect, but are so different in many other aspects of their biochemistry as to be classified as an entirely separate domain of life. Archean microorganisms are of particular interest because they include many of the inhabitants found in extreme habitats such as sub-aqueous hydrothermal vents and hot springs. Furthermore, biochemical and genetic dating techniques suggest that archean life is closer than any other to the proposed original ancestor from which all other life on Earth arose.

While there is clearly enormous diversity in life, many important attributes are common to all life. For example, all life is characterized by at least five key functions: reproduction, metabolic activity, growth, evolution, and sensory response to stimulus from the surrounding environment. Such commonality points to a general definition of life according to function, independent of how those functions are carried out. And while rudimentary, such a definition is none the less of great value when investigating the legacy of life on Earth and when searching for life elsewhere in the Universe, because it gives indications of what to look for without having to know the fine detail of any such life, or its ecosystem, in advance.

We also find many striking common features in the underlying morphology and biochemistry of all life on Earth. As already indicated, all organisms, from the microscopic bacterium to the largest plants and animals, are built upon the cellular morphological unit. Furthermore, despite the enormous diversity and specialities in the roles of cells throughout life, all share many underlying characteristics, such as a fatty-protein semi-permeable outer membrane, and, among many others, an internal aqueous solution that concentrates the materials needed for (and which enables) life activity. Indeed, as we traverse the microscopic to the molecular levels at which the biochemistry of life operates, we again see striking commonality. For example, with the exception of some viruses, replication in all life on Earth occurs though the biochemical symbiotic relationship of deoxyribonucleic acid (DNA), ribonucleic acid (RNA), and protein-based enzymes.

DNA is an enormous molecule composed of billions of molecular building blocks called nucleotides. A nucleotide is composed of three molecules linked together—a pentose (5-carbon) sugar molecule, one or more phosphate groups and one of four types of organic bases: adenine, cytosine, guanine, and thymine. Nucleotides join together into enormous chains called polynucleotides. DNA comprises two polynucleotide chains connected together along their length, with adjacent bases linked together into base-pairs similar to the rungs of a ladder, which then twists into the famous "double helix" shape. While DNA contains the relevant genetic information needed for the various functions of life, it cannot carry out its own operations without the aid of RNA and enzymes. RNA, of which there are many types, is composed of a single polynucleotide chain, with the organic base uracil replacing thymine. Enzymes are composed of proteins that are themselves made of amino acids linked together into polypeptide chains.

DNA, RNA, and enzymes work together in the most exquisite of symbioses found in nature. First, DNA synthesizes RNA using information encoded along its nucleotide chains. During RNA synthesis, part of the DNA double helix temporarily unwinds and, with the help of enzymes, manufactures a new strand of RNA by aligning along its length freely available nucleotides within the cell aqueous solution. The pattern of DNA nucleotides determines the sequencing of the new RNA molecule, with enzymes assisting in both the positioning and linkage of the nucleotides. When complete, the new RNA molecule is released, with DNA rewinding into a double helix, ready to repeat the process.

Among RNA's roles is the manufacture of enzymes. It does this by linking together individual amino acids into peptide chains in a way not dissimilar to the DNA construction of RNA. Finally, DNA contains the genetic information necessary for its own replication, although once again it depends upon enzymes to carry out the process. During replication, the DNA double helix unwinds completely, producing two separate single polynucleotide strands. With the help of enzymes, free nucleotides within the cell are collected and linked to each of the separate strands, producing the two DNA double helixes required in replication. Hence in the most sophisticated of interdependencies at the core of the life process, DNA manufactures RNA, and RNA manufactures the enzymes on which DNA depends to carry out its various functions.

Another important biochemical function in life is how energy from the outside environment is acquired for use in metabolic activity. There are only two sources of external energy harnessed by all life—sunlight and chemical energy in the form of redox chemical reactions with the surrounding environment. A redox reaction is one involving the transfer of an electron from an "electron donor" such as a metal, to an "electron acceptor" such as oxygen, while in the process releasing a small amount of usable energy. Organisms such as plants that derive their energy from sunlight are called phototrophs, while microorganisms that derive their energy by redox reactions, often with inorganic materials in the environment, are called chemotrophs.

After acquiring energy, all phototrophs and virtually all chemotrophs then activate a sophisticated "electron transfer chain" to move the acquired energy across their cell membranes. While doing so, they also move protons (hydrogen atoms with their single electron removed) from one side of the membrane to the other. This is crucial, because the accumulation of protons across a membrane acts like a small electrical battery, storing the energy harnessed from the outside environment for future use. Subsequently, in what is called a "proton pump," the stored energetic protons release their energy to manufacture the enzyme ATP-synthase, which itself makes adenine triphosphate (ATP), the vital "energy currency" molecule used within all living organisms. Even in organisms where an electron transfer chain is not activated, as in some archean microorganisms, ATP is still manufactured using the energy from protons acquired directly from their immediate environment. Hence at the core of energy production within all life, we find proton pumps used to manufacture ATP-synthase and subsequently ATP.

Echoes of our Origins

Given an origin for planet Earth itself, there must also have been a moment when life first emerged. Whether that involved an actual origin to life on the planet, or its arrival from elsewhere in the Universe during Earth's early history, is currently unknown. Nevertheless, what we have so far learned about the natural history of our planet and the evolution and biochemistry of life provide valuable clues to the first organisms to inhabit our world, and may also help to uncover an actual origin to life here. Within the fabric of life today reside echoes of our origins.

For example, while currently DNA is crucial to virtually all life, its complexity and dependence upon enzymes constitute a highly developed biochemical system, suggesting a precursor means of preserving genetic information and of replication and, hence, a time-line of development in the biochemistry of life. Other biochemical processes suggest similar development over time. Photosynthesis, for example, constitutes a sophisticated mechanism for acquiring energy, suggesting that it too has evolved from precursor methods of acquiring metabolic energy from sunlight. And aerobic respiration—the process of releasing biochemically stored energy in the presence of oxygen—could only have become dominant subsequent to oxygen becoming available in our atmosphere, indicating that anaerobic respiration emerged at a prior stage. Many other examples reveal similar complexities, symbioses, and dependencies that point to an evolution in the structure and biochemical operation of life over time.

Overall, the developmental path in life indicated both by Darwinian evolution and within biochemistry suggests that the first organisms on Earth were simpler in morphology and biochemical operation. Indeed, the fact that all living organisms consists of one or more living cells, and are underpinned by the five basic functions of life outlined above, perhaps points to a minimum requirement for life. For example, fossilized evidence of the earliest known life on Earth, dating back to approximately 3.5 billion years ago, is of microbial life only, while RNA-only viruses and chemotroph microorganisms both point to alternative modus operandi in life separate from the more sophisticated biochemistry associated with DNA and photosynthesis. Such evidence points to the first life on Earth being single-celled entities capable of at least the basic functions of life, and having biochemical properties that were perhaps related to RNA-only viruses and archean chemotrophs found in extreme environments.

Even with such insight into the possible nature of the very first organisms, as stated above, we cannot yet tell whether the first organisms to inhabit our planet originated here, or whether they came from elsewhere in the Universe. Hence, a search for an actual origin to life must include a widespread and thorough investigation of many of the other bodies of the Solar System. Such a search, however arduous, should eventually reveal whether ancient microbial life emerged elsewhere in our Solar System, or perhaps even came from the original nebula from which our Solar System emerged.

If, on the other hand, life originated on Earth itself, clues to its actual origin may be found from our improving grasp of the natural history of our planet and the nature of life. For example, the longstanding connection between life and the environment points to a similarly close coupling between the processes of any origin and Earth's earliest environment. Indeed, given the ever-changing and evolving biosphere, the conditions within which an origin to life occurred would have been very different to those of today. However seemingly toxic by our standards, Earth's earliest environment was conducive to life and could well have brought about its origin. Certainly the discovery of archean extremophile microorganisms at sub-oceanic hydrothermal vents demonstrates that life is possible in the most hostile environments, while chemoautotrophic microorganisms demonstrate that basic life can harness even inorganic materials both for energy and organic nutrients, suggesting that Earth's early and relatively hostile environment could have supported an origin to life.

Hence, we envisage an origin process involving a previous and separate era to life itself—an era of prebiotic chemical evolution, dependent on the prevailing planetary conditions and natural resources, from which emerged single-celled organisms exhibiting at least the basic functions of life, perhaps simpler in biochemical operation yet with a capacity to evolve into ever more capable forms.

The search for origins is reduced to an intricate investigation of our planet's earliest history, to determine the natural resources that were available, the prebiotic chemistry that was possible (and actually occurred), and how it led increasingly toward biochemical systems that isolated themselves from their surroundings in cellular structures and developed the basic essentials of life.

The Origin and Early History of Earth

Five thousand million years ago, our Sun formed within a vast nebulous cloud of gas and dust. The material of the cloud originated inside pervious stars long since deceased and from gases created during the formation of the Universe. Over millions of years the cloud coalesced into a number of slowly rotating clumps. Eventually, our Sun began to take shape from one clump as a dense swirling globe more than a million kilometers across, with its remaining material condensing into a thick rotating disk stretching for billions of kilometers beyond. Pressure at the center became so great as to trigger nuclear fusion, converting hydrogen into helium and releasing vast amounts of energy into space. Our Sun had begun its life as a new star within the immense Milky Way galaxy.

As the surrounding debris coalesced about the Sun, complex chemical and mechanical interactions were triggered by solar energy, cosmic radiation, and electrostatic and magnetic activity within the field itself. Many new materials were synthesized, including volatile and organic materials such as water and carbon dioxide, among others, to at least the complexity of amino acids and nucleic acids. Most of the remaining hydrogen and helium found its way to the outer region of the disk, eventually to become the giant planets Jupiter, Saturn, Uranus, and Neptune. Closer in, the remaining light gases mixed with heavier materials to form dust and ice particles that, over millions of years, accumulated into rocks, boulders, and colossal planetoids, eventually becoming the protoplanets of the inner Solar System.

The emerging planets all jostled and struggled with one another. Chaos ensued as each settled into orbit while enduring cataclysmic collisions with other forming planets as well as relentless bombardments from the millions of planetary remnants. Eventually only four rocky planets survived the battle for the inner Solar System—Mercury, Venus, Earth, and Mars. Each would endure further pounding for millions of years, but they would survive. They had grown to hundreds of times the mass of even the largest remaining planetary invader, and although regularly inflicted with serious surface damage, each planet could by now hold together and maintain its orbit. The inner Solar System as we know it was taking shape.

Having emerged from a single debris field, three of the four inner planets—Venus, Earth, and Mars—may have begun their existence quite similarly. All enjoyed, to various degrees, a soothing heat from the Sun. All three were by now substantial rocky planets, settling in the same broad region of the inner Solar System, and all were quite similar in material composition, internal planetary dynamics, and surface conditions. Significant differences would eventually emerge among the three, but in the beginning they would have been broadly similar.

As Earth grew toward its present size, its overwhelming gravity drew all of its accumulated material into a near-perfect globe. Any mountain too high would crush under its own weight and any rift too deep would fill with debris falling from above. A process of material and chemical differentiation drew heavier materials toward the center of the planet, forcing lighter material toward the surface. The core of the planet became hotter, powered both from the gravitational contraction and radioactive decay of heavy metals. The initial heat may have been sufficient to melt up to 60 percent of the interior of the planet, with the core softening and a magma ocean hundreds of kilometers thick forming around it. Hot molten rock and iron circulated within colossal convection currents moving from the depths of the planet toward the surface, where they cooled and flowed inward once again. An electric-dynamo generated by the molten iron created a planetary magnetic field that extended for thousands of kilometers into space, protecting the surface from harmful solar and cosmic radiation.

Even with such inner turmoil, the surface began to cool and solidify. Convective cycling had quickly dissipated much of the internal energy and sorted the materials of the planet according to their density and chemistry. Silicon and oxygen, comprising most of the surface, solidified and formed a relatively low density crust that floated on the more dense material below, maintaining a coherent surface despite the continuing inner activity. Over time, the inner core solidified, although the outer core has remained molten to the present day. The mantle also mostly solidified, though again the outer asthenosphere has remained soft and malleable, churning through slow convection cycles that take hundreds of millions of years to complete.

In the final stage of differentiation about 4.3 billion years ago, in what is called the Hadean Period, hot volatile gasses were ejected in a process of outgassing via the planet's tectonic rifts and volcanoes. In perhaps only a million years, 80 percent of Earth's original atmosphere was produced, with the oceans appearing shortly thereafter. From a central iron core to the upper atmosphere, the materials of the planet had been separated though a process that we suspect is common to many rocky planets.

Along with silicon and oxygen, the new crust consisted of iron, aluminum, calcium, magnesium, and phosphorus, among other materials, all of which chemically reacted to form minerals. The most common minerals on Earth are silicates—minerals made from both silicon and oxygen—including feldspar and pyrite, which are low in metal content, and basalts, which contain more iron and magnesium. The formation of those early igneous rocks became the foundations upon which further surface activity could occur.

Earth's earliest atmosphere, composed primarily of hydrogen, was quickly lost to space. Subsequently, a more stable atmosphere was created from outgassing and from the condensation reactions of water upon the newly formed igneous rocks, producing an atmosphere of carbon dioxide, nitrogen, methane, ammonia, hydrogen chloride, hydrogen sulfide, and sulfur dioxide. Although toxic when compared to today's atmosphere, that early atmosphere acted as an important stabilizing influence. Carbon dioxide and methane would have brought about a greenhouse effect that helped to raise the surface temperature above the freezing point of water, while the increased surface pressure allowed liquid water to persist on the surface for the first time. And with continuing supplies from the outgassing of steam and from impacting comets, the surface became increasingly dominated by water. Eventually vast oceans covered much of the surface and, from then on, all that happened on Earth would be intimately connected to liquid water.

Toward the Origin of Life

Earth had taken about 100 million years to form, attaining its present size approximately 4.4 billion years ago. Along with the turmoil of its early indigenous activity, the planet endured repeated bombardment from space for hundreds of millions of years, but which rapidly ceased about 3.9 billion years ago. With firm evidence of microorganisms as far back as 3.5 billion years, and tentative evidence that life existed 3.8 billion years ago, it seems that life emerged quite rapidly in Earth's early history.

Hence, in attempting to determine how life emerged, it is to that tumultuous young planet and its immediate space environment that we must turn our attention. We must understand the effect of mass bombardments, the nature of our early Sun and Moon, and the indigenous events, such as tectonic, volcanic, hydrothermal, and water-based activity, the formation of the atmosphere, and the resulting climate. We must also determine the planetary environments that could have led to the synthesis of organic compounds and then to their assembly into ever more complex systems, toward prebiotic chemistry and, finally, life itself.

We must begin by considering the origin of organic materials—whether synthesized in space to arrive on Earth from impacting celestial bodies, or synthesized on Earth itself, and, if so, by what processes. We must also determine how biochemically functional blocks, such as nucleic acids and amino acids, originated. Here also we must address one of the most significant features of the biochemistry—that of chirality. Chirality in living systems refers to an asymmetry in the use of biochemical molecules. For example, each amino acid occurs in nature in two forms called left-handed (L) amino acids and right-handed (D) amino acids. While both are found in equal quantities in nature, living organisms use mostly L-amino acids, indicating a required mechanism that either preferentially *synthesized* L-amino acids or preferentially *selected* L-amino acids during the origin of life.

Next we must determine possible methods of energy transductance within the first biochemical systems. The processes leading toward life could not have occurred in an unchanging or stagnant environment. Rather, they must have arisen in an environment of thermodynamic disequilibrium (that is, an environment whose thermal and material characteristics were in flux) and capable of providing energy of a type useful to chemistry and ultimately for metabolic activity while simultaneously providing meta-stable (temporarily stable) conditions that allowed for organization and complexity in form and function to emerge.

Also of importance is how such energy mechanisms became coupled to polymerization chemistry and specifically how the polymerization of nucleic acids into polynucleotides and amino acids into proteins occurred. Here many fundamental issues arise due to the complexity within genetic and protein materials, and their interaction among one another. There is general consensus that such complexity could not have emerged in nature all at once, but was the outcome of a chemical evolution over many stages. Here, several scenarios present themselves. In one, polynucleotides capable of retaining genetic information may have arisen first, subsequently aiding the polymerization of amino acids into proteins akin to how RNA synthesizes proteins in life today. Alternatively, proteins and organic membranes could have emerged first, providing cell-like environments within which concentrated solutions of nucleic acids could then polymerize.

While far from certain, many think it less likely that protein could arise in nature without genetic assistance and that it is more likely that genetic polymers would have arisen first, subsequently aiding the emergence of proteins and cellular membranes. Indeed, many now think that the emergence of RNA in particular was a pivotal event in the origin of life; and that, for a time in Earth's early history, all life may have been based on RNA (or some other related genetic polymer) rather than DNA. Certainly the dependence of DNA on enzymes makes it difficult to see how DNA could have emerged first and have been capable of self-replication before the emergence of proteins and enzymes. Similarly, it is difficult to see how proteins could have arisen without genetic assistance. RNA, on the other hand, has been identified in the laboratory as being capable of self-replicating, while, as already mentioned, RNA-only viruses show that it is possible to have life that is independent of DNA.

Although tentative, such a scenario at least offers a starting point for further investigation. If we accept that life originated in nature, then some sequence of events currently unknown to us led to its emergence and we now consider the emergence of RNA to have been important in this stage.

Possible Pathways—Organic Synthesis

There are two broad scenarios for the origin of organic material. As discussed in Chapter 1, organic synthesis would have occurred throughout the nebula from which Earth formed. We know this because we can currently observe organic synthesis taking place throughout vast molecular clouds between the stars, deep within star-forming nebulae, and within the flattened disks surrounding newly formed stars. Furthermore, evidence from a particular class of meteorites called carbonaceous chondrites, which formed during the origin of the Solar System, reveals a number of organic materials to the complexity of amino acids.

Several astronomical studies have even provided new insights into the issue of chirality. First, a particular meteorite called the Murchison Meteorite (discovered in Murchison, Australia, in 1969) not only contains amino acids but also shows an excess of L-amino acids, indicating a celestial process that originally synthesized more left-handed amino acids within the original solar nebula. Furthermore, recent observations of a region of the Orion nebula, called OMC1, even suggest a possible process—that of circularly polarized light from nearby stars bathing the region and preferentially triggering the formation of L-amino acids. If such mechanisms occurred within the original solar nebula, it is plausible that organic material suitable to life could have been delivered to the surface of our planet during its early history.

A second scenario considers organic synthesis having taken place on the surface of the planet itself. Here, several different mechanisms are envisaged. One mechanism, demonstrated in a famous experiment by Miller and Urey in the 1950s, leads to the production of organic compounds when volatile gases such as carbon dioxide, methane, and ammonia are energized by solar radiation or lightning discharges. In was realized in the 1980s, however, that methane and ammonia in particular would not have remained in Earth's early atmosphere long enough to produce significant quantities of organic material, casting doubt over the validity of this process. More recent studies have shown, however, that even without gases such as methane and ammonia, substantial organic synthesis can occur within a carbon-dioxide-rich atmosphere in the absence of oxygen, indicating that Earth's early atmosphere would have been conducive to organic synthesis.

An alternative and potentially significant source of organic synthesis has presented itself with the discovery of hydrothermal vents on the ocean floor. Hydrothermal systems arise when volcanic materials solidify on or near the surface and then cool, contract, and crack. Any water that circulates through the cracks heats up and flows more rapidly over the newly formed igneous rocks, releasing hydrogen gas which then reacts with carbon dioxide in the atmosphere to produce a range of simple organic materials. Given the favorable conditions during Earth's early history—a carbon-dioxide-rich atmosphere, planetary oceans, and widespread tectonic and volcanic activity—hydrothermal systems may have been an important contributing factor to the supply of organic material, as well as a range of other life-related activity.

Possible Pathways—Polymers and Membranes

The synthesis of organic compounds, even to the complexity of amino and nucleic acids, represented but the first of many steps toward the origin of life. Even the smallest single-celled organism contains more than 100 billion atoms organized in extraordinarily sophisticated ways. Nevertheless, numerous plausible mechanisms for the polymerization of amino and nucleic acids, and for the emergence of cellular-like structures, are now postulated and are the subject of intense and ongoing investigation.

In one scenario, polymerization of nucleic acids into genetic polynucleotides could have occurred with the aid of particular mineral clays found in sedimentary basins, dried lakes, and retreating coastlines during periods of warming climate, or from dry-heat at volcanic settings. For example, in one experiment conducted by James P. Ferris of the Rensselaer Polytechnic Institute, the mineral-clay montmorillonite (hydrated aluminum silicate) was shown to catalyze the synthesis of RNA oligomers (partial polymers). Composed of regular charged layers of aluminum and silicate, montmorillonite can catalyze the production of RNA oligomers by aligning and linking individual sections (called monomers) between adjacent sheets, holding them in place by electric charge. Although far from revealing how RNA emerged in nature, this experiment at least demonstrates a possible process leading toward basic genetic polynucleotides that are capable of catalyzing their own replication. Hence, it is speculated that through such a "naked-genes" scenario, the rise of RNA in nature may have been supported initially by minerals, and that this subsequently brought about the synthesis of organic membranes.

Another scenario considers Earth's early organic-rich oceans giving rise in the first instance to membranes and cellular-like entities, which subsequently facilitated the chemical evolution of genetic polymers. For example, when surrounded by water, many organic compounds, such as amino acids and proteins, coalesce into microscopic spherical structures called coacervates. These organic-rich colloids even allow other organic materials from the surrounding medium to enter, suggesting that coacervates may have provided the first isolated environments within which complex prebiotic chemistry occurred.

Yet another scenario considers hydrothermal systems as a possible seat for the origin of life. As originally considered by Gunter Wachterschauser in the 1980s, and subsequently by many others, life may have emerged within black smoker hydrothermal vents on the ocean floor. In such settings, tiny caverns coated with iron sulfide (pyrite) inorganic membranes could have acted as micro-environments within which life's first steps began. The hydrothermal vent itself could have provided both organic raw material and chemical energy; while within each tiny cavern, polymerization of nucleic acids toward RNA could have been supported by the unique geometry of pyrite crystals, which are now considered to be capable of catalyzing such reactions. Subsequently, synthesized lipids and proteins could have gradually replaced the inorganic iron sulfide membranes, leading increasingly toward organic-based cellular entities. Given such a range of intriguing possibilities, many now regard hydrothermal systems as a serious contender for the origin of life.

However rudimentary, all current scenarios point to a geochemical context for the origin of life, with mineral rocks and clays potentially as significant as organic synthesis and the presence of water. Indeed, minerals could have performed numerous important roles during the origin of life. First, they could have acted as containers and scaffolds—as with montmorillonite—supporting and aligning organic monomers and enabling polymerization. They could also have acted as templates for particular reactions and perhaps even offered a planetary context to the chirality issue. For example, in experiments by Robert Hazen, Timothy Filley, and Glenn Goodfriend in 2001, crystals of calcite immersed in a solution of both left- and right-handed amino acids were coated with a surface layer of mostly L-amino acids *and* also assisted in their polymerization, representing a plausible geochemical mechanism for the production of protein from L-amino acids on the young Earth. Minerals could also have acted as catalysts, providing an intermediary role in organic synthesis—as in the example above, where dissolved iron sulfide may have provided the initial framework for the formation of membranes and for genetic polymerization in sub-aqueous hydrothermal vents. Finally, inorganic minerals, which participate in the biochemistry of life today, could equally have been involved in the chemistry of life from the beginning. Overall, a geochemical context for the origin of life is now recognized, through which otherwise impossible chemical processes relevant to the origin of life may have been enabled.

A Way Forward

Uncovering the origin of life on Earth may seem to be an enormous if not insurmountable task. But if our origin has a basis in nature, there is no better means at our disposal than the relentlessness of modern science to uncover the processes involved. Nevertheless, it is unlikely that we will discover a singular piece of evidence that quickly and resolutely reveals "the origin of life." Rather, we now recognize that the search for origins is a process of elimination and refinement through experimentation and exploration, leading us to ever more elusive yet vital clues that will gradually improve our understanding of the processes involved.

As shown throughout these opening chapters, we have already made an excellent start—identifying an ancient and planetary context for the emergence of life, realizing the likely nature of the first organisms, and even identifying a range of possible scenarios through which the steps toward life may have occurred. But it is just a start. We cannot yet even conclude that life actually originated on Earth. For this we must obtain a statistically significant sample set from across the entire Solar System that conclusively reveals either the presence or absence of microbial life elsewhere in the early Solar System. And while we pursue that agenda, it is also prudent to consider an origin of life on Earth itself, arguably as good a place as any for life to have originated. In this we must continue with our in-depth investigations into the many contributing factors on our planet and in its immediate vicinity in space. We must determine the organic synthesis that occurred within the original solar nebula, as well as the precise stocks of volatile and organic materials that were delivered to the planet's surface through mass bombardment. We must comprehensively understand Earth's earliest intrinsic tectonic, volcanic, and hydrothermal systems, its early atmospheric composition, and the character of its early water systems—their acidity, redox potential, water activity and salinity, and so on. The greatest challenge, however, will be in determining the actual environmental settings in which organic synthesis, polymerization, and other complex prebiotic chemistry could have led toward the first single-celled organisms.

While the search for the origin of life will always be pursued on Earth, the discovery of new evidence can be extraordinarily difficult. In perhaps the greatest of ironies, the relentless evolution of, and ancient link between, our planet and life itself means that virtually no trace of Earth's earliest history persists (although ever-improving techniques allows for ever more elusive evidence to be uncovered). But given the significant planetary context for origins, an important new opportunity now presents itself with the planet Mars. With broadly similar early histories, many of the external and intrinsic planetary factors contributing to the origin of life on Earth may also have affected Mars. From our robotic Mars missions to date we can already see that, unlike on Earth, Mars actually retains a substantial record of activity from its earliest history, presenting a significant new avenue for directly probing planetary and Solar System activity from that era. Indeed, such is the mass of evidence awaiting us and such are the gaps in our knowledge, that Mars may well provide significant new insights into Earth's early history and the origin of life. In both challenging our current ideas and in stimulating brand new thinking, Mars can provide opportunities that are simply not available on Earth, enhancing our understanding of the circumstances through which life originated. And if, as we now think plausible, life actually emerged in Mars' early history, that planet may even provide far-reaching answers both on the origin of life and its broader, cosmological context.

From Antiquity to the Canals

Human perception can be a curious faculty. There was nothing to prepare the first listeners to Edison's acoustic wax phonograph for what they were about to hear—no point of reference—and so they were literally unable to distinguish the recording from the real thing. So fantastic was that first recording that it mesmerized the minds of those who heard it.

And so it must have been for the first civilizations to look skyward with the curiosity of minds capable of eventually taking them there. At a time when the night was illuminated by starlight alone, it must have been the most amazing spectacle. The mesmerizing power of such a realm must have been as substantial as any virtual reality experience today. The spectacle of a clear, crisp night sky was potent enough, but the untimely arrival of a shooting star streaking across the sky would have been dazzling. The slow arrival of a comet in the sky, becoming slowly brighter with each passing week, must have been a challenge—a source of apprehensive and perhaps even of anguish.

A change in the sky was regarded as significant, and shooting stars and comets evoked surprise and apprehension, but the regular motions of the Sun, Moon, planets, and fixed stars signified regularity and certainty. They were beyond Earthly troubles and everyday experience, and in time they were regarded as eternal and god-like. Many cultures built myths about the sky gods, and those myths were our first way of deciphering the meaning of it all. Humanity was building a perception of the Universe and at the same time attributing purpose to it. The ancient sky myths were a reality to generations of smart, discerning people.

The Sun, Moon, and planets all had their place among the gods. But the planets were peculiar. While the Sun, Moon, and stars circled about the center of the Universe (Earth) without interruption, the planets would regularly stop, move backwards for a while, then return to their regular journey in the same direction as the other heavenly bodies. Of the planets, Mars' *retrograde motion* was noticeably greater than the others. And while Jupiter, for example, commanded great presence in the sky on an almost

K. Nolan, *Mars, A Cosmic Stepping Stone*,
DOI: 10.1007/978-0-387-49981-9_3, © Praxis Publishing, Ltd. 2008

yearly basis, Mars would fade to insignificance and then brighten about every two years. At its brightest it would rival Jupiter, but with a striking deep red color, it became a symbol of war to the Greeks and Romans who worshipped it.

The retrograde motion of the planets was a conundrum to the brilliant minds of ancient Greece. A fabulous and enduring model, first proposed by the exceptional mathematician and astronomer Hipparchus in the second century BC—and later developed by Ptolemy, the last of the great Greek astronomers—was that the planets moved in small circles called epicycles as they traveled around the Earth. The planets would sometimes appear to move backwards on their epicycles, explaining the temporary retrograde motion seen in the sky. Despite being ultimately incorrect, Hipparchus' and Ptolemy's model gave unprecedented accuracy in predicting the position of the planets for the next 1,500 years.

Over the centuries the Greek and Roman empires faded and Christianity took hold across Europe. By AD 400 there was scarcely an individual in Western civilization who thought the planets were gods. The one true God had been revealed and the planets, though still not understood, were relegated to the status of heavenly bodies circling the Earth, now seen as the seat of God's chosen people. But the curious motion of the planets gnawed at the purity of such an Earth-centered, God-created Universe. In the final analysis, Ptolemy's model was not complete. It was clear to astronomers and priests alike that the epicycle model did not adequately describe the motion of the planets as well as we could map them. It was also cumbersome, and was not sitting properly with a notion of the *harmony of the heavens* as created by God.

In 1543, the Polish astronomer and priest, Nicolaus Copernicus, while supposedly on his deathbed, published his life's work regarding a new and revolutionary model for the Universe—one that put the Sun at the center of all things. The motivations for the Copernicus model were the limitations of Ptolemy's model and a striving for a purer explanation of God's Universe. He had no special insight into the future he was about to unfold and had absolutely no desire to attack his church. Despite the brilliance of his idea, it was just that—an idea. Copernicus had started the debate, but he had not provided the means for a satisfactory conclusion.

Fifteen hundred years before, the substantial retrograde motion of Mars had helped Ptolemy to develop his model, and now Mars would once again assume a key role. It was the nature of Mars' actual orbit—being more elliptical than that of any other planet—and its proximity to Earth that would provide the means for the truth of the motion of the planets, and the underlying laws of gravity, to be finally uncovered.

Not long after Copernicus' death, a brilliant astronomer, mathematician and observer called Tycho Brahe decided to examine Copernicus' ideas. As telescopes had not yet been invented, there was no means by which astronomers could look closely at the planets and examine their true nature. Nevertheless, Brahe's exceptional talent at astronomical instrument making, and his keen insight into the limitations of his predecessors' efforts, provided him with the means to make observations of the motion of planets that were no less than five times more accurate than those of even Hipparchus. Brahe is still widely regarded as the greatest natural observer who has ever lived.

But it was not Brahe who was to confirm Copernicus' ideas. Brahe's measurements were so good, but so revolutionary, that they led him to a conclusion he could not accept. The argument goes like this. The stars were thought to be fixed to a crystal sphere beyond the furthest planet, Saturn. If Earth truly moves around the Sun, Brahe should have been able to see an apparent movement in the position of the fixed stars over the course of a year. But he saw no such movement. For years he checked and rechecked, but not once did he see any of the stars apparently move. There could only be two conclusions: either the stars are much further away than previously thought and the Earth does indeed circle the Sun, or the Earth is at the center of the Universe and Copernicus is wrong. Brahe's observations were so good that he could say with confidence that if any star was closer than 200 times the distance from the Earth to Saturn, he would see its apparent motion in response to the Earth's true motion. But he witnessed no such apparent stellar motion. As he could not accept that the stars were so far away, he rejected Copernicus' hypothesis.

Brahe was unable to accept the profound conclusion exposed by his own remarkable observations—that the Earth and the other planets are part of an isolated system and that the stars are very far away. With the unaided eye, Brahe had shown that the stars were at least 300 billion kilometers away. The closest star to the Sun, Proxima Centauri, is actually 42,000 billion kilometers away, but in Brahe's time it was thought that the stars were just beyond the planets, at perhaps at a few billion kilometers. Brahe had made an astounding discovery regarding the nature of the Solar System and the distances of the stars, and didn't realize it.

Despite Brahe's rejection of Copernicus' idea, the flawed theory of Ptolemy still remained. A contemporary of Brahe, called Johannes Kepler, worked with Brahe's observations to see if he could unlock the secrets of Copernicus' universe. But Brahe and Kepler did not get along, and it was not until Brahe's death in 1601 that Kepler gained access to all of Brahe's astounding observations.

It was Brahe's meticulous and comprehensive observations of Mars that were to provide the key. By 1604, Kepler had uncovered the only possible conclusion: not only must Mars orbit the Sun, but it must do so in an elliptical manner, not a circular one. Brahe's observations of Mars were so good that they gave Kepler the means to discover the true laws of planetary motion, and in so doing reveal a fundamental truth about the nature of the Universe. Mars was the tool with which humanity unlocked the nature of planetary motion and ultimately of the nature of gravity. Kepler himself stated: "In order to be able to arrive at an understanding, it was absolutely necessary to take the motion of Mars as the basis, otherwise these secrets would have remained eternally hidden." The true proximity of Mars and its particularly eccentric orbit (far more so than that of any other planet apart from Mercury) led Kepler to the inevitable conclusion that we are not at the center of the Universe.

Worlds Revealed

For ever more, the Universe was a different place. No longer was Earth situated at the privileged position of the center of all things. Instead it was revealed to be like the other worlds—a planet, a wanderer, and perhaps relegated even more than the mythical planetary gods of old. For thousands of years our ancestors had not known the nature of the Universe they occupied, but finally the efforts of Copernicus, Brahe, and Kepler had uncovered the nature of Earth and some measure of its place in the Universe. It took only six more years for the Italian astronomer Galileo Galilei, with his new invention of the telescope, to observe the changing phases of Venus, a gibbous Mars and four moons circling the planet Jupiter, adding significant evidence to the hypothesis of a Sun-centered Universe.

Though these findings were revolutionary, they were not to achieve general acceptance. The Roman Catholic hierarchy would not allow a challenge to the source of its authority—Earth's privileged place in God's Universe. The notion of Earth as just another planet would cause unrest and a questioning of Rome's supreme authority and so Galileo's findings were not to be widely publicized or embraced. That the planets were entire worlds and that Earth was one of them remained unknown to ordinary people of the time and the significance of the moment went largely unnoticed. Over the next 250 years the new Universe gained slow acceptance, but the impact it might have once had was lost. Only a handful of talented scientists and astronomers would explore our planetary neighbors in earnest. What they did observe laid the foundation for much of the development of astronomy for hundreds of years and for much the general perception of the Universe even today.

One of the first of that small band of seventeenth-century observers came from the then more relaxed and enlightened Netherlands. His name was Christiaan Huygens, a brilliant mathematician and scientist whose improved telescopes revolutionized astronomy. With his rudimentary telescopes, Galileo could identify the planets as tiny and mostly featureless globes, but Huygens' telescopes revealed details of their surfaces. In 1656 he discovered a large triangular dark region on Mars, now called Syrtis Major. In just three consecutive nights he determined that Mars' day is about 24 hours long. He discovered that Mars has polar ice caps just like those of Earth (though he offered no explanation of their nature). A contemporary of Huygens called Cassini, improved his findings, verifying that Huygens' measurements of Mars were real and not illusionary.

From 1672 to 1719, the astronomer Giacomo Maraldi (Cassini's nephew) conducted a detailed and lengthy study of Mars. Maraldi noticed that Mars' polar ice caps changed in size over the course of its 667-day year. He surmised that Mars might pass through the four seasons of spring, summer, autumn, and winter over the course of that time. The British astronomer William Herschel later verified Maraldi's conjectures when he spectacularly measured Mars' axial tilt to be 28 degrees (Herschel was only about 4 degrees out—Mars' tilt is close to 24 degrees). Since it is the 24-degree axial tilt of Earth that gives rise to our seasons, Herschel reasoned that Mars' almost identical tilt could also result in seasons, with each being twice as long as those on Earth because Mars' year is twice as long.

Time passed and the more that was learned about Mars, the more Earth-like it seemed. Its 24-hour day, polar ice caps, 28-degree axial tilt, annual seasons and surface detail that apparently revealed seas and land, all led astronomers to believe that they had a reasonable understanding of Mars, and that it was not too dissimilar to Earth. By the end of the eighteenth century, it was even considered that Mars might be inhabited. The first steps had been taken toward considering life elsewhere in the Universe.

Martian Canals

Every two years or so Earth and Mars line up with the Sun. We call this *opposition* because Earth is then between the Sun and Mars, with Mars directly opposite the Sun in our sky and most favorable for observation at night. But because Mars' orbit is elliptical, some oppositions bring Earth and Mars closer together. If, at opposition, Mars is at its furthest point from the Sun (aphelion), then both planets cannot come closer than about 100 million kilometers. But if, at opposition, Mars is at its closest point to the Sun (perihelion), then Earth and Mars can be as close as 56 million kilometers. The latter scenario occurred on August 27, 2003, when Mars was the closest it had been to Earth for 57,000 years. For several weeks during August and September 2003 Mars was more striking in the sky than it had ever been in recorded history. In 1877 there was also an extremely good perihelic opposition, which heralded one of the most important periods of Mars observation.

By the mid-nineteenth century, telescope building was a sophisticated affair. William Herschel had constructed a giant 36-inch reflecting telescope, and the Third Earl of Ross built an even larger 72-inch telescope at Birr in County Offaly, Ireland. Both telescopes were larger than anything that had been built previously, and signified a time of major advancement in the construction of lens-based refracting telescopes and curved-mirror reflecting telescopes. For the first time, such capable instruments revealed the nature and material make-up of the stars, and hitherto unseen distant objects categorized as *island universes* together with deeper insights into the nature of the Sun, Moon, and planets.

Mars, too, seemed to reveal itself in all its glory, and astronomers felt that they were finally developing an understanding of the planet. The first detailed maps of Mars were produced from drawings made by the leading astronomers—Wilhelm Beer, Johann Heinrich von Madler, Angelo Secchi, Richard Anthony Proctor, and the brilliant Reverend William Rutter Dawes. Those astronomers were even called *Areographers*; they were mapping an entire world and had created a new field of study—Areography.[1]

Observational standards and practices were also developed during that time. Serious observers had to be meticulous and follow a protocol aimed at reproducing factual drawings of real features on the planets. Adequate sketches were to be made at the telescope in pencil, with copious notes regarding clarity of seeing, precise coloring and shading, contrast, and so on. As soon as the observing session was complete, the observer would produce a high-quality color drawing under good light conditions, based on the sketches and notes taken at the telescope. In this way, little was left to memory or chance and accurate and consistent observations could be produced and relied upon. Through such practice, nineteenth-century telescopic observing by eye reached new heights.[2]

[1] The original Greek god of war was Ares, from whom the Roman god of vegetation, Mars, inherited his attributes of a warrior.

[2] Such practices are still in operation today and many serious observers still produce hand drawings of the planets and comets. Indeed, at fleeting moments of supreme atmospheric clarity, the human eye can witness details on the surfaces of the planets that can still be too difficult to capture on a photograph.

However, from the mid-nineteenth century to the first decades of the twentieth, telescopic observing by eye was to reach its limit. The limiting factor is our own turbulent atmosphere, which, through the largest telescopes becomes critically important and blurs our view of celestial objects to devastating effect. Significantly, that limit was reached during the period of great telescope building, and most poignantly regarding observations of Mars, which helped to catalyze some of the most controversial episodes ever to involve science and society. The outcome of that controversy, which still impacts on how we regard Mars and life in the Universe today, arguably set back the study of Mars by 60 years and exposed human errors perhaps still to be acknowledged.

The Areographers of the nineteenth century were pushing the boundary of observation. As they stretched the limits of their telescopes and of their observing practice, it seemed that in moments of extreme clarity in Earth's atmosphere, linear features could be seen on Mars. As early as 1830 Beer and Madler claimed that they saw such features, and Secchi claimed to see *canali* or channels on Mars in 1858. There was, however, uncertainty regarding the nature of such features, and even their existence: first, the observed features were at the limit of perception even with the best telescopes then available, and only when Earth's atmosphere was extremely stable; second, even features that were thought to really exist seemed to fade then reappear over periods of weeks and months.

The close perihelic opposition of 1877 brought Mars within 57 million kilometers of Earth and presented a unique opportunity to observe the planet. By then even better telescopes were at the disposal of the world's leading astronomers and Mars was at one of its best oppositions ever. The most famous observations of that year were by the Italian astronomer Giovanni Schiaparelli (Figure 12). He, too, was an excellent observer, and with his fabulous 8.6-inch refracting telescope produced the most detailed drawings of Mars of his time. They were so good that they rendered R.A. Proctor's excellent map out of date. Schiaparelli observed vastly more detail, seeing many new features for the first time. He was discovering a more detailed Mars than had been seen before, but he, too, was often perplexed by what he saw. Features seemed to fade and reappear over periods of several months. To Schiaparelli it was clear that there was more to Mars than had been hitherto considered, and there seemed to be activity there.

Schiaparelli noted many linear features, which he also termed *canali*, meaning channels in his native Italian. Moments of extreme clarity revealed minute details on Mars that appeared to be linear. The closer Schiaparelli looked, the more linear features he saw. To be certain of their locations, he devised a 62-point reference system on the globe of Mars by which he could

Figure 12: Giovanni Schiaparelli, who in 1877 identified canali (channels) on Mars. [Source: http://en.wikipedia.org/wiki/Image: GiovanniSchiaparelli.jpg. Credit: Public Domain]

precisely chart their position. However, his drawings of 1877 were presented to a skeptical world, as many other observers could not see the reported channels. None the less, as Schiaparelli was a reputable and trusted scientist, his observations were taken seriously. In the opposition of 1879, he devised a more extensive 114-point reference system, and even invented new methods of observing to maximize the contrast of feature on the Martian surface. This time he claimed to see even more channels that appeared even sharper than before. Furthermore, during the opposition of 1881, he actually claimed to witness as many as 20 different canals splitting in two—a process he compared to germination. Where he had previously observed a single straight line, he could now clearly see two. He checked and rechecked his observations to be absolutely sure that he was seeing real features, and convinced himself of their accuracy (Figure 13).

Other astronomers could not see Schiaparelli's channels, but he explained that they were not easy to see. Only under supreme seeing conditions and through exceptional telescopes could they be seen, and then only for fleeting moments at any given time. He also explained that the extravagant maps he produced were not true representations of what an observer would see—rather they were technical charts showing the positioning of the channels (Schiaparelli was a skilled draftsman). He also said that he did not know the nature of the channels, but suggested that they were probably natural features such as valleys and gorges. He proposed that the surface of Mars consisted of desert-type land and shallow seas, hence the vast array of channels that he witnessed.

An error in translation into English radically heightened interest in the

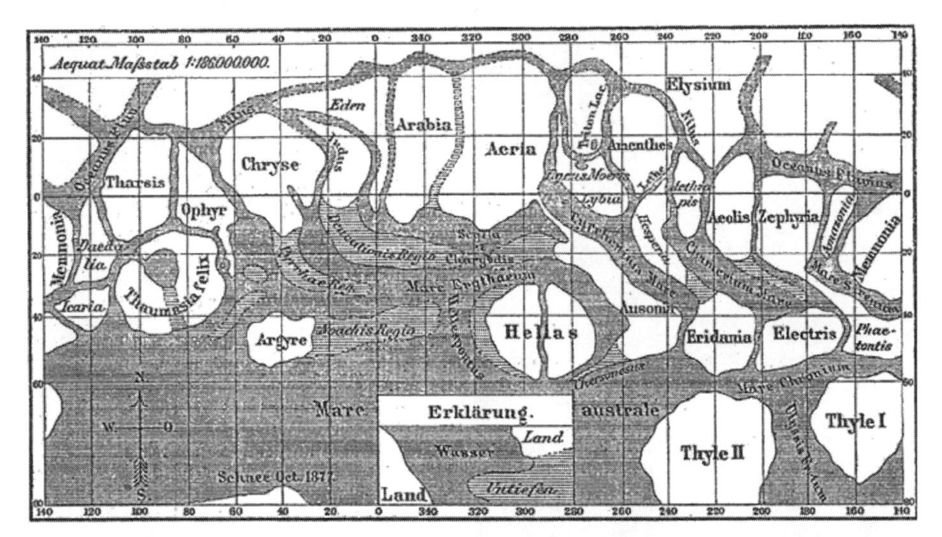

Figure 13: A map of Mars by Schiaparelli from 1888 showing his famous channels. [Source: http://en.wikipedia.org/wiki/Image:Karte_Mars_Schiaparelli_MKL1888.png//tgc. Credit: Public Domain]

affair. Secchi's and Schiaparelli's Italian term *canali* was translated into English as canals, implying the existence of artificially created waterways. The notion spread that Schiaparelli had seen canals built by a race of intelligent Martians. Schiaparelli insisted that this was unlikely and that they were most likely of natural origin. Whatever their nature, the case for canals on Mars gained substantial credibility in the mid-1880s when two major observatories—the Nice Observatory in France and the Lick Observatory in the USA—both confirmed sightings of canals precisely where Schiaparelli had said they would be. When the Lick Observatory also reported that they saw several bright patches on the terminator of Mars (the region between day and night), the press sensationally misrepresented their statement, printing that Martians were signaling Earth by giant flashing lights.[3] Interest among the public in Martin canals was intense.

The controversy of the Martian canals deepened during the opposition in 1894. A wealthy US businessman called Percival Lowell had been following the Mars story with huge interest (Figure 14). He teamed up with a controversial astronomer called William Henry Pickering with a view to verifying the Martian canals. Pickering had been dismissed from his role as a Harvard astronomer in 1892, partly for observing Mars (and making sensationalist claims) instead of observing Saturn as directed during an

[3] Those patches were actually the giant volcanoes of the Tharsis region of Mars, first revealed in all their glory by Mariner 9 in 1971.

Figure 14: Percival Lowell, who was instrumental in popularizing and perpetuating the notion of canals built by intelligent Martians.

expedition to Peru. For the opposition of late 1894, Lowell and Pickering set up a temporary observatory at Flagstaff, Arizona, with 12- and 18-inch refracting telescopes. In a lecture to the Boston Scientific Society in 1894 (several months before he had actually carried out observational work of his own) Lowell stated:

> ... Nevertheless, the most self-evident explanation from the markings themselves is probably the true one; namely, that in them we are looking upon the result of the work of some sort of intelligent beings ... The amazing blue network on Mars hints that one planet besides our own is actually inhabited now.

After only several months of observations during the 1894 opposition, Lowell drew far-reaching conclusions about Mars that differed even from Schiaparelli's idea of a planet consisting of seas and deserts. In his book *Mars* (1895) he considers the universal context for life, the historical context for the observation of the canali, published his own observations of the canal network, and provided a detailed explanation as to what he could infer of their origin and nature. He proposed that Mars was a dying planet and suggested that the dark regions on Mars were marshes while most of the remainder of the planet, being lighter in shade, was desert. Lowell proposed that since Mars was a dying planet, it was older than Earth; and since the observed canals come and go, he proposed that during summer in the southern hemisphere, for example, its polar ice cap would partially melt and provide running water to the rest of the planet through an intricate network of canals build by an ancient, intelligent, and technologically advanced race

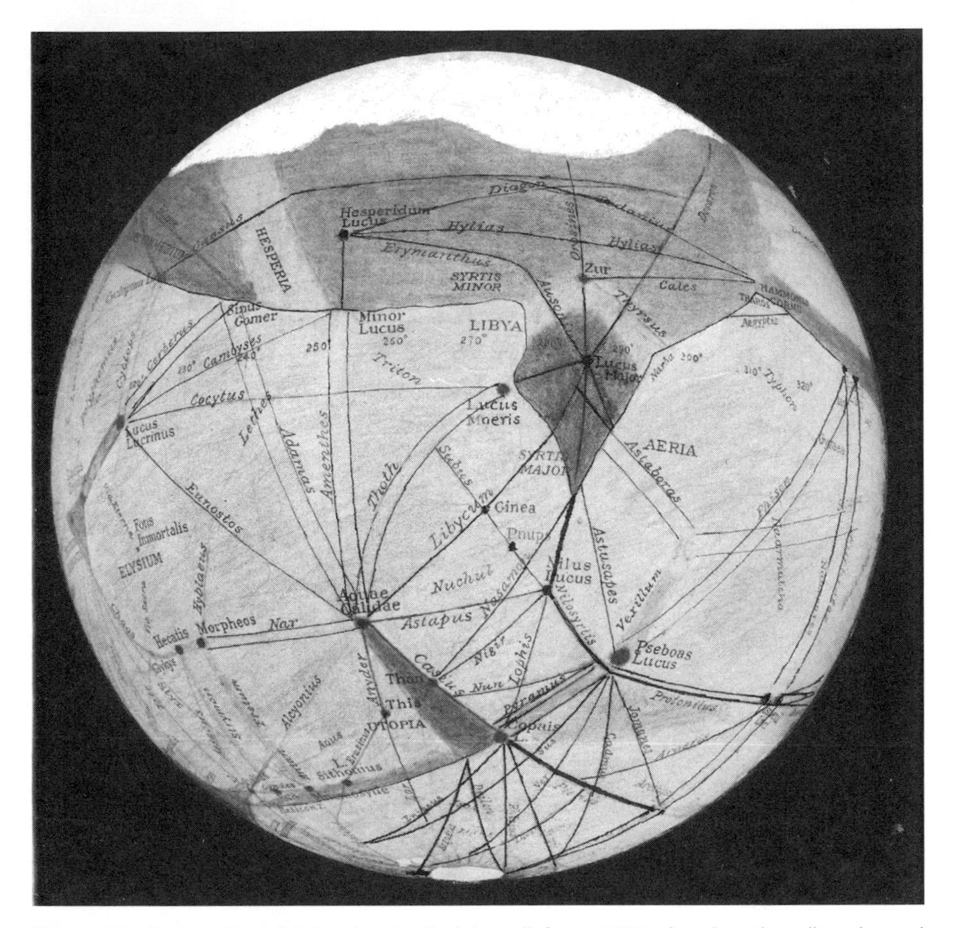

Figure 15: A drawing of Mars by Percival Lowell from 1905 showing the alleged canal network. See Plate 5 in the color section. [Credit: Lowell Observatory Archives]

of Martians. He did not propose that they were necessarily alive to day as they may have built this network in the distant past. He suggested that some of the canals were over 2,000 kilometers long, and that water would flow through them at a rate of about 3 kilometers per hour. As a Martian summer progressed, the canals would fill, and vegetation for miles on either side would be watered and blossoming, making them visible to Earth observers (Figure 15).

Lowell's conclusions were printed in newspapers across the USA, while he also traveled and lectured widely. The combination of his efforts and Schiaparelli's continued observations, as well as sightings of the canals by other astronomers, meant that by 1895 doubters were in the minority. It seemed that the Martian canals, whatever their origin and nature, were real.

Incredibly, during the oppositions of 1892 and 1894, the astronomer

Edward E. Barnard carried out observations of Mars using the Lick 36-inch refractor. Using magnifications in excess of 1,000 (usually observations were made at magnifications of 300–600), Barnard saw an altogether different Mars. He actually saw, as he stated, "a vast amount of detail—spots, patches, irregular and broken up," and where Schiaparelli had said there were canals he saw "a bewildering amount of natural features, so many that they were impossible to commit to drawing." He further stated:

> Under the best conditions these dark regions, which are always shown with smaller telescopes as of nearly uniform shade, broke up into a vast amount of very fine details. I hardly know how to describe the appearance of these "seas" under these conditions. To those, however, who have looked down upon a mountainous country from a considerable elevation, perhaps some conception of the appearance presented by these dark regions may be had. From what I know of the appearance of the country about Mount Hamilton as seen from the observatory, I can imagine that, as viewed from a very great elevation, this region, broken by canyon and slope and ridge, would look just like the surface of these Martian "seas."

Perhaps for fear of adversely affecting his career, Barnard did not publish his findings and only referred to them in passing in a paper several years later (Figure 16). What he saw, now known to be of the true Mars, went virtually unnoticed.

Furthermore, in 1894 the English astronomer Edward Walter Maunder carried out experiments that demonstrated that the appearance of the canals could be an optical illusion. He also pointed out that just because the canals were the finest details we could see with our telescopes, there was no reason to believe that those were the finest actual features on the surface of Mars. Surely, he proposed, there are even finer details that our telescopes cannot yet see, and that perhaps we see features that simply look like canals through our telescopes. Such was the intense interest in canals, however, that virtually nobody took notice of either Barnard or Maunder. They had actually succeeded in presenting a substantial explanation for the appearance of the canals *and* observational evidence of the real Mars, yet it was all ignored. Indeed, a book on Mars by French astronomer Camille Flammarion, *La Planète Mars*—among the most respected books on the subject at the time—supported the idea of canals, that Mars comprised deserts, marshes and shallow seas and possibly even red vegetation. H.G. Wells, on writing the epic *The War of the Worlds* in 1898, far from being based on fanciful notions, was based on popular scientific thinking of the time.

A change was about to take place, however. In 1896 Lowell claimed to see

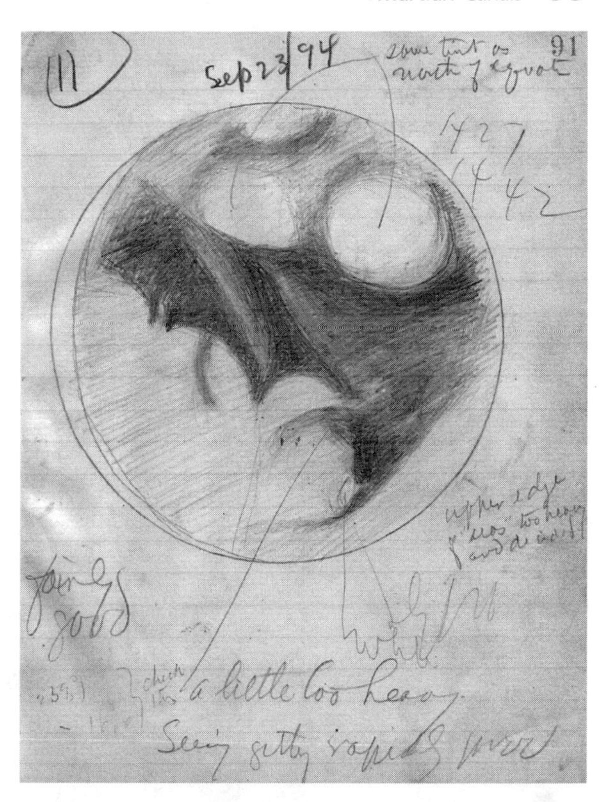

Figure 16: A drawing of Mars by E.E. Barnard from the opposition of 1894, showing no evidence of canals. [Credit: Lick Observatory]

spoke-like features on Venus. Virtually all astronomers knew that whatever the case for canals on Mars, there were no such features on Venus. Many in the astronomical community recoiled from their acceptance of canals, at least as seen by Lowell. Then, in 1897 the Italian astronomer Vincenzo Cerulli, observing Mars with an exquisite 15-inch refractor and under excellent seeing conditions, literally witnessed one of Schiaparelli's canals—Canal Lethes—break into a myriad of minute detail. Before his very eyes the illusion of the canals was exposed and the surface of Mars revealed itself as a myriad of natural features. As the nineteenth century closed, the British Astronomical Association's Mars section, under the directorship of the great astronomer Eugène Michael Antoniadi (Figure 17), rejected the notion of canals and in 1903 published a Mars chart devoid of canals altogether (Figure 18).

But the controversy was far from over. Through to 1909, Lowell claimed to have photographed the canals of Mars. Magazines and newspapers bid furiously for the right to publish, and once again there was widespread interest (the images were never good enough to print and their supposed "proof" was never confirmed). The astronomical world remained divided. Antoniadi, using the great 33-inch refractor in Meudon near Paris in 1909, studied Mars as Lowell had photographed it. Antoniadi reported seeing "a

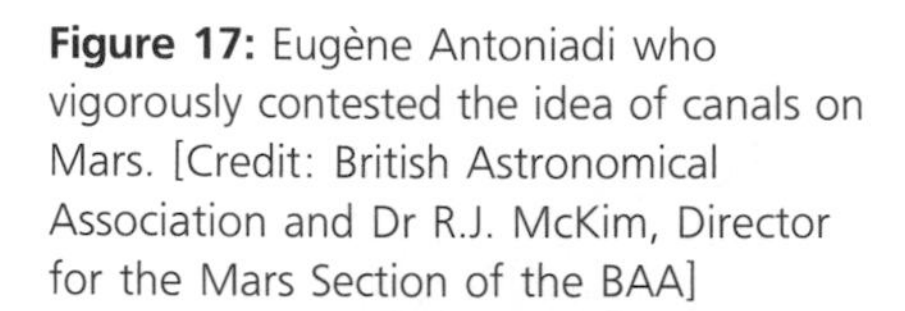

Figure 17: Eugène Antoniadi who vigorously contested the idea of canals on Mars. [Credit: British Astronomical Association and Dr R.J. McKim, Director for the Mars Section of the BAA]

Figure 18: A drawing of Mars by Antoniadi, revealing no trace of canals. [Credit: British Astronomical Association and Dr R.J. McKim, Director for the Mars Section of the BAA]]

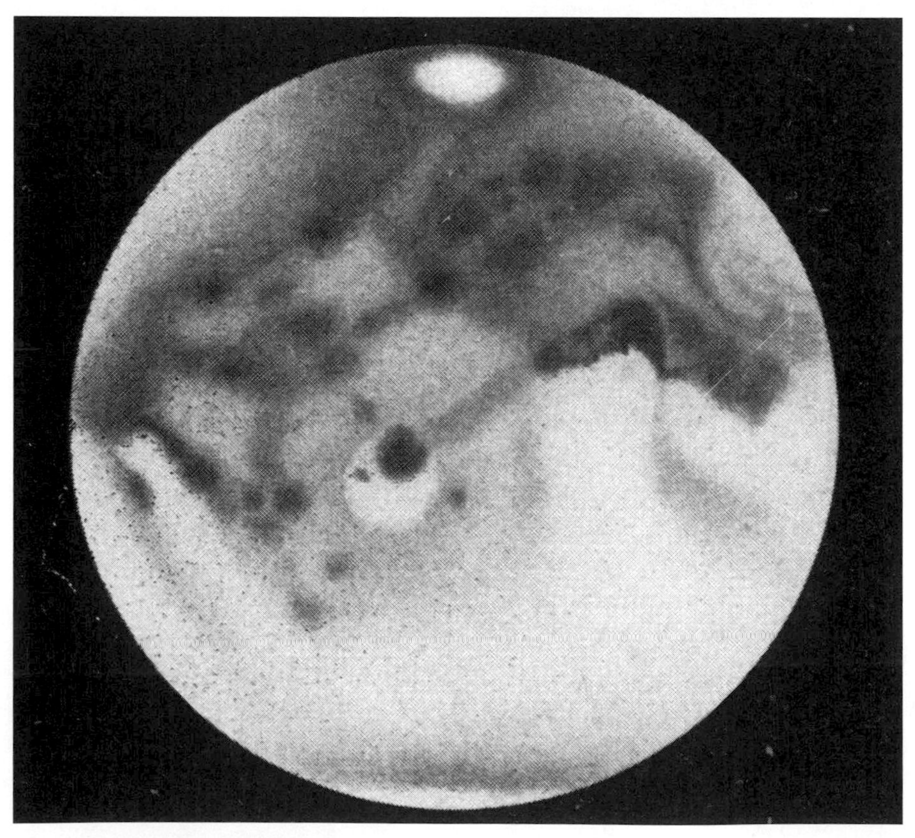

detailed, natural Mars" during moments of supreme seeing, testimony to the natural features previously seen by Barnard and Cerulli. He even charted many of them, which have been found to match the Mariner and Viking space probe images of the 1960s and 1970s.

But even this evidence could not end the controversy, and Lowell stood by his findings until his death in 1916. In fact, the noted planetary astronomer Earl Slipher, who worked at Lowell Observatory from 1908, shared Lowell's belief in canals, and reported seeing them as late as the 1950s, as did others, and even provided charts showing canals to the US Air Force when initially considering Mars exploration.

Opportunities Missed

Over the first half of the twentieth century interest in Mars faded. There were some efforts to determine the composition of Mars' atmosphere and polar ice caps, but by and large Mars was no longer a priority. Apart from a general consensus that there was little more that could be learned about Mars from visual observations alone, most astronomers in the early twentieth century turned their attention to the emerging science of Astrophysics and the myriad of exciting problem it engaged—with little development taking place in our understanding of the Solar System or of Mars for the next 50 years. Even so, the legacy of Schiaparelli and Lowell lingered. Many astronomers would not have been surprised if vegetation were found on Mars by the first space probes to visit the planet in the 1960s, for perhaps no other reason than the notion of vegetation on Mars that was proposed during the "canali" episode over half a century earlier. Why a planet should contain only vegetation and not a more realistic ecosystem was not considered until the emergence of a new generation in planetary scientists of the late 1950s, signifying a general lack of serious thought given to Mars in the first half of the century.

Even today we cannot claim immunity. It can be argued that much of our motivation for studying Mars, our view of aliens as portrayed in popular culture, and some of our motivations for searching for life elsewhere in the Universe are in part underpinned by the Mars canals episode. And, too often, references to the canals affair are extremely casual, ignoring the intricacies, treating it as a curiosity or side note, or as events that happened to a band of misguided astronomers who somehow deluded themselves. Careful consideration of the affair, however, suggests otherwise—that those astronomers were no less capable than any who came after, that today we would fair no better, and, indeed, that later complacency regarding the episode contributed to the subsequent setback in the interest in Mars.

To set the events of the time in context, we must consider a number of factors. First, we must remember that in the earlier part of the nineteenth century even the best telescopes were incapable of showing fine details on Mars. By 1877, however, better telescopes, *and* the exquisite opposition that year, presented an unprecedented opportunity to observe Mars and for the illusion of the canals to present itself most prominently. Even through Schiaparelli's excellent telescope, the ever-present atmospheric turbulence demanded extreme patience, requiring that he observe the frustratingly shimmering image of Mars for hours on end just to avail of an occasional moment when the air settled for a second or two and the features of Mars became clearer. Even then, any details revealed were on the verge of perceptibility. The smallest features discernible on the planet would have been 50 kilometers or more across and no doubt subject to the perceptual confusion Maunder had pointed out, i.e. separate features on Mars connected in the minds-eye as linear features. And we cannot ignore the subjectivity that Schiaparelli himself brought to the situation. Schiaparelli, who was severely color-blind, was observing contrast features on a red–green globe, which surely contributed to the pitfall. Furthermore, as a trained draftsman and meticulous record keeper, his maps and charts of Mars were more representative of architectural diagrams than true representations of what was actually seen while observing Mars—a point he highlighted on many occasions. Therefore, in pushing the boundary of observational astronomy, Schiaparelli inadvertently witnessed real optical effects as lineations on Mars. This, coupled to his own style of observing as well as subsequent misinterpretation of his maps, all contributed to the popularization of the idea of canals on Mars.

It is often noted that Schiaparelli's lineations do not correspond to actual features on Mars, but there need not necessarily have been any correlation between his first sketches of Mars and the actual features on the planet. The features that he committed to paper in 1877 were a selection that were randomly affected by the atmospheric conditions during observation, coupled with his unique "mind's-eye" preference toward linearity. Once Schiaparelli had committed linear markings to paper, however, a bias toward those same features (irrespective of their correlation with actual Martian features) then predominated for all subsequent observations. As soon as Schiaparelli's charts were produced, they became a template for most observers of that era. It is worth noting that the 1895 confirmation of Schiaparelli's canals by both the Nice and Lick observatories was made while the astronomers in question actually used Schiaparelli's charts at the telescope, repeatedly cross-checking his charts with their telescopic view of the planet. While they previously could see no canals, they could see them with Schiaparelli's charts available at the telescope. The fact that in 1895 many regarded the canals as reality was based on mounting (albeit flawed) evidence, not on whimsical belief or delusion.

Had this remained the case, the fallacy behind the illusion and misinterpretation of Schiaparelli's charts might have quickly resolved itself. But several other factors confused the issue and ensured perpetuation of the canals myth. First, there was the involvement of Percival Lowell. In retrospect it is clear that Lowell's science was poor even for the time. The rapidity and certainty with which he personally observed canals, and the sweeping conclusions arrived at within just a matter of months, were dubious even for their time. None the less, many of Lowell's other ideas on considering the nature of Mars as an ancient and dying planet and on the rationale of life elsewhere in the Universe were quite substantial. Lowell was a smart man, and even considered the possibility of optical illusion, but rejected it on what he felt was reasoned grounds. Although misguided, his commitment to the idea of canals on Mars, coupled to his phenomenal success at popularizing his work, ensured that their perceived existence went far beyond optical illusion and into popular culture.

There were other contributing factors. As the nineteenth century drew to a close, larger telescopes became available, such as the Lick 36-inch and Meudon 33-inch refractors that were far more capable than Schiaparelli's 8.6-inch refractor of 1877. These great telescopes were capable of revealing far more detail. While the resolution of linearity on the surface was at the limit of perceptibility for Schiaparelli, the great telescopes under exceptional seeing could apparently expose in bewildering detail the actual natural features on the surface. Yet the debate initiated in 1877 remained largely unchanged well into the early twentieth century and the progression in the capability of telescopes, and what they could reveal of Mars, seemed to be largely ignored. Barnard, Cerulli, and Antoniadi all witnessed a myriad of natural features on Mars, possibly including craters circled by wind-blown dust, streaks, gorges, and valleys. There for all to see was the true Mars, yet it went almost unnoticed. This was in part due to the dominance of Schiaparelli and Lowell in the field and also because, even for the great telescopes and at the best of times, observations of Mars were so incredibly difficult.

We must also remember that practices and standards in science and astronomy were not as rigorous at that time as they are today. As late as 1909 Lowell and Antoniadi argued over the differences between their observations from Lowell's 24-inch refractor and Antoniadi's 33-inch refractor. What might the outcome have been had both telescopes been identical, or had Lowell and Antoniadi worked on observing sessions together? Today, any comparisons would require a normalization of all conditions, yet this did not occur at the time. In essence many of the debates regarding the visibility of canals were at cross purposes: as individual astronomers observed Mars at different latitudes, at sites of different seeing, and using different telescopes, there could only be disagreement on the observations. Irrespective of the perceptive dimension to seeing canals, the physical circumstances in which the various observations were made varied so widely that they perpetuated the myth for longer than would otherwise have been accepted had standardization of work practices and observing conditions been implemented.

And we must also consider that, perhaps, the idea of life on Mars was no more far-fetched than many of the stories from new and undiscovered regions of our own planet at the time. Certainly the idea of life elsewhere in the Universe was not seen as pivotal to explaining our origin or revealing something fundamental of our true nature. The motivations for questioning the validity of the canals were different, and the importance of canals on Mars, as important as they were, did not fundamentally impact upon humanity's perception of itself in the greater scheme.

Another important dimension to the canals episode should perhaps still be acknowledged today. Barnard, Cerulli, and Antoniadi presented their evidence of Mars as a bewildering and natural environment as far back as 1892, yet their findings went unnoticed until the beginning of the space age. None of the notes of those scientists was given serious consideration by any astronomer during the first half of the twentieth century. While it is perhaps excusable for astronomers of the time to have ignored their findings in the confusion that reigned, there can be little excuse for later scientists of the twentieth century who chose to sustain the myth of the canals, or who decided that Mars could not be better understood from ground-based observations. Common astronomy texts up to 1965 make no reference to an ancient, natural landscape with craters, streaks, and valleys, yet those discoveries had been made. Astronomers of the first half of the twentieth century chose, incorrectly, to believe that Mars could not be examined properly even through the great telescopes, yet Barnard had done so as early as 1892. The available evidence was ignored and, as a result, a great opportunity was lost—to know the real Mars somewhat as we are coming to know it through our robotic space explorations. Mars was never going to be an easy object to study, but it was possible, as Barnard had shown, to identify something of its true nature some 60 to 70 years before the arrival of our first space probes. The near ridicule of studying Mars during the first half of the twentieth century could perhaps have been avoided had scientists shown a better and more realistic understanding of all the evidence. That would surely have enhanced our first space explorations of the planet, which was so dogged by controversy during the 1960s and 1970s and still reverberates today.

We now acknowledge the role that Mars has to play in answering the universal question of life. It is one of our few substantial test-beds and we are only now maturing in our views. Who knows where we would be now and what experiments we would be conducting on Mars had we taken on board what Barnard, Cerulli, and Antoniadi discovered between 1895 and 1909. They were the unsung masters of their time. In perpetuating the "myth" and "mystery" of the Mars canals episode, all who came after are as guilty as those first astronomers who could not decipher the real evidence from the bogus claims.

4 Viking

Modern Science

With the passing of Schiaparelli in 1910 and Lowell in 1916, the fervor of the Mars canals episode began to fade. The ancient, barren Mars as witnessed by Barnard, Curelli, and Antoniadi went unnoticed and the notion of canals lingered in some quarters even until the 1960s. The effect on the public psyche was, however, substantial and persists even to this day. The idea of life on another planet had been given serious thought over a prolonged period of time, setting both the Earth and humanity in a new and broader context. Despite the setback regarding interest in Mars over the first half of the twentieth century, the possibility that we may not be alone in the Universe had been seriously contemplated and is arguably a true legacy of the whole affair.

Mars received relatively little attention after 1916. There were other priorities occupying science, with no less than two revolutions in physics taking place at that time. The first involved the work of Albert Einstein in finally explaining a conundrum that had been an issue in physics for decades—an inconsistency between the laws of mechanics and the laws of electricity. With his *Special* and *General Theories of Relativity* of 1905 and 1915 respectively, Einstein solved the conundrum and in so doing rewrote the laws of nature as they were then known. Prior to Einstein it had been thought that the dimensions of both space and time were fixed and constant everywhere in the Universe. A meter was a meter to everyone measuring it, and a minute was the same for all experiencing the passage of time. But Einstein showed that this is not the case. Instead he showed that neither space nor time is fixed in any particular way, anywhere in the Universe, and that the distance between two points and the passage of time are different for different observers, depending on how they travel with respect to one another as well as their distance from massive objects such as stars and black holes. His ideas also revealed that the countless galaxies that make up the

K. Nolan, *Mars, A Cosmic Stepping Stone*,
DOI: 10.1007/978-0-387-49981-9_4, © Praxis Publishing, Ltd. 2008

Universe cannot be stationary in space, but are all in perpetual motion on a universal scale.

The astronomer Edwin Hubble, using the great 100-inch telescope at Mt Wilson in California, confirmed Einstein's "moving universe"' in 1923 by showing that our Universe is in fact expanding and growing larger moment by moment. He showed that the further away a galaxy is from us, the greater is its velocity and that each galaxy is racing away from every other. Each galaxy sees exactly the same picture—all other galaxies across the Universe are accelerating away from it. Given such universal expansion, astronomers wondered if the Universe was smaller in the past and for how long it had been expanding. Was there a time when the entire Universe was contained entirely within an unimaginably small volume? Might there have been an actual start to the expansion—a beginning to the Universe? With astronomy and physics confronted with such new, fundamental questions regarding the origin and nature of the Universe, both became inextricably linked in a new endeavor known as Astrophysics. Astronomy, a generally observational activity, now needed fundamental physics to explain its new and profound concept of the Universe. As those new astrophysicists set out on the quest for answers, few gave serious thought to the planets of our Solar System.

At the same time, a revolution through Quantum Theory was also under way. Quantum theory explains the nature of space, time, matter, and energy on the scale of the infinitesimally small. From around 1900 through to 1937, it helped to explain the hitherto unsolved mysteries of the tiniest building blocks of nature. It revealed that there are limitations in the Universe—there is a smallest amount of space, a smallest chunk of matter, and a smallest quantity of energy. Quantum theory also revealed that, at the tiny dimensions of atoms, nature is uncertain and fuzzy and obeys a set of curious, somewhat indeterminate natural laws on the scale of the infinitesimally small. Yet, for all their elusiveness, the quantum laws of nature manifest themselves on a grand scale as the world we see around us and are so comfortable with. Atoms and molecules, light and sound, water and air—all exist in their various forms because of the elusive rules of quantum mechanics. Everything we are and everything we know is built on uncertain foundations. Quantum theory even gave shape to twentieth-century technology. From being mostly an esoteric set of principles up to 1937, quantum laws were then rapidly exploited in the most ingenious ways across science and engineering, leading directly to the invention of the transistor only 10 years later. The computer—built upon the transistor—is the grandchild of quantum theory.

The Emergence of Exobiology

Quantum theory also brought about a deeper understanding of chemistry and biochemistry, both theoretical and experimental. Through our developing grasp of solid-state physics and new investigative techniques such as X-ray crystallography, the way was paved for the discovery of the general structure of DNA, the foundation molecule of virtually all life. To understand its structure is to substantively understand the nature of life on Earth, and even today we are just embarking on a determination of the precise make-up of DNA for different organisms—a task that will occupy biology for decades or more. But the determination of the basic structure of DNA in 1953 was in itself hugely important because it provided the explanation as to *how* Darwinian evolution actually happens. Evolution explained that all life originated from a single organism—a common ancestor—on Earth billions of years ago. But Darwin could not explain how evolution worked; all he could do was show that it happened. The discovery of the structure of DNA changed that, however, providing clues to the mechanism that actually drives evolution. DNA was found to be a *self-replicating* molecule, containing *genes* that manifest themselves as *traits* in organisms. All species have DNA but they have different genes and so possess different traits. Reproduction could now be understood through DNA replication and the origin and diversity of organisms explained as a consequence of random changes in copies of DNA giving rise to changes in the traits of offspring. Over billions of years, and in an ever-changing Earth, such random changes produced widespread differences in organisms and hence the origin and diversity of species. Our understanding of the structure of DNA had finally explained a key mechanism in evolution, and unified evolution with the entire biology of our planet.

There were other developments in life sciences during the first half of the twentieth century. Even as far back as the 1920s the Soviet scientist Aleksander Oparin and the British scientist J.B.S. Haldane independently suggested that ultraviolet light from the Sun or lightning discharges might have caused the constituents of Earth's original atmosphere (carbon dioxide, methane, ammonia, and water among others) to react chemically and form organic compounds even to the complexity of amino acids, nucleic acids, and sugars. If this was the case, then this process of organic synthesis might have been the first important step in the origin of life on Earth. In 1953 Stanley Miller and Harold Urey (Nobel prize winner in 1934 for the discovery of heavy hydrogen) actually verified Oparin's and Haldane's hypothesis by simulating such an atmosphere and subjecting it to an electric discharge, resulting in the production of significant amounts of water-soluble organic compounds including amino acids. In 1959 Miller and Urey further realized that with organic synthesis possible in any similar atmosphere devoid of oxygen, such as those of Venus and Mars, life was also conceivable. Further, Joshua Lederberg, the 1958 Nobel prize winner for Medicine (for generic recombination and the organization of genetic material in bacteria), led the way in emphasizing the connection between microbiology and biochemistry, stating that "biochemistry consummates the unification of biology revitalized by Darwin." He also pointed to common structural units of all life on Earth—amino and nucleic acids, carbohydrates, etc.—suggesting that with such underlying global commonality to all life and a common chemical origin of all the planets, the existence of life on some of them was possible.

Almost as soon as the underlying structure of life began to reveal itself, leading figures in conventional chemistry, biology, and medicine quickly recognized a natural context for the origin of life on Earth and a link with a universal context for all life. They also realized that both might be knowable and explorable, in particular by searching for and studying life elsewhere in our Solar System. If, for example, independently originating life on Mars was seen to be similar to life on Earth, then it would verify the prevailing ideas regarding the origin of life. If, on the other hand, life on both planets was seen to be very different, it could provide insight into advancing our theories on origins and reveal a more general context for the operation of all life.

By the late 1950s and early 1960s, a sophisticated set of ideas regarding the nature of the Universe, Earth, and life were emerging. We had a broad grasp of the origin and evolution of the Universe through Einstein's relativity and were beginning to understand how galaxies, stars, and planets formed, through pioneering observational astronomy. We had uncovered the building blocks of matter and energy through quantum theory and were also developing an integrated picture on how the continents, oceans, and atmosphere of our own planet began, as well as taking our first steps toward an understanding of the origin and nature of life itself. At that time, however, each area of science operated largely on its own, but as the twentieth century progressed many sciences began to work together. Astrophysics was a combination of physics and astronomy; geophysics provided a deeper insight into the mechanical working of our planet through geology and physics; and the field of exobiology started to make an appearance through a combination of astronomy, biology, chemistry, geology, and medical science, among others.

The discovery of DNA and a biochemical context for life allowed the question of the origin of life on Earth to step beyond hypothesis and become a true experimental science. How organic material combines to form DNA and how it happened on Earth billions of years ago were questions that could now at least be considered for investigation, exploration, and experimentation. However, for meaningful answers we would also need to know about Earth as a planet: how and when it formed, the composition of its surface and atmosphere in the beginning, how it interacted with the Moon and its planetary neighbors, what our Sun was like when life formed, and so on. Biological and chemical questions regarding the origin of life increasingly called upon astronomy and geology when formulating questions, designing experiments, and in seeking meaningful answers.

Serendipitously, astronomy was making similar strides forward. Of course Mars brought with it its historical legacy. Even in the 1960s Barnard's and Antoniadi's descriptions of a complex and intricate Mars remained largely unknown and the general perception of Mars was still overshadowed by the legacy of the canals episode, the existence of which had still not been fully refuted. So, while the emerging science-based quest for the origin of life on Earth and search for life elsewhere brought Mars once again to the fore, the legacy of ill-conceived ideas about that planet prevailed. Such were the advances in other areas compared to planetary astronomy that by the 1950s, of the 1,000 or so professional astronomers living in the USA, only 20 were engaged in planetary science. Most considered that all that could be learned about the planets from telescopic observations had already been learned. Indeed, even in the 1950s there were only 14 telescopes larger than 1 meter in diameter, with six of those made prior to 1920, but there was some progress none the less. Increasingly scientists did not take the canals seriously, and Antoniadi's exquisite Mars global map gained wide acceptance among serious astronomers. Among the few serious planetary astronomers in all of the USA was G.P. Kuiper, a formidable scientist who pioneered the area of Planetary Science through the application of rigorous modern scientific methodologies when observing the planets through the greatest telescopes of the time. Interestingly, his observations, and those of others, continued to reveal coloration changes on Mars. Some proposed that they could be due to vegetation, where changes in color indicate seasonal changes in growth. But, while Kuiper pursued a quite conventional path—examining planetary surfaces and atmospheres—a small number of younger astronomers began to make the connection between planetary science and the question of life elsewhere in the Universe, as well as realizing a deep and ancient connection between life on Earth and its astronomical context.

Frank Drake, for example, pondered the possibility of detecting other intelligent civilizations in our Milky Way galaxy. He even devised experiments to search for such extraterrestrial civilizations via revolutionary new giant radio telescopes, which could readily detect transmissions by another civilization. Carl Sagan, Iosif S. Shklovskii and others considered the question of origins and a universal context for life, and also realized that there may be an ancient commonality among Venus, Earth, and Mars. They noted that all three planets share a similar place and history in the Solar System and yet have become very different worlds. They also questioned why Venus became a boiling planet with a runaway greenhouse effect; why Mars became a cold barren world; and why Earth became a haven for life. These pioneering planetary scientists brought into sharp focus just how little was known about any of the planets, including the Earth, and they realized that there was a deep-rooted link between the origin and abundance of life and its planetary context. Soon, planetary science became (in part) the quest for understanding the cosmic nature of life, with Earth's planetary processes as well as the evolution and make-up of the other planets now recognized as of central important. The questions of the origin of life on Earth and of its cosmic abundance were increasingly seen to be the same question approached from different perspectives. But with such poor facilities and virtually nobody involved, any hope of realizing answers would require a seismic shift in approach.

Space Race

In tandem with our ability to pursue fundamental questions about the Universe, the Cold War gathered strength through the 1950s with the USA and the USSR applying the very same science to develop nuclear weapons capable of eradicating all life from our planet, as they applied in a ferocious space race. We thus became a space-faring civilization on October 4, 1957, when the USSR launched and successfully placed into Earth orbit the Sputnik-1 satellite. From then until the end of the 1980s, both nations repeatedly tried to outperform each other with ever more daring ventures—the first man in space, the first spacewalk, the first man on the Moon, and so on. And although vast billions of dollars were spent and motivations were often questionable, truly astounding feats in space flight and exploration were achieved.

Of course, as soon as space flight became a reality, a new possibility for studying the planets emerged—that of actually traveling to them. Such possibility was lost on no one. Eisenhower saw the sociological significance of a civilian space program; and with the emergence of exobiology, conventional and planetary scientists recognized that investigations could now be carried out by sending robotic space probes to the planets. So convinced were the scientific community of the value of robotic planetary exploration that a new Space Science Advisory Board was established to present the case and to act as a unified voice. Further advocacy came from W.H. Pickering (*not* Lowell's compatriot!—but the first Director of the Jet Propulsion Laboratory or JPL), who pointed out to NASA the central importance of exploring the planets—not only because great feats in planetary exploration would showcase US technical superiority over the USSR, but also because the civilian space program, and even NASA itself, might be jeopardized in the face of military space priorities if no planetary exploration was undertaken. Urey and Lederberg also advised NASA on the importance of studying the Moon regarding the origin of the Earth, and of planetary studies regarding the origin of life. The vision of Eisenhower and an overwhelming validation from JPL, Nobel Laureates, and planetary scientists was enough. Within just a few years of its formation, NASA acknowledged the importance of both space science and planetary exploration and in 1960 set up its Office of Life Sciences, launched a 10-year program of planetary exploration, and made Mars exploration of prime importance.

Of course, exobiology and space exploration were both still in their infancy and in an effort to legitimize both areas, NASA sponsored an exobiology and planetary exploration *Summer Study Camp* in 1964. The camp served two purposes: (1) to sharply define imminent exobiological and planetary exploration issues and priorities, and (2) to provide respectability to the subject in order to attract the new talent. The outcomes of the camp were substantial, not only because key priorities for planetary exploration in general were identified but also because specific objectives for the upcoming Mars program were also determined. While, it was argued, there was no firm basis to assume life on Mars, the possibility provided a mechanism to explore the uniqueness of life on Earth—stating "At stake is our place in the Universe and hence the exploration of our close neighbors is of immense importance." It was also argued that although we knew little more about Mars than had been known in the early part of the twentieth century, conditions there were not totally non-conducive to life as we know it. Despite our legacy with misgauging Mars, life there remained a possibility. But the camp also recognized that to fully understand Mars there would have to be a far-reaching program of exploration to fully characterize its surface and chart its evolution, as well as to search for life past and present. It was also emphasized that the rewards of the search were not dependent on detecting life, but simply on understanding the planet to completion, and that would be of immense value in its own right.

Voyager

From the beginning it was realized that a search for life elsewhere required a robust definition of what life is—and not one based on "Earth Chauvinism," as Carl Sagan put it. Further, such a definition would also need to point to a basis for the exploration on Mars—that is, experiments must be devised to search for such life on Mars. In general there were six different scenarios to consider regarding the possibility of life on Mars:

- life similar in biochemistry and evolution to that on Earth
- life of similar biochemistry but not having evolved
- life based on carbon but of different biochemical structure
- life based upon different chemistry (such as silicon)
- extinct life
- no life at all.

While it was impossible to verify, the leading exobiologist Wolf Vishniac pointed out that there were good grounds to consider that any life on Mars would at least be based on carbon. Of all the elements, carbon is the most capable at reacting with itself as well as with other elements to form complex structures capable of storing information and preserving continuity over generations. Even if based on carbon, however, any devised tests would have to be able to look for certain attributes associated with life. A broad yet robust definition of life was therefore proposed (and still stands today). Life was declared to be that which is capable of reproduction, metabolism, growth, irritability (response to stimuli), and evolution.

For Mars, in particular, it was decided that two key attributes should be sought out: growth and metabolic activity. It was also decided that detection systems should look for microbial life with a cellular and genetic character, given the global nature of cellular structure to all life on Earth. And so, with a cautiously generalized definition of life based on five key attributes, as well as a chemical basis for life on Mars, the foundation upon which to design biological experiments had finally emerged. NASA immediately put out a request for proposals for organic chemistry and biological experiments, with the very first being appropriately awarded to Wolf Vishniac. Eventually, five experiments were selected and designed with a view to sending a sophisticated exobiological lander mission to Mars:

1. *The Wolf-Trap*—designed by Wolf Vishniac. Here, a sample of Martian soil would be dropped into a solution of known acidity. If growth occurred, the acidity would change causing the solution to cloud, thus verifying the presence of living organisms.

2. *The Pyrolytic Release experiment*—designed by Norman Horowitz. This experiment involved an enclosed chamber containing an artificial Martian atmosphere with the addition of small amounts of radioactive carbon. A soil sample would then be added and if organisms were present and used the air, the radioactive carbon would act as a tracer and allow for their detection.
3. *Gulliver* (Labeled Release experiment). Similar to the Pyrolytic Release experiment, nutrients containing radioactive carbon added to a Martian soil sample, if used in metabolism by microorganisms, could be detected when released into the enclosed atmosphere as gaseous by-products.
3. *Gas Exchange experiment* (GEX). This worked in two modes—first, by providing moisture and carbon dioxide to a dry soil sample to see if metabolic activity occurred; and second, by adding moisture and even more nutrients than with Gulliver and again monitoring the atmosphere for traces of radioactive carbon released from the soil.
4. *Gas Chromatography Mass Spectrometer* (GCMS). This highly sensitive organic chemistry experiment would not test for life but instead detect organic molecules to one part per billion, conclusively verifying the presence of even trace amounts of organic matter in the soil.

With major exobiological priorities defined and even key issues of planetary contamination and protection being addressed though stringent space probe sterilization techniques, a serious Mars program was finally started. Given the name Voyager, the aim of the program was to set one and possibly two landers onto the Martian surface by 1970 with a view to directly searching for life there. In the meantime, however, simpler spacecraft would be sent to Mars to conduct planetary surveys and to develop the required technology to get to Mars in the first place. Incredibly the Jet Propulsion Laboratory, with just a few years prior experience, could by then, in under a year, turn out a new space probe that was capable of traveling to either Venus or Mars. Thus, Mariner 4, the first space probe to successfully reach Mars, duly arrived at the planet on July 14, 1965.

Flying within 13,000 kilometers of the planet and transmitting just 21 low-quality images back to Earth, the images received from Mariner 4 immediately dealt a near fatal blow not only to the Mars program, but also to NASA's planetary program, and indeed the entire field of exobiology. The images comprehensively eradicated the previous hundred years' legacy of Mars observations, including virtually all contemporary ideas. Although they covered just 1% of the planet, the images clearly revealed a surface littered with impact craters similar to those on our Moon (Figure 19). Mars,

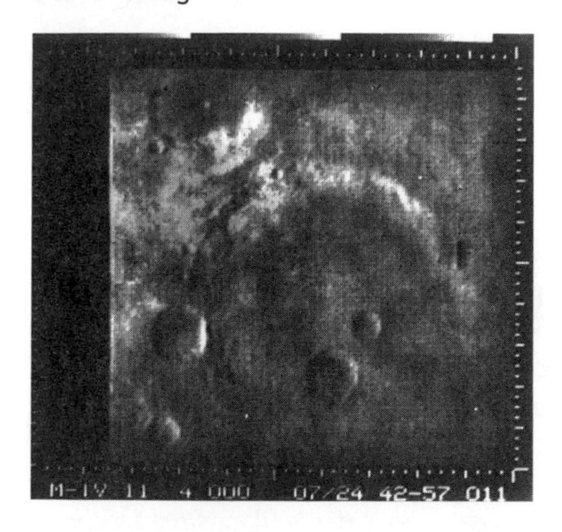

Figure 19: The eleventh picture of Mars from Mariner 4 (in "raw" state) taken through a green filter from 13,000 kilometers and showing a crater 120 kilometers in diameter in the Atlantis region. [Credit: NASA/JPL]

according to Mariner 4, was similar to our ancient and desolate Moon and nothing like Earth. The chaotic and ancient terrain observed by Barnard, Cerulli, and Antoniadi 70 years previously was now devastatingly revealed to all. And with craters in excess of 100 kilometers in diameter, NASA scientists realized that Mars' surface had probably remained unchanged for hundreds of millions, if not billions, of years. There never had been a race of canal-building Martians. There could never have been vegetation, and it seemed that there never was any running water, let alone any life on the surface of Mars. It was a completely dead planet. The Mariner 4 results were a devastating blow both to canal supporters and exobiologists. Major criticism came from within NASA as well as from the wider public regarding the expectations set by the exobiological community. But, in response, the exobiologists reminded all concerned that the rewards of Mars exploration were not dependent on the hypothesis of life on that planet, as they had stated from the beginning. Further, even with the Mariner 4 results, life *was* still possible on Mars, albeit in more basic form. With significant unknowns about the planet remaining, Mars exploration was still of immense value. The exobiological and planetary science community were robust and also correct in their reasoning; and having weighted the situation NASA agreed to continue with its Mars program, and the Voyager program in particular.

Many issues dogged Voyager by then, however. First, there were major concerns about being able to adequately sterilize the proposed exobiology payload. Second, there were significant problems with actually building a rocket powerful enough to send such a heavy payload to Mars. The financial burdens of the Vietnam War also added to NASA's woes, leading to budget cuts from the government that placed a great strain on all its programs. And even the validity of the Voyager program itself was called into question. Bruce Murray, then Chairman to the Caltech–JPL Planetary Studies Group, pointed to the huge risks of sending a sophisticated and costly payload to a surface that had not even been mapped; suggesting that a phased program of detailed planetary reconnaissance followed by lander missions was more appropriate. All of these issues, combined with the bitterly disappointing results from Mariner 4, led to the demise and cancellation of the Voyager program in 1967.

Despite the setback, NASA administrator James E. Webb immediately pulled out all the stops to save NASA's overall planetary program. He reminded government that to shelve the planetary program would be to jettison 30,000 man-years of experience by then gained from sending space probes to Venus and Mars. Webb was so convincing that President Johnson personally assured him that the planetary program would not be abandoned; and within just several weeks of the cancellation of Voyager a *Plan 5* extension to the planetary program was announced, leading to the launch of two more Mariner-class flyby missions of Mars in 1969, two Mariner orbiters in 1971, and even a brand new exobiological lander program called Viking.

Mariners 6 and 7

The Mariner 6 and Mariner 7 flybys of Mars in 1969 were almost as dramatic as Mariner 4. As with Mariner 4, both probes flew past the planet and did not enter orbit. They were substantially more capable than Mariner 4 however, returning more than 1,000 photographs. But they provided the final reality check regarding any possibility for advanced life on Mars or of finding a current Earth-like planet. With widespread cratering confirmed and absolutely no sign of plate-tectonic activity across the planet, it was clear that Mars' surface had remained motionless and unchanged for billions of years (Figure 20). Also, with an atmosphere measured at around 10 millibars (about 1% as dense as Earth's atmosphere) there could be no hope of finding liquid water on its surface today. Carl Sagan and one of his students, James Pollack, demonstrated that the changes in coloration thought to be caused by vegetation could equally be explained by the movement of dust in the Martian atmosphere. Most conclusive of all, however, was the verification that the majority of the ice at Mars' north and south poles is made of frozen carbon dioxide, not frozen water, and if carbon dioxide exists as ice, then the conditions on Mars are simply too cold for life.

There were other intriguing findings. Images revealed a gigantic object on the surface at the exact location traditionally called Nix Olympica, but whose character could not be fully discerned (Figure 21). Also, vast tracts of what could only be described as chaotic terrain were spotted in the northern

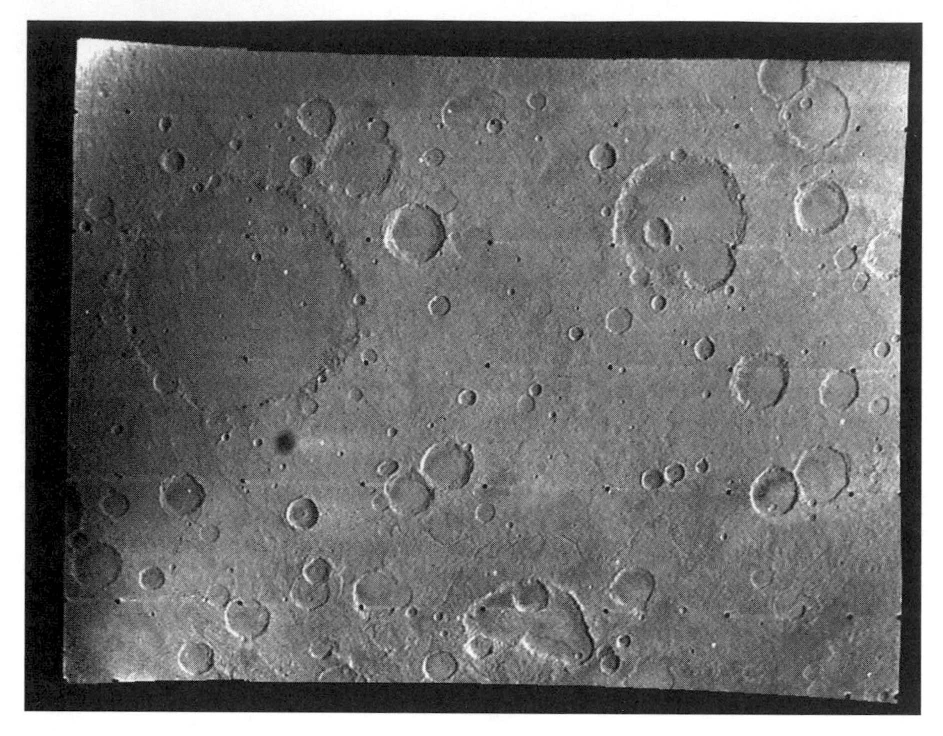

Figure 20: Mariner 6 confirms, with great clarity, an ancient cratered Martian surface as indicated by Mariner 4. [Credit: NASA/JPL/Piotr. A. Masek]

hemisphere, perhaps hinting at some past indigenous planetary activity. The superior quality images also showed that the edges of many craters were smooth as if eroded; and a giant crater basin, called Hellas, in the southern hemisphere was seen to be the smoothest place known in the Solar System. Mariners 6 and 7 revealed an ancient and desolate planet, but they not only suggested activity on Mars throughout its history, but that it is not as dead as had been inferred from Mariner 4 images.

The overriding realization from Mariners 6 and 7, however, was one of resignation that any hope of finding substantial habitats was now completely gone. The entire canals saga—and even the (perhaps) more realistic hope of finding vegetation—were never more than wildly over-optimistic aspirations. With such resignation emerged a new maturity in approach, but neither planetary scientists nor exobiologists were ready to give up. They could now start to explore the real Mars—a planet almost completely unknown to us, a planet with a great deal to teach us, and a planet still close enough in character to Earth, despite their differences, to perhaps have something to say about microbial life.

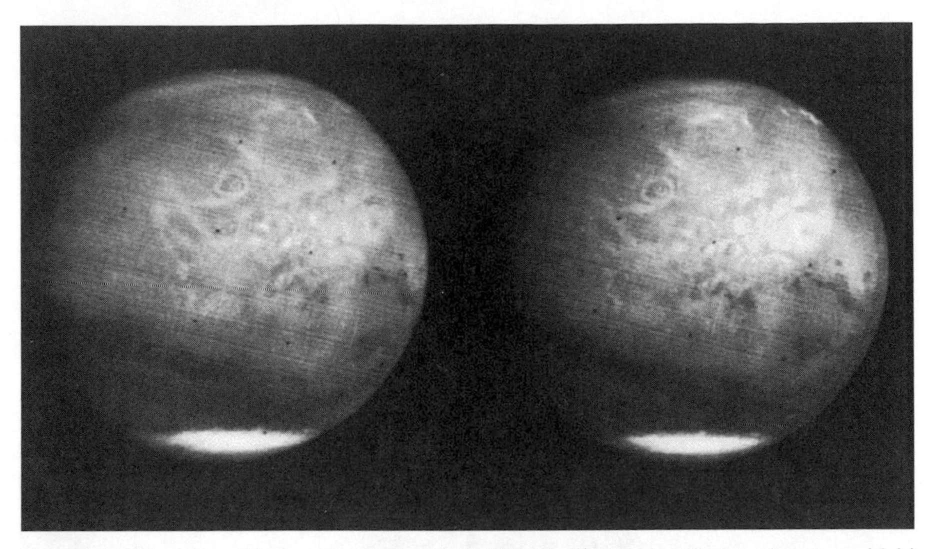

Figure 21: Mars from Mariner 7, showing Nix Olympica (later identified as the giant shield volcano Olympus Mons); and Polar Caps. [Photographed from 300,000 kilometers. Credit: NASA/JPL]

Mariner 9

Despite all that had been learned about Mars from Mariners 4, 6, and 7, the next space probe to visit Mars—Mariner 9 in November 1971—presented yet another startling new perspective on Mars. This time a Mariner probe would orbit Mars for several years and carry out a detailed analysis of the planet; and thus Mariner 9 became the probe that changed our entire understanding of the planet.

Even as Mariner 9 arrived at Mars it confirmed that Mars was far from dead. The first images sent back showed a dust storm that encased the entire planet. It lasted for over a year, demonstrating a Mars with an active climate. The mysterious changes in features seen on Mars over a century before were finally explained—vast dust storms that occur regularly with seasonal changes on the planet. Peering above the vast dust storm, a number of enormous volcanoes were seen, the largest of which was identified as Nix Olympica. Renamed as Olympus Mons, it revealed itself to Mariner 9 as a truly gigantic volcano—the largest in the Solar System. It is so big that its base would cover all of France, while its summit rises 27 kilometers above the Martian surface—over three times the height of Mount Everest. The discovery of Olympica Mons and the other giant volcanoes of the Tharsis region were significant because they verified that Mars had been volcanically active in its past.

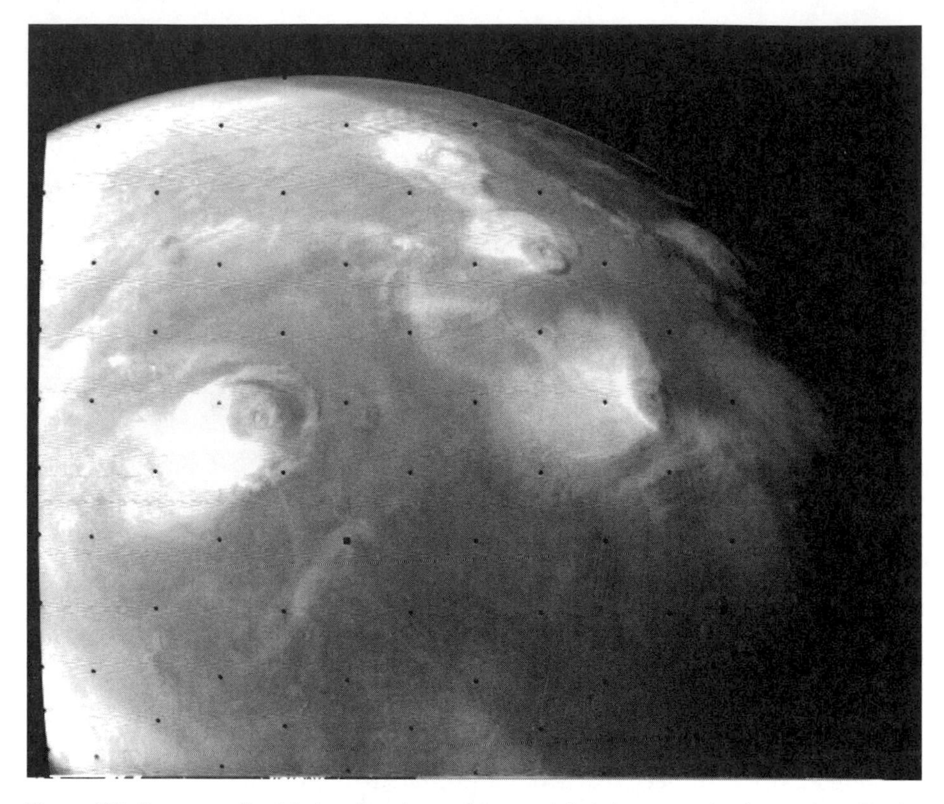

Figure 22: One year after Mariner 9 arrives at Mars a global dust storm begins to clear. To the left is the gigantic volcano Olympus Mons, while further right are the three volcanoes sitting upon the Tharsis Bulge. [Credit: NASA/JPL/Piotr. A. Masek]

A full year passed before the planet-wide dust storm finally settled, and Mariner 9 began to properly photograph the planet from orbit (Figure 22). The images sent back were worth the wait and were once again a revelation, showing layered structure near the polar regions indicating recent climatic change and, critically, countless winding channels that looked just like dried river valleys (Figures 23 and 24). The implications were far-reaching: if water had flowed across the surface of Mars at one time, the planet must have also possessed a much thicker atmosphere. In total over 7,000 images were transmitted back to Earth from Mariner 9 and literally transformed our view of the planet from the barren, lifeless landscape of Mariner 4 to a planet that looked as if it had had a very active past for upwards of a billion years—where volcanoes created and replenished a dense atmosphere and where, perhaps, even liquid water carved countless river channels and filled ancient lakes and seas. Also discovered was the most breath-taking of all feature on Mars—an enormous canyon over 4,000 kilometers long that looks like a gigantic

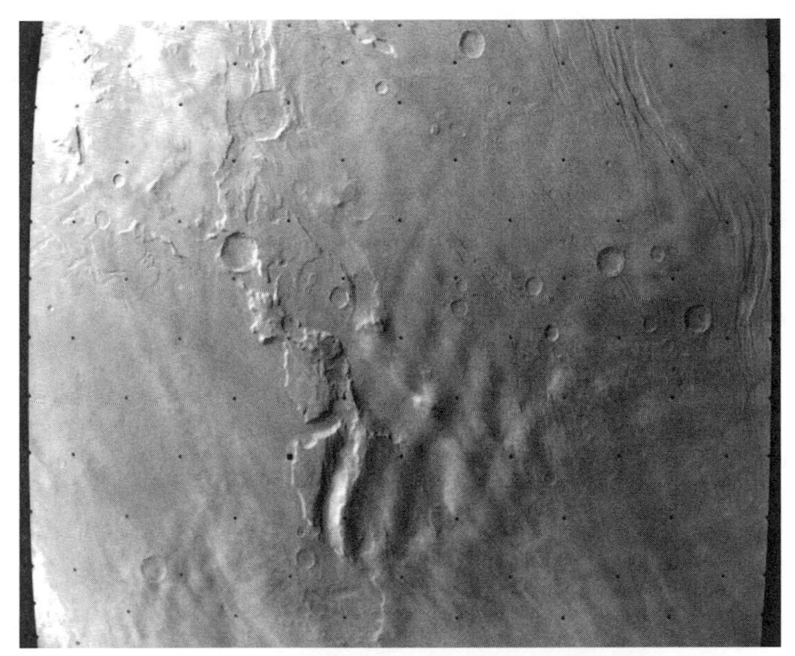

Figure 23: Mariner 9 images the Mareotis Fossae and Tempe Terra region of Mars, showing evidence of ancient tectonic and volcanic activity. [Credit: NASA/JPL/Piotr. A. Masek]

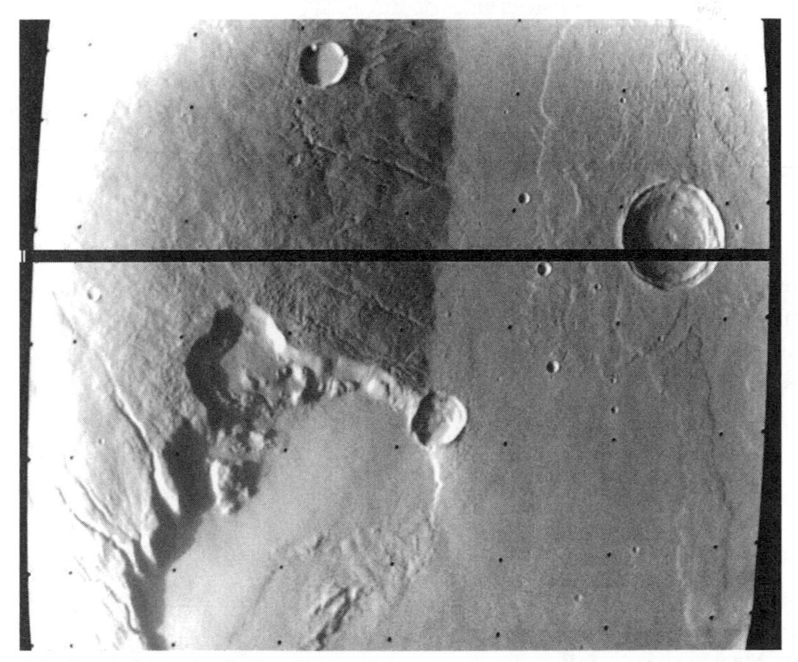

Figure 24: Channels on the flanks of the volcano Hecates Tholus as seen by Mariner 9. Various radial channels have been interpreted as erosional ash channels, lava channels or channels eroded by fluvial processes. [Credit: NASA/JPL/Piotr. A. Masek]

rip in the planet. It was unanimously agreed that that this continent-sized valley network should be called Valles Marineris, in honor of the astounding achievements of Mariner 9 and indeed the entire Mariner program.

With wind and dust storm activity, huge variations in temperature, perpetual bombardment from solar radiation, and large variations in atmospheric pressure all occurring today, Mars had finally revealed itself as a planet with a rich and detailed past and a complex and energetic present. The possibility of life having existed there in its distant past, and perhaps even of microbial life today, were possibilities that once again could not be ignored. Despite a century of conjecture and a decade of turmoil, some semblance of the true Mars was finally realized and it pointed squarely at a possibility that few dared hope for—that of far-reaching answers from Mars regarding the origin of life and the possible presence of microbial life there today.

Viking

Even as Mariner 9 was on its way to Mars, the Viking program was in full development under the leadership of Jon Martin. One of the major issues to be resolved was that of the selection of a landing site, placing great demand on Mariner 9 to deliver images that would make that possible. In this, Carl Sagan provided a significant contribution, chairing a brain-storming session in 1970 from which 35 candidate sites were selected. However, as Mariner 9 images proved to be of too low a resolution, Sagan urged for more extensive studies of the Martian surface using Earth-based radar systems. As Sagan pointed out, a surface resolution to 100 meters tells you nothing about smaller scale features that could inhibit a safe landing. And so through extensive radar analyses of the Martian surface in 1973, two sites were announced for two Viking lander missions: Chryse, a valley at the mouth of Valles Marineris, and Cydonia, a low-lying smooth plain upon which there would be a better chance of detecting any present atmospheric water.

At any other time in history, it might have taken many years, perhaps even decades, to make the leap from planetary orbiter to planetary robotic lander. But such were the prior preparations during the Voyager program—the ferocity of technological development of the era and the vast budgets available during the Cold War years—that NASA accomplished the task in just five years. In less than 19 years since a bleeping Sputnik satellite circled Earth, NASA successfully sent four robots—two orbiters and two landers, each over half a tonne in mass—to Mars in 1975. By all standards, it was a significant achievement.

Vikings 1 and 2 launched on August 20 and September 9, 1975, respectively, but as Viking 1 arrived at the planet in the summer of 1976, landing site selection became a major concern once again. Upon settling into orbit the Viking 1 orbiter photographed the surface and revealed, *yet again*, an entirely different world to that imaged by Mariner 9 (Figure 25). Areas of the surface appearing featureless, flat, and smooth to Mariner 9 were now often seen through Viking's superior cameras to be of rugged surface relief and containing a myriad of small craters (Figure 26). Both of the chosen sites were clearly too treacherous and too risky to be used as landing sites. With only a month to set Viking Lander 1 down before Viking 2 arrived (and demand the full attention of the team), Jon Martin took the brave decision to delay the first landing (scheduled for July 4, 1976) while the team rapidly decided upon two new sites. With the superior images from the Viking 1 orbiter, two new sites were chosen within just two weeks: Chryse Planitia (the Plain of Cold), further north of the original Chryse site, and Utopia

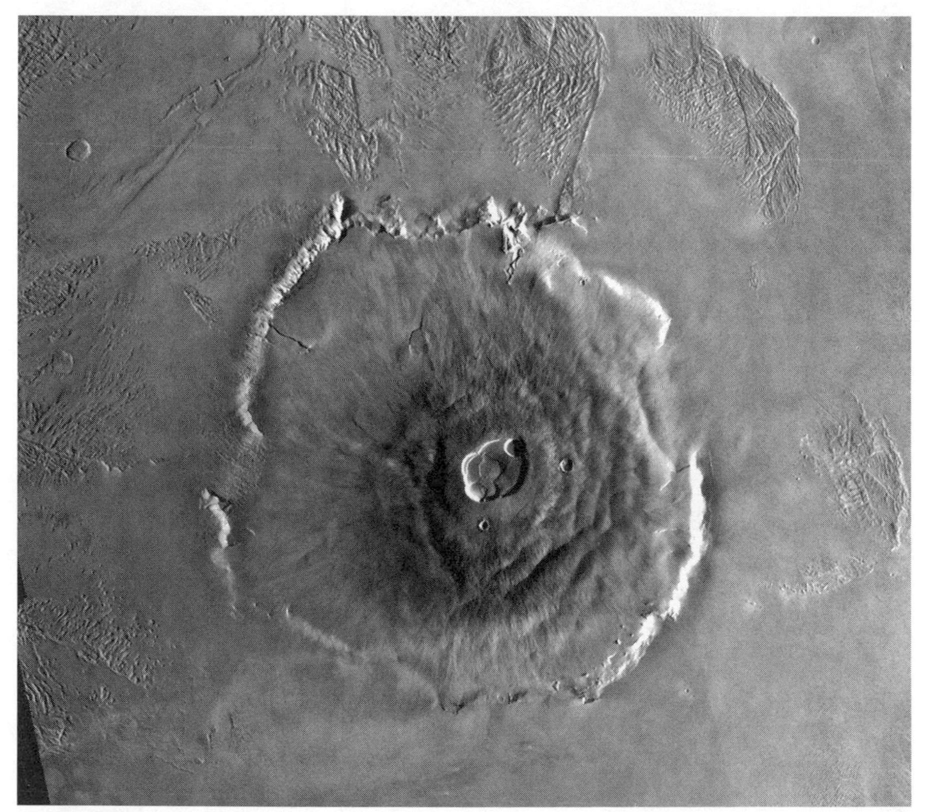

Figure 25: Viking 1 orbiter photo-mosaic of Olympus Mons, which is 600 kilometers in diameter and 27 kilometers high. The entire volcano is flanked at its base by 6-kilometer-high cliffs. See Plate 6 in the color section. [Credit: NASA/JPL]

Figure 26: Mars as seen by the Viking 1 orbiter. The center of the scene shows the entire Valles Marineris canyon system over 4,000 kilometers long and up to 8 kilometers deep, extending from Noctis Labyrinthus system of grabens in the west, to the chaotic terrain in the east. See Plate 7 in the color section. [Credit: NASA/USGS]

Planitia, half way around the planet and further north at 48 degrees. After a massive flurry of activity, Viking 1 landed perfectly on Mars on July 20 (Figures 27 and 28), with Viking 2 following suit on September 3. Over the next six years (until 1982) the Vikings carried out a monumental study of the entire planet that was to revolutionize our entire perception of the planet—all over again!

Viking Surface Science

By the time the Viking landers had set down, Bruce Murray was Director of the Jet Propulsion Laboratory (JPL), replacing W.H. Pickering (who had served in that position from 1954 to 1976). NASA science and exobiology

Figure 27: The first photograph of the surface of Mars, taken by Viking 1 just minutes after the spacecraft landed. The center of the image is about 1.4 meters from Viking lander's camera. [Credit: NASA/JPL]

Figure 28: Sand dunes and large rocks characterize this panorama view of Mars by Viking I. The horizon is approximately 3 kilometers away. [Credit: NASA/JPL]

were also by then sophisticated affairs with no less than 72 scientists organized into 13 teams working on Viking mission science. Apart from equipping the Viking orbiters with sophisticated high-resolution color cameras that sent back over 57,000 images of the planet, each Viking lander was equipped with a dazzling array of equipment to photograph the surface in color, determine the chemical make-up of the atmosphere, monitor the weather, and carry out chemical and biological analyses of the soil and rocks. No less than five onboard mini-laboratories—two chemistry and three biology—were placed on each lander, designed specifically to look for evidence of microbial life.

Through budgetary cutbacks, however, Wolf Vishniac's Wolf-Trap had been dropped from the Viking project. Vishniac was devastated, not only because of his passion for the endeavor but also because so much of his time and dedication had been committed to this one-off project over the past decade. True to the end, Vishniac continued with his life-science research, even traveling to the coldest and driest deserts of Antarctica to search for evidence of microbial life in such hostile environments. Ironically, while there he actually used his Wolf-Trap to verify that life did indeed exist in Earth's toughest environments. Tragically, however, Vishniac never returned from his expedition to Antarctica, apparently having fallen down a ravine to

Figure 29: A boulder-strewn plain stretches to the horizon at Viking 2's northern hemisphere landing site in Utopia Planitia. Superficially it resembles the Viking 1 landing site in Chryse Planitia. See Plate 8 in the color section. [Credit: NASA/JPL]

his death. It was an extremely sad if most meaningful end for one of the great pioneers of the twentieth century in exobiology.

Despite having dropped the Wolf-Trap, the Viking landers brought to Mars the other experiments designed for the Voyager program—the Pyrolytic Release, Labeled Release and Gas Exchange biological experiments as well as sophisticated organic chemistry detectors. Both landers also provided an unprecedented scene to humanity—that of the surface of another planet (Figures 29, 30, 31). Strip by strip, the first color pictures of the surface unfolded—first to the NASA scientists in JPL and within just hours to the world through TV news broadcasts. What was revealed was a red, barren landscape that looked intriguingly like an ordinary desert scene from Earth, with one exception—it had an orange sky! The Viking landers transformed our view of Mars forever. It would no longer be a place of elusiveness and mystery. Through the Vikings it became a real, tangible place—somewhere we could relate to and a place of realistic possibilities. While the results from the lander cameras were enthralling, the same could not be said for the results from the biological experiments. At first all three

Figure 30: Viking 2 landing site at Utopia Planitia, showing a thin coating (less than 1/100 millimeter thick) of water-ice on the rocks and soil. Atmospheric dust picks up solid water and carbon dioxide, becoming heavy enough to sink to ground and form the visible layer. See Plate 9 in the color section. [Credit: NASA/JPL]

Figure 31: A Martian sunrise captured by the Viking 2 lander on June 14, 1978, on the lander's 631st sol (Martian solar day). [Credit: NASA/JPL]

experiments seemed to actually provide positive results, suggesting the existence of organisms in the Martian soil. The Gas Exchange experiment, for example, detected changes in its chamber, suggesting that something in the soil was breaking down the provided moisture and nutrients. The Labeled Release experiment showed similar positive results, as if something in the soil was also using up its provided nutrients. Finally, several weeks later, on August 7, 1976, Norman Horowitz also announced that the Pyrolytic Release experiment had yielded a positive result, although not a conclusive one.

While the results were there to be seen, drawing a conclusion that life had been detected was a different matter. First, the Gas Exchange experiment released about 200-fold more oxygen than could be accounted for, suggesting that the oxygen was coming from the soil itself rather than from biological behavior, although biological behavior could not be discarded. The Labeled Release experiment suggested that the gas released was similar to that released by terrestrial bacteria absorbing the supplied nutrients; but again reactions with the soil could not be discarded. Also, while the Pyrolytic Release experiment suggested biological activity, a simpler chemical reaction could not be ignored. Although the biological experiments were inconclusive, the GCMS organic detector dealt a massive blow to the prospect of finding any biology at all. It found, conclusively, that the Martian soil contained less organic matter than our lifeless, desolate Moon (which retains trace levels of organic matter from impacting comets and asteroids). The GCMS, capable of detecting organic matter to one part per billion, found absolutely none at all; suggesting that not only did organic material not exist on the surface today, but it also appeared that the billions of years of ultraviolet radiation from the Sun had converted the soil into a super-oxidizing material that actively destroys organic matter.

With such devastating data, Horowitz in particular became convinced that the results from the biological experiments were simply chemical in nature. He pointed out that, in any case, with the Gas Release experiment, the interesting reactions observed were due only to the added moisture, and the later added nutrients in part quench the reaction. He also pointed out that the reactions observed with the Labeled Release experiment could be accounted for if there were peroxides on the soil, as was suggested from the extent of sterilization from the GCMS. Despite all of this, however, the Pyrolytic Release experiment from Lander 2 did not reveal the same results as from Lander 1. Indeed its results were more indicative of biological activity—and critically—at levels that could not be discounted by the GCMS organic detector at that site. Even here, however, Klaus Biemann, molecular analysis team leader with the GCMS experiment, pointed out that typically hundreds or thousands of times more organic matter is found during typical tests and that, overall, with a parts per billion lack of organic matter at either site, it was difficult to envisage how microorganisms could inhabit the Martian soil.

Overall, the results were both inconclusive and hugely disappointing, to the point of leaving the Viking team somewhat bewildered as to their next move. It had taken 16 years or more to develop the exobiological arguments and experiments for Viking, and the results provided answers neither one way nor the other nor a clear path forward. Gerald Soffen, Carl Sagan, and Joshua Lederberg remained open to the possibilities of life on Mars because of the continuing unknowns, while Norman Horowitz became convinced that there could now be little chance of any life existing anywhere else in the Solar System, and that there was no point in any further exploration of Mars.

There were, however, other intriguing results from Viking. The orbiters revealed the true extent of the dried river channels covering the entire planet. They also revealed what looked like remnants of vast flash-flood channels and plains, while some the basins on Mars looked like the remains of ancient lakes and seas. The landers also made a pivotal discovery: locked in a permafrost below the surface of the entire planet is a gargantuan reservoir of water.

Through the late 1970s and early 1980s, the Viking mission presented complex and perplexing perspectives on Mars. On the one hand, the entire surface of the planet is currently hostile to life, yet on the other there may exist a vast frozen reservoir of water and substantial evidence of a past that could almost have been Earth-like. Such diverse perspectives were reflected in the lack on consensus across the scientific community at the time. Yet, despite the jadedness of many of the exobiologists from the Viking effort, others saw the possibilities that Viking had presented. Carl Sagan and Bruce Murray, for example, although originally approaching Mars from very different standpoints, both now realized that Mars had yet to reveal most of its secrets and that life on Mars was not just a question of life on the planet at present but also a question of life in its distant past. They pointed out that it is not possible to draw conclusions regarding an entire world from just two sample locations and monitoring just the top half of one meter of soil. There was a lot more work to be done before any conclusions could be drawn.

Perhaps the legacy of Viking is that it finally brought humanity, after 100 years of tortuous effort, to the realization that Mars is not a simple place. It is complex; but it is a place, and we can go there. Viking pushed us into a realization that questions regarding life are not easy. From here on we would have to be more objective and more sophisticated on how we think about life—and Mars exploration. From that perspective, Viking was more of a success than we could have imagined from the outset. Viking revealed a Mars that is a vast and complex world with a four-billion-year history and dynamic past. Through Viking, Mars finally pointed the way for the right types of questions about the actual evolution of life and the Solar System, perhaps in a way not even possible with Venus and Earth. With both those planets having long since wiped clean from their surfaces any evidence of their early history, Mars still offered an ancient territory, a laboratory and a stepping stone for humanity to tackle the profound question of the origin of life. While Carl Sagan, for example, could see such opportunity, others could not, and 16 years would pass before the next successful mission to the Red Planet was launched.

5 The Legacy of Viking

The direct search for biology on Mars had ended in controversy. With Mars looking utterly hostile to life on the surface, there was persistent confusion over the biological experiments, and with no imminent possibility of firm conclusions, both the exobiological basis for exploring Mars and the prospect of further exploration ran aground.

The vulnerability of pursuing exobiological experiments based on little supporting evidence had been exposed. No severe decisions were made to halt Mars exploration or to castigate its exobiological basis. On the contrary, many now recognized what an interesting yet still largely unknown planet Mars had revealed itself to be. But it had taken 20 years to develop the exobiological arguments and experiments to search for life, and it was now simply impossible for even the (by-then) developed exobiological community to respond quickly enough to the confusing Viking findings to chart the way forward. With such ambiguity among the Viking community and no NASA capability to follow up, the pursuit of life on Mars and the origin of life through planetary exploration were off the agenda for the foreseeable future. It may be that Earth and Mars shared a common chemistry in the beginning; it may also be that carbon-based life with a not dissimilar genetic structure and microbiological morphology arose on Mars; but for now there could be no way of pursuing any of these questions to discover if they had any basis in reality.

Although the exobiological strategy for Mars had encountered an impasse, the Viking orbiters (whose imaging cameras were at one stage under serious consideration for cancellation) almost immediately began to suggest an entirely fresh beginning for Mars exploration. In their four years of operation, until Viking 1 orbiter ceased operating in 1980, no less than 57,000 images were relayed back to Earth, providing an exquisite and unprecedented view of the entire planet. The Viking image set represented our first comprehensive study of the planet with sufficient quality to determine something of the history and true character of the planet. With images of the entire globe, and at various resolutions from global to

K. Nolan, *Mars, A Cosmic Stepping Stone*,
DOI: 10.1007/978-0-387-49981-9_5, © Praxis Publishing, Ltd. 2008

kilometer *regional* scale (and in some cases even down to 10-meter scales), we could finally begin to study and analyze the planet, relating its surface features to the planetary processes that created them.

Mars Revealed

From the very first Viking orbiter images it was clear that Mars was more varied and complex than had been revealed through Mariner 9 (Figures 32 and 33). Craters, dried river valleys, ancient flood channels, volcanoes and rift valleys, vast low-lying plains, ancient highlands, and seasonally varying polar ice caps all pointed to a planet we did not yet understand and had not envisaged. One of the most striking characteristics to emerge was a great north–south divide or dichotomy, with each of the northern and southern hemispheres showing radically different geological character to the other, as though two entirely different planets had been joined together at the equator (Figure 34). The southern hemisphere was seen to be dominated by heavily cratered highlands, clearly an ancient landscape from the period of heavy

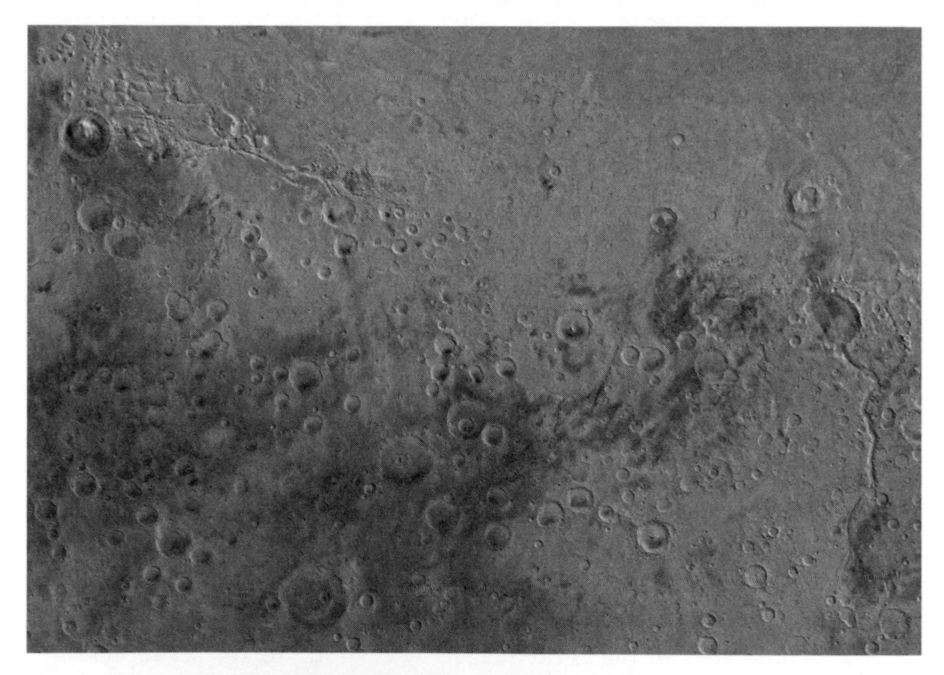

Figure 32: The Aeolis region of Mars. The southern highlands are separated from the northern plains of Elysium Planitia by a highly dissected scarp. Ma'adim Vallis, an ancient channel, is seen running into Gusev Crater on the far right. See Plate 10 in the color section. [Credit: NASA/JPL/USGS]

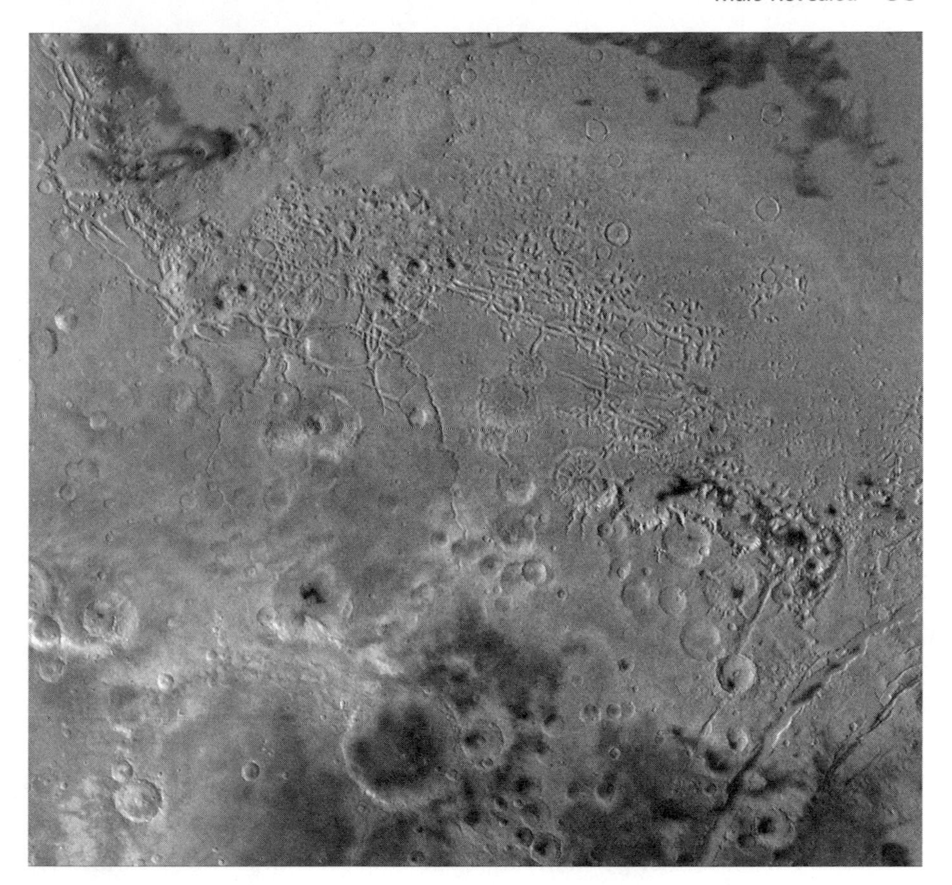

Figure 33: The Nilosyrtis Mensae region of Mars showing heavily cratered highlands in the south separated from the northern lowland plains by a belt of flat-floored valleys, mesas, buttes, and the channels. See Plate 11 in the color section. [Credit: NASA/JPL/USGS]

bombardment. Excavated from the highlands is a gigantic impact basin called Hellas. At about 2,000 kilometers across, its floor was seen by Viking to be one of the smoothest surfaces in the entire Solar System. From its sheer size, there could be little doubt that the impact that created Hellas must have been a significant event in the history and evolution of Mars.

By contrast the northern hemisphere was seen to be dominated by flat, low-lying plains devoid of significant cratering and apparently lying at several kilometers below the datum (Mars' mean radius, or, its sea level—if it had seas). With such a scarcity of craters, the northern hemisphere must be younger, perhaps created where indigenous planetary processes either eroded or filled its ancient craters. Also evident across the north were vast tracts of *chaotic* terrain with no analogous feature on either the Earth or the Moon.

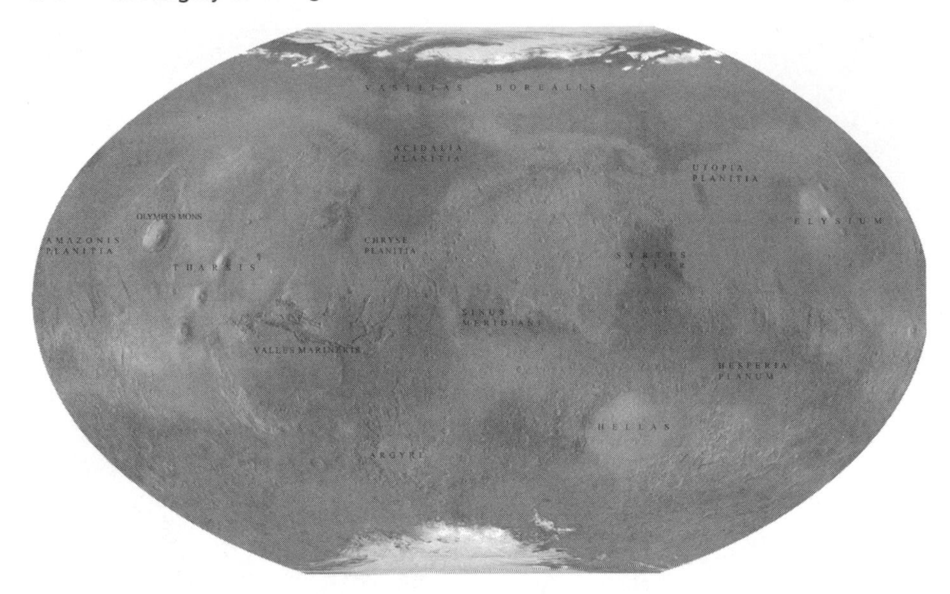

Figure 34: A map of Mars based on MGS MOLA data, showing the major regions and features. North is at top. Notice in particular the heavily cratered southern hemisphere, and the smoother northern lowlands. See Plate 12 in the color section. [Credit: NASA/JPL/MGS MOLA Group]

Although initially unexplainable, one possibility is that such terrain may have emerged slowly over the eons with a slow, creeping movement of planetwide near-surface permafrost, suggesting great quantities of water-ice bound into the surface.

Further evidence of Martian permafrost came from *lobate* craters. Unlike craters on the Moon which show extensive radial distributions of debris blasted from the impact site, many craters on Mars show surrounding layers of deposition, with each layer flanked by a ridge as if formed by debris flowing smoothly from the point of impact; again suggesting a near-surface permafrost that flows smoothly when heated by an impacting comet or asteroid.

Viking also revealed widespread evidence of ancient features possibly created by the surface water. Across the ancient southern highlands we can see what look superficially like dried river channels. Some meander for hundreds of kilometers while others show extensive dendritic features. Some even cut into crater walls on one side and exit through the other, suggesting crater ponding. Even more enigmatic are ancient and gigantic flood channels across the equatorial regions of the planet, which largely point northward (Figure 35). The most extensive of the flood channels emanate from the eastern end of Valles Marineris where they turn northward toward

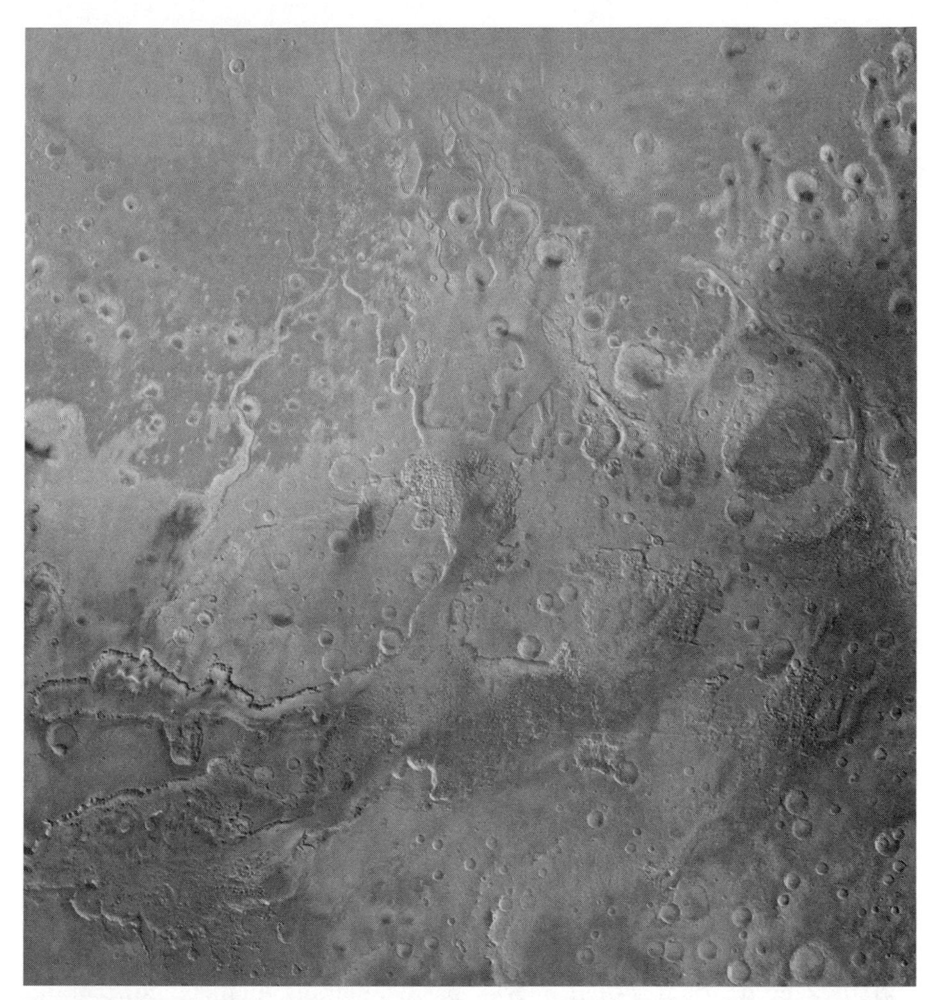

Figure 35: Chaotic terrain at east Valles Marineris from which outflow channels Shalbatana, Simud, Tiu, and Ares Valles (left to right) drained north into the Chryse Basin billions of years ago. [Credit: NASA/JPL/USGS]

Chryse Planitia. Some of the flood channels are truly colossal, suggesting flooding on a scale perhaps 10 times greater than any occurring on Earth. And some single floods may have involved a discharge of 100 times the annual discharge from the Mississippi river.

Viking orbital data also showed significant evidence of ancient tectonic activity as well as past and recent volcanic activity. To the west of the planet and straddling the north–south dichotomy is Tharsis—a great bulge in the planet stretching over 4,000 kilometers across and 10 kilometers in height. Sitting upon Tharsis are the culprits—three gigantic volcanoes arising from internal stresses on the crust and whose lava outflows helped to create the bulge. Also, with a notable absence of large craters over the entire region, it appears that Mars has remained volcanically active even to the current geological era. On the northern rim of Tharsis is the vast volcano called Alba Patera. Unlike any volcano on Earth, Alba Patera rises just 5 kilometers above the surrounding landscape yet, having ejected as much as 10 times more lava than any volcano on Earth, stretches over 1,500 kilometers from side to side. West of Tharsis is the most spectacular volcano of all—Olympus Mons. At over 27 kilometers in height (over three times the height of Mount Everest) and over 500 kilometers across, it is the most colossal volcano in the entire Solar System. Its sheer size and lack of craters suggest that it has been active throughout all of Mars' history and perhaps even to the present day. While the Vikings showed no evidence of tectonic plates as on Earth, they did reveal widespread evidence of tectonic activity in Mars' distant past. For example, emanating radially from all sides of Tharsis is a network of tectonic rifts reaching out to over a third of the planet. The eastward fault, Valles Marineris, is over 4,000 kilometers long, over 7 kilometers deep, and hundreds of kilometers wide, dwarfing Earth's Grand Canyon by comparison (Figure 36). On the eastern hemisphere and also straddling the north–south dichotomy, we find another bulge, Elysium, also thousands of kilometers wide and dominated by shield volcanoes.

Other evidence suggested that Mars had an active past. Sedimentary layers seen within the walls of Valles Marineris indicate an episodic mass movement of material—whether eolian (wind), aqueous, or volcanic in nature—over hundreds of millions of years. Also, Mars craters were not like those on the Moon: while Moon craters that were made even three or four billion years ago remain perfectly preserved, those on Mars are heavily eroded. Their walls are often worn down or rounded, their central peaks are missing, and debris fills their interior basins, making them shallow; all this points to substantial eolian based erosion from a denser atmosphere in Mars' distant past.

The Viking orbital images showed a planet that was very different to the one we see today—a once active planet where tectonic and volcanic activity may have driven planetary systems, including running water, and a dense atmosphere comprising carbon dioxide and other volatile materials. Intriguingly, much of that activity may even have been similar to that occurring on its sister Earth at the same time. The Viking orbiter images suggested that Mars was perhaps not a total mystery after all—and may not have always been hostile to life. Perhaps it had been similar in important ways to Earth in the very beginning—two planets of similar material composition, in similar regions of the Solar System, and undergoing similar planetary processes from material differentiation to surface water and

Figure 36: Part of Candor Chasma in Valles Marineris. The geomorphology is complex, having been shaped by tectonics, volcanism, and wind and perhaps by water. See Plate 13 in the color section. [Credit: NASA/JPL/USGS]

atmospheric activity. Perhaps they are different today not because they have always been so, but because two planets with some similarities and some differences started out broadly similar, but followed ever divergent evolutionary paths through history?

Immediately after Viking, some inertia seems to have set into the space community. There were several attempts to send more orbiters back to Mars—the Soviet *Phobos* probes in the 1980s and the US *Mars Observer* in the early 1990s—but all failed. And so, with a lack of an integrated program and real impetus from the space-faring nations or space agencies, it was not until the arrival of Pathfinder in 1997 that we successfully returned to Mars with real and long-term intent. This, however, did not mean that post-Viking there was a lack of interest in Mars from planetary scientists and exobiologists; on the contrary, Viking only increased their appetite to investigate the planet. From the huge archive of data made available by the

Vikings (including literally millions of atmospheric and climatic measurements made by both landers), the planetary science community slowly began to piece together a broad outline of the history and evolution of Mars. Careful geological analysis of a multitude of surface features revealed the broad nature and extent of past tectonic, volcanic, water, and atmospheric activity. Also, because of the unprecedented quality and coverage of the planet, they could conduct a stratigrapic analysis of the surface for the very first time, revealing the relative ages of the various terrains and features. By the early 1990s, we had extracted as much from the Viking dataset as was possible; and although not achieving definitive answers about Mars' early history and evolution, the results pointed firmly to a vibrant and enigmatic past. It could now be seen that there had been three great ages in Mars' history. The first, called the Noachian period, lasted perhaps a billion years after Mars' birth and was characterized by heavy bombardment as well as significant planetwide indigenous processes, perhaps leading to a dense atmosphere and surface water activity. Subsequently, it seems that Mars underwent a devastating transition during an age called the Hesperian period, where over hundreds of millions of years Mars' interior cooled and collapsed, virtually curtailing all activity on the surface. Finally, over the past two to three billion years and up to the present day in what we call the Amazonian period, Mars became an increasingly dormant world, belying its ancient legacy.

The Viking orbiter results brought about a new realism in our perception of Mars. And with such intriguing evidence of a rich and vibrant past, they also brought about brand new scenarios about the origin of life there that could not have been conceived previously. The possibility of life arising on Mars in its distant past had been given a new, yet more grounded, basis—and one that could not be ignored by the scientific community. To examine this basis, let us consider the history of Mars as suggested by the Viking orbital investigations.

The Noachian Period

About four and a half billion years ago Mars began to take form from the same protoplanetary disk as did its sister Earth. Over a period of 100 million years or so Mars gathered material from the debris field, slowly emerging as a rocky planet capable of gravitationally binding all its material into a near perfect globe about 7,000 kilometers across. With a similar origin to Earth, both planets formed with essentially the same materials—carbon, oxygen, silicon, and metals such as iron and aluminum, with smaller amounts of lighter elements such as hydrogen and helium.

During Mars' formation, a process of material differentiation occurred—

separating heavy from light materials, resulting in a dense core, a less dense mantle, and a crust of lighter silicate-based minerals. Internal heat generated by gravitational contraction and from radioactivity may have triggered internal convection currents, where molten material deep within the planet heated, moved toward the surface, cooled, and slowly descended back into the bowels of the planet. An electrical dynamo, caused by moving metals within the convective currents, may have produced a planetary magnetic field that extended into space and protected the surface from harmful cosmic and solar rays. Meanwhile comets and asteroids pounded the surface for millions of years, covering its entire surface with impact craters. One particular asteroid, perhaps 100 kilometers or more across, pounded the southern hemisphere, creating the giant Hellas Basin. It blasted enough material to cover the entire planet and raise surrounding regions by several kilometers.

Two sites in particular became major seats of tectonic and volcanic activity on early Mars—Tharsis to the west and Elysium to the east. Over the ages both have become continent-sized bulges on the surface, created perhaps by internal tectonic pressure and also from accumulating lava flows. Extensive tectonic rift valleys emanate from Tharsis in particular, reaching out to over a third of the planet. The enigmatic Valles Marineris is a spectacular example of one such rift valley lying along the eastern Tharsis fault. Without evidence of tectonic plate movement from the Viking orbital data, it appears that volcanoes at both sites remained at their original locations and grew larger with every eruption throughout history, eventually to become the colossal volcanoes we see today. This is also testimony to repeated stress from within the planet and to volcanic lava flows over billions of years. With relatively few craters at Tharsis and Elysium, both regions appear to have remained geologically active until recent times, suggesting that Mars may still be evolving.

Volcanoes also acted as pressure vents for the final phase of material differentiation—the pressurized release of the lightest materials of all, water, carbon dioxide, and other volatile materials. Outgassing for perhaps millions of years may have produced a dense atmosphere comprising carbon dioxide, nitrogen, ammonia and methane, among others. Ice and possibly liquid water could have arisen on and near the surface for long periods, supplied by water ejected from within the planet and from comet bombardment. Tectonic activity, volcanic rejuvenation, and comet and asteroid impacts, coupled to geothermal and solar energy, could have contributed to surface conditions that supported a significant cycling of volatiles through planetary reservoir interacting with one another and with the crust, compensating for and adjusting to changes in the planetary environment.

Tantalizingly, the orbital data returned by Viking was not sufficient to allow us to determine the precise nature, extent, and duration of such early activity. We could not, for example, determine the density and make-up of its early atmosphere or the length of time it persisted. Nor could we determine the actual details of surface water activity or of the cycling of volatiles. The precise planetary governance of Mars' ancient activity would remain indeterminate from Viking data alone. Critically, however, the analyses did suggest scenarios for Mars with new and far-reaching implications to our understanding of the origin of planets and of life itself.

In one scenario, for example, a carbon-dioxide atmosphere may have emerged on Mars that was dense enough to create a mild greenhouse effect, raising surface temperatures to those capable of sustaining liquid water on the surface. A resulting planetary water cycle might even have supported precipitation levels similar to Earth's average annual rainfall today. River networks would have arisen across the planet, carving out valleys and even creating a great northern sea. For millions of years in its early history, Mars might have been a spectacular watery world, just like its sister Earth.

Alternatively, Mars may have had a dense atmosphere, but one incapable of generating a greenhouse effect sufficient to allow for liquid water. In this scenario the surface would have remained cold and dry from the beginning. Ice sheets may have covered parts of the planet, with wind-based erosion and deposition dominating the exposed landscape. Surface and subsurface activity could still have been widespread however, where geothermal energy coupled to volcanic and impact replenishment of volatiles could have sustained an atmosphere, subsurface aqueous and hydrothermal systems, and even subglacial rivers, lakes, and seas. Even on such a relatively cold Mars, cycling of the volatile materials through their natural reservoirs and complex planetary chemistry could still have occurred.

Other scenarios were also conceivable. Mars may, for example, have periodically warmed in response to external factors such as occasional large asteroid or comet impacts, or to changes in its axial tilt and orbit about the Sun. These, coupled to tectonic, volcanic, and geothermal activity, may have periodically tipped Mars from being a cold and dormant world to a warmer, active one.

The Hesperian Period

Whatever the original scenario, Viking data suggested that the interior of Mars began to cool after about a billion years, spelling disaster for any chance of a vibrant future. At only one-tenth the mass of Earth, Mars could not sustain the internal heat needed to drive convection, tectonic, and volcanic activity. With ever fewer volcanoes, the atmosphere would no longer be replenished, consigning it to become ever more tenuous. If Mars had originally been warm, rain would now quickly extract much of the atmospheric carbon dioxide, dissolving it in seas and lakes and permanently precipitated out as carbonate sediments. Any greenhouse effect would now collapse, freezing all surface water into the crust as permafrost or causing it to sublimate (change from ice directly to vapor) into the ever more tenuous atmosphere.

Other factors could also have contributed to the removal of the atmosphere. Impact blasts would have ejected significant amounts of the atmosphere into space; and with no internal dynamo driving a magnetic field, ultraviolet and particle radiation from space could now penetrate the atmosphere, breaking its constituent molecular gases to their atomized state, and allowing them to rapidly react with the crust or escape into space.

Even in a scenario where Mars had always been a relatively cold world, a similar fate awaited it. Here, the cooling of the planetary interior and cessation of volcanic activity would likewise have stopped atmospheric replenishing, geothermal energy and volatiles that drove subsurface or subglacial activity. However Mars had lived out its early years, it was destined to become ever more quiescent with the collapse of its internal heat source.

As we have come to expect from Mars, yet another Martian puzzle was revealed by the Viking orbital data. While the supposed river channels to the south of the planet formed during the Noachian period, the giant flood channels leading to the northern lowlands were determined to be much younger—perhaps by as much as a billion years—suggesting that even as Mars was undergoing a devastating and downward transition, forces on a planetary scale triggered gargantuan floods across the planet. Today we still do not know the cause of the flooding, but evidence clearly shows that Mars has seen significant activity throughout much of its history and not just in its earliest years.

Even the great floods of the Hesperian period came to an end, however, and in the millions of years that followed, virtually the entire atmosphere dissipated through chemical bonding with the crust and by leaking into space. Today the Martian atmosphere is perhaps less than 1% as dense as it was in the beginning—equivalent in density to Earth's atmosphere at an altitude of 30 kilometers. With such a catastrophic loss of atmosphere, the surface would for ever more endure extremes of temperature, from highs of 20°C during summer days down to −140°C during winter nights. Over millions of years, contraction and expansion caused surface rocks to shatter into a fine dust that quickly dispersed across the planet on seasonal winds. The Sun's ultraviolet radiation energized oxygen to bond with the constituents of the dusty iron-rich soil, forming iron-oxides and sterilizing peroxide and superoxide compounds, giving Mars its distinctive red color. The naturally emerging sterilizing agents of the Martian soil soon obliterated any persisting organic material on the surface. Today, at least where examined by Viking, there is less organic material on Mars than on our own lifeless Moon. By the end of the Hesperian period, somewhere between two and three billion years ago, Mars succumbed to the inevitable—becoming a cool, dormant world.

The Amazonian Period

As the countless millennia passed from the late Hesperian period to the present day—in what we call the Amazonian era—the characteristics of Mars changed completely. The atmospheric pressure dropped to the point where even water-ice could no longer exist on the surface except at the poles. Virtually all of the planet's water became chemically bonded to the soil, locked into subsurface permafrost or sublimated into the atmosphere. From the equator to about 40 degrees north and south, all ice to a depth of 100 meters sublimated and was lost to space or redeposited at the poles. At latitudes above 60 degrees, however, ice may reside as close as 1 meter beneath the surface, and at the poles it can still reside on the surface.

Despite its current relative quiescence, Viking also revealed that Mars is far from being completely dormant. Surface layers seen at high latitudes suggest periodic climate change on timescales of hundreds of thousands of years, persisting even until today. Such climate change could, for example, result from the numerous periodic changes in Mars' axial tilt and orbit. First, there are large changes in Mars' obliquity. While, currently, Mars' obliquity is about 24 degrees and gives rise to seasons similar to those on Earth, over 100,000 years the obliquity varies from just 13 degrees up to 42 degrees, severely changing the climate as well as the extent of seasonal variations. Superimposed upon this is a precession of the planet itself. The direction of the planet's axis changes over a 55,000-year cycle, altering the seasons of the Martian year. Added to this are periodic changes to the eccentricity of Mars' orbit on timescales of 100,000 *and* 200,000 years. Although we cannot yet determine the true effect of such axial and orbital variations, and even though they are somewhat chaotic, layering seen by the Viking orbiters near the polar regions suggests associated climate change. Along with occasional tectonic and volcanic activity and comet and/or asteroid impacts onto a surface containing vast quantities of frozen water and other volatiles, opportunities for periods of higher activity—extending for months, years, decades or even thousands of years—may on occasion arise.

There are also less extreme cycles to the Martian planetary environment that bring change to the planet. The Martian year of 687 Earth days, or 669 sols, is about twice as long as that of Earth, and the planet enjoys four seasons—spring, summer, autumn, and winter—that last twice as long as the seasons on Earth. Coincidentally, a Martian day, called a "sol," is just 39 minutes longer than an Earth day. The extreme temperature changes that come with a tenuous atmosphere, and the day and night and seasonal variations, all contribute to winds of up to 300 kilometers per hour across the planet. Seasonal temperature variations can produce atmospheric pressure changes so severe as to induce dust storms that cover the entire planet for months, as encountered by Mariner 9. Intriguingly, Mars' northern summer is warmer than the southern summer even though it is further from the Sun. This is caused by smaller temperature gradients that cause less dust to rise into the atmosphere and therefore greater exposure to the Sun.

Other recent activity is also evident from the Viking orbiters. For example, many craters are worn down and encased in dust streaks and fluvial deposits, suggesting that erosion and deposition have occurred for long periods in Mars' history. Most intriguing are the north and south polar ice caps, composed of carbon dioxide and water-ice. During winter the caps grow substantially as carbon dioxide from the atmosphere condenses to form a thin layer of ice. When summer arrives the carbon dioxide sublimates and returns to the atmosphere, revealing smaller but permanent water-ice caps. Emanating from the poles in spiral patterns within the caps are curved valleys hundreds of kilometers long, each separated by perhaps 50 kilometers from the next. These once again suggest variations in deposition caused by circulating wind patterns, or perhaps by periodic climatic change over intervals of thousands of years. Red soil and dust dominate the landscapes of Mars today, but the variety of landscapes is staggering. Vast flat plains from horizon to horizon, giant craters, sand-dune-filled deserts, dried lakes and river valleys, mountain ranges and gigantic rift valleys all reveal a planet shaped by a legacy of activity and change.

As revealed by the Viking landers, Mars is an accessible world to humans—given adequate protection. The daytime sky is red, not blue as on Earth, and is caused by a fine suspension of dust that permanently resides in the atmosphere. The Sun appears smaller in the sky because it is further away from Mars, but at night, an observer on Mars will see exactly the same constellations as are seen from Earth. A striking difference does exist, however. Depending on the time of year, but always at dawn or dusk, a brilliant blue star appears regularly, accompanied by a tiny companion. The Earth and Moon are Mars' morning and evening stars, as Venus is ours.

A New Beginning

The Viking lander results had been both inconclusive and controversial. Indeed, by the 1980s even the exobiological basis for the lander experiments—carbon-based life emerged from organic synthesis fueled in Mars' atmosphere—was severely challenged. By then it had been realized that methane and ammonia (needed in the chemical *reduction* process that would give rise to organic compounds) would not have remained in either Earth's or Mars' original atmosphere long enough for significant organic synthesis to occur. The entire basis for organic synthesis upon which life might emerge on rocky planets, including Mars, was in doubt; and with it the basis for assuming that life existed on Mars was equally shaky. Instead, new ideas on the emergence of life on Earth were taking shape—involving hydrothermal vents where the presence of mineral rocks, hot fast-flowing water, and a carbon-dioxide atmosphere may have led to organic synthesis and other chemical pathways toward life. These were pathways that were never considered for Mars.

Serendipitously, the Viking landers revealed Mars as it really is—a planet shaped by natural planetary processes throughout its history and especially in the Noachian period. Intriguingly, with emerging evidence of tectonic, volcanic, atmospheric, and water activity also occurring on Mars, the possibility of hydrothermal systems in the presence of a carbon-dioxide atmosphere in Mars' distant past now also seemed a distinct possibility. Although not conducive to life on its surface today, the Viking orbital data suggested the possibility of activity on an ancient Mars similar to that increasingly considered central to the emergence of life on Earth! Both planets were probably never identical and followed different evolutionary paths, but the Viking orbiters suggested that they may have been close *enough* in many early planetary processes relevant to the origin of life. By the early 1990s it was clear, once again, that Mars was hugely relevant to the quest for origins and the question of life elsewhere in the Universe. The exobiological basis of Earth and Mars sharing an ancient commonality in chemistry was no longer speculation. The particulars of chemistry common to both planets may have turned out to be different to that surmised by the pioneering exobiologists, but the Viking orbiter findings none the less suggested a commonality in chemistry between both planets in their early histories. Although the original hypotheses for organic synthesis and the origin of life had changed, the argument for any Martian life being based on carbon (so eloquently reasoned by Vishniac) was now supported by the Viking observations. With similar composition and early planetary processes, any life emerging on Mars would have probably been both carbon based and microbial. Incredibly, contemporary thinking on the origin of life on Earth matched our new conjectures about activity on early Mars, to the extent that the prospect of life emerging on Mars seemed once again plausible.

The Viking orbiters revolutionized our thinking about Mars. The landers exposed the limitations of pursuing exobiology blind while the orbiters immediately stepped in and revealed, for the first time, the broad nature and evolution of the planet, including a vibrant early history that slowed down to the dormant and toxic world we see today.

A new beginning for exploring our origins and the question of life elsewhere had emerged. This time we would start from a firm basis and take on the task of fully understanding Mars in a mature manner. The Viking lander experiments had been limited in their scope and frustrating in their results, but they were absolutely necessary. They had been developed from the best possible stance and by the best minds at that time, and no one today could have done it better. And even from an almost blind starting position, those involved in Voyager and Viking had developed a broad hypothesis regarding a planetary context for the origin of life that still holds. Most of all, however, the Viking project was a risky venture that has given rich rewards in ways not envisaged from the outset, and on which we can now build a new effort to comprehensively explore the Red Planet.

ALH84001

Although Viking set the scene for a legitimate return to Mars, one other event in 1996 provided added impetus. Over the past 200 years, several dozen of a class of meteorites called the SNC meteorites have been discovered, some of which originated on Mars. They are called SNC meteorites because they represent three subgroups of meteorite—Shergottites, Nakhlites, and Chassigny—that all contain minerals uniquely characteristic of rocky planets that have undergone differentiation. Detailed analyses of gas pockets in some of them show a striking similarity to Mars' atmosphere as examined by the Viking landers, leading to broad agreement that they originated on Mars.

In 1984 an expedition to Antarctica discovered one such meteorite, labeled ALH84001, which has turned out to be particularly important to the question of life on Mars. Many meteorites that collide with our planet survive the journey through the atmosphere, landing in tact. Most of those that land remain undetected, but one place on Earth where meteorites can be found more easily than anywhere else is at the Allen Hills ice field in

Antarctica. As with all locations on Earth, the glaciers of Antarctica are subject to occasional bombardment from small rocky meteorites. Movement of the glaciers in the region toward the Allen Hilis however, coupled to a unique weather pattern that heavily erodes the glaciers, causes all debris, including meteorites, to accumulate at the foot of the hills. The Allen Hills act, therefore, as a natural repository for debris falling to Earth from across our Solar System.

Analysis of minute amounts of trapped gas—including of very specific *isotopic markers* unique to Mars—revealed that the ALH84001 meteorite (the first meteorite of 1984 discovered at Allen Hills) originated on Mars, that it formed during Mars' early history approximately four and a half billion years ago, was blasted off the surface about 16 million years ago by a comet or asteroid impact and subsequently drifted between the planets until it landed on Earth 13,000 years ago. Yet further analysis created a sensation. In 1996, a joint UK Open University/NASA research team led by David McKay and Everett Gibson announced that they had uncovered evidence of fossilized microbial life from Mars within the meteorite. Powerful electron-microscope images show what appear to be fossilized bacteria within the meteorite. Near-by detected magnetite mineral is consistent with metabolic waste from bacteria and there is even organic material within the supposed fossil—also consistent with the presence of once-living organisms within the rock. While there is little debate on the presence and origin of those constituent materials within the meteorite, intense debate continues as to their interpretation. Many scientists have subsequently offered alternative non-living and inorganic chemistry explanations for the observed features. The evidence is intriguing, but it is not conclusive (Figures 37 and 38).

Despite the continuing debate, ALH84001 and other SNC meteorites are pivotal to the broader question of a cosmic context for life. First, they demonstrate that material interactions among the planets take place—a fact that in itself is of significance regarding the spread of life throughout the Solar System and beyond. Second, ALH84001 has prompted new life-science studies on Earth that have even led to the discovery of microbial life an order of magnitude smaller than had previously been considered possible, showing that life is more diverse than previously envisaged even on our own world. Furthermore, chemical analyses of the SNC meteorites point to cycling of material between the Martian crust and atmosphere, possibly at ancient hydrothermal sites and suggesting the presence of both water and hydrothermal systems on early Mars. Finally, the SNCs unequivocally verify that organic material existed on Mars in its early history. Hence, while the SNC meteorites might not provide conclusive evidence for life on Mars, they

Figure 37: Meteorite ALH84001 is a piece of Martian rock that found its way to Earth. Analysis of the meteorite suggests activity on ancient Mars relevant to the origin of life. [Credit: NASA/Johnson Space Center]

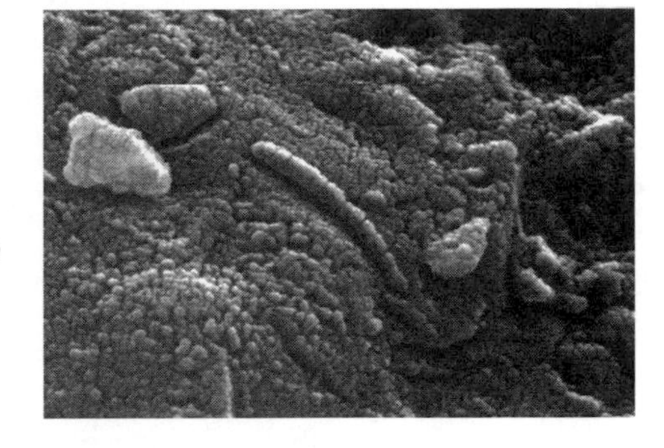

Figure 38: An electron microscope image of a piece, ALH84001, revealing what may be fossilized evidence of ancient microbial life on Mars. [Credit: NASA/ Johnson Space Center]

suggest that processes now considered to be important to the emergence of life were taking place on Mars during its early history.

When combined, the emerging picture of Mars from Viking and from SNC meteorites was both intriguing and strong enough that, by the mid-1990s, a major new exploration program was once again set in place to uncover the true nature of the Red Planet and what it might tell us about the nature of life in general.

Mars: A Cosmic Stepping Stone

6

We have come to a realization regarding the nature of life: there is a connection between the origin of life on Earth and its cosmic context. If there was a natural origin to life on Earth, then it came about from the natural environments and resources available at that time.

The question of our origins is therefore not just about Earth in isolation, it is also about all that occurred before—the grand evolution of the Universe, the origin of our Sun and Solar System, as well as Earth's formation, make-up and early surface environment. With the emergence of life on an Earth that was initially very different to the one we inhabit today, the nature of life must have been more robust and adaptable than we have traditionally considered, and perhaps can still imagine.

A universal context for life is also suggested not only because of the astronomical context for life here, but also because life is so adaptable and the processes giving rise to it may be widespread, given the vast number of places in the Universe and the commonality of nature at each and every one. Although fundamentally different types of life may be possible, commonality suggests that if life occurs elsewhere in the Universe, then organic life on rocky planets may be relatively common. Certainly the mounting evidence of extrasolar planets, newly forming protoplanetary disks, and the ubiquity of organic matter suggests such potential.

We may not yet know the likelihood of life arising on any given planet, or the precise steps by which it originated on Earth, but we now know that there are planets throughout our galaxy and that similar planets may have similar chances of life. We regard life as a potentially widespread phenomenon in nature. It may be possible to determine the actual steps that led to life on Earth, because if they are based in nature then we can attempt to understand them through science. Also, if we can achieve this, we will also gain significant insight into the origin and abundance of life elsewhere.

Such hard-earned perspectives have brought maturity to our questions about the origin and cosmic abundance of life. A previously speculative

K. Nolan, *Mars, A Cosmic Stepping Stone*,
DOI: 10.1007/978-0-387-49981-9_6, © Praxis Publishing, Ltd. 2008

endeavor that was often crowned by ad-hoc opinions and premature conclusions, it is now tangible and earnest science, not just because our questions and hypotheses are based in science, but because we can pursue answers through observation, exploration, experimentation, and analysis. We need no longer wait for future generations to find answers—we can look for them ourselves. Nor is it time for conclusion or opinion on such a grand context for life or on our ability to find defining answers. We are only now starting on this quest and it is therefore a time for new questions and new exploration. It is time to find out what precisely "Earth-like" meant for our planet at the time life emerged; and it is time to determine whether the potential for life across a cosmos filled by countless ordinary places has actually been realized.

Such an undertaking is huge and complex however. We must deal with the enormity of the Universe, yet determine how it works on a molecular level and the complex ways that bring about life. Despite the hurdles, we are already making headway. There is, for example, the intricate study of Earth itself during its earliest history. By looking at the earliest rocks and crystalline minerals using sophisticated isotopic and radioactive techniques, we can infer a great deal about Earth's earliest environment. But there are huge challenges with this line of investigation because little of Earth's earliest environment remains, having been radically transformed by the activity of the past four billion years—and because life on Earth today reveals little about the prebiotic processes that led to its origin. Earth studies are of paramount importance, but they are far from straightforward.

A second approach is the continued observation of the galaxy in ever-increasing detail for evidence of organic materials, newly forming solar systems, extrasolar planets, and even life. As previously mentioned, the Hubble and Spitzer space telescopes continue to deliver unprecedented insight into the formation and make-up of protoplanetary disks around young stars, hugely improving our understanding of precursor activity relevant to the origin of life. Giant new ground-based telescopes and networks of smaller yet sophisticated robotic telescopes are becoming increasingly adept at finding new planets around other stars. And with emerging revolutionary instruments such as the proposed Terrestrial Planet Finder, it may only be a matter of 10 to 20 years until we detect other Earth-like planets and even life upon them. We have a long way to go before we exhaust what we can achieve with telescopes and the future decades promise to confound and dazzle us with new views on the Universe—perhaps even unveiling a picture of an abundance of life across our galaxy. However, as far as telescopes can take us, definitive answers about life on such distant worlds will ultimately require closer examination, demanding that we actually travel to them. Today we can envisage ways of sending tiny automated space probes to nearby stars, but it will be well beyond our lifetime before we have the capability of carrying out even one manned journey to our nearest star. We will soon be able to look at the stars and planets across our galaxy in unprecedented clarity, but traveling to them is too difficult a feat for the foreseeable future.

While we accept all the shrouding of Earth's ancient past and the prohibitive vastness of the Cosmos, there resides on our cosmic doorstep the most exceptional opportunity to explore our origins and the cosmic dimension to life. That opportunity is the planet Mars. Historically we may have looked for Earthly qualities in Mars that simply are not there, but we have recently begun to recognize real and ancient parallels between both worlds that suggest that Mars may hold far-reaching answers to our questions.

We reach this conclusion not only because of a new and better understanding of the planet itself, but also because of our maturing perspectives on the nature of life, on our own planet, and on the Solar System in general. Set within such a broad context, we suspect that in its distant past Mars may have been (and may continue to be) similar *enough* to Earth in ways that are important for the emergence of life. Mars may be a stepping stone to understanding our own origin and the cosmic nature of life.

To see how this might be, let us begin by looking at Earth and Mars from their broadest context—as worlds that are part of a system composed of many distinct *classes* of world. From our colossal central star to the swarm of comets occupying the outermost region of the Solar System and from the magnificent gas giants to their innumerable icy moons, our Solar System is organized into clear and distinct *classes* or *groupings* of worlds. Yet, although one group may be radically different from the others, within each we find individual and separate worlds bearing remarkable and revealing similarities. For example, each of the four gas giants shows many bulk planetary characteristics that closely resemble the other three. Indeed, so alike are all four that we are sure they have shared a common legacy and origin. Likewise all comets, despite their variation in size, shape, and ice content, are all members of a single class. Their similar composition and general size (always much smaller than any planet or moon) and the nature of their orbits about the Sun, suggest a commonality shared since their origin billions of years ago. In general, each class of object is defined not just by similar characteristics among its constituent worlds, but also by a common origin and history that brought about a similarity in their physical attributes.

It is within this context that we see the first connection between Earth and

Mars. Among the various categories we also find that the four inner planets—Mercury, Venus, Earth, and Mars—all share broadly similar characteristics and also constitute a single category of celestial object. Although their surface conditions are very different today, they share sufficient bulk characteristics—size, material make-up, internal structure, and location in the Solar System—to suggest a common origin. All four constitute a single class of world. Despite their obvious differences today, they are more similar to one another than to any other world within our system, and have been since the outset.

Further, when we look at the surface environments of Earth and Mars within the broad context of the Solar System, we again see interesting commonality. No clearer can this be illustrated than by considering the diversity in behavior of volatile materials. On Venus, for example, the surface temperature is so high that water—a volatile that readily exists on Earth as solid, liquid, and gas—cannot even be sustained in its vapor state. Venus is so hot that its water has long since boiled away and dissipated into space. Mars, on the other hand, still possesses substantial water supplies, but because it is further from the Sun and possesses only a thin atmosphere, it cannot support surface liquid water, though it can support water-ice at its poles and trace amounts in its atmosphere. However, as we travel beyond Mars toward the icy worlds of the outer Solar System, we find that all surface water is permanently frozen, never to be released from a perpetual ice-age endured since the beginning of the Solar System. Other volatiles show similar trends. Carbon dioxide, for example, exists only as a gas on Venus and Earth. On Mars it exists as a gas, but it also freezes at the Martian polar caps as "dry-ice" during winter, returning to the atmosphere in summer. Once we travel to the outer Solar System, however, we find that, as with water, any carbon dioxide existing on the outer icy-rock worlds is permanently frozen.

These examples only partially demonstrate the full diversity in volatile behavior encountered across our Solar System. With every new outer world visited by our space probes, we encounter entirely different realms of volatile activity hitherto unimagined. Consider Saturn's giant moon Titan, for example. At almost the same size as Mars, and possessing an atmosphere far denser than even our own, Titan has a mean temperature of -178°C and exhibits spectacular and unique volatile behavior. It is so cold that neither water nor carbon dioxide plays any part in volatile activity, being completely bound into the surface as rock-hard ices. Instead, entirely different materials such as methane and ethane are found in abundance as solid, liquid, and gas—perhaps creating and shaping an enigmatic and lurid planetary landscape that may curiously mimic how water has shaped our planet. Volatile activity on Titan constitutes an entirely different domain of planetary activity to that encountered in the inner Solar System. But the various activities do not end there. As we journey even further to the edge of the Solar System we find perhaps stranger activity. On Neptune's giant moon Triton, for example, the surface temperature is $-235°C$ which is the coldest natural environment ever encountered. The result is that even methane—so vibrant on Titan—is itself frozen into the surface and nitrogen, capable of existing as a gas on worlds all the way from Venus to Neptune, is itself cooled on Triton to its liquid and solid-ice states. Triton is so much colder than all of the other worlds closer to the Sun that it is capable of freezing virtually all volatile materials, except for a few of the very lightest elements of all.

In a context of such staggering diversity of planetary environments and volatile behavior, a similarity in volatile behavior between Earth and Mars has revealed itself. As different as they may be—and there are indeed significant differences—they are far closer *and* similar in surface environment and volatile behavior than any other world found across the Solar System. Indeed, such a broad perspective on volatile behavior across our Solar System also suggests at least part of the reason for life having arisen on Earth, because our region of the Solar System is significantly more conducive to the volatile behavior associated with life as we know it.

While the required environment and volatile activity necessary for life may be exclusive to Earth today, we are increasingly confident that this was not always the case. Inasmuch as the history and evolution of the Solar System can be characterized by bombardments and impacts, it can also be tracked through the history of volatile behavior, especially on the inner planets. In the beginning Earth was a planet characterized by planetary differentiation, tectonic and volcanic activity, and outgassing of biogenic and volatile materials to the surface, but we now suspect that it was not alone and that Venus and Mars were also active in these same ways. Today Earth is very different from Venus and Mars. However, while Earth has always been fundamentally different from, say, Jupiter, the same is not true when compared to Venus and Mars. Rather, with all three having emerged within the one region of the Sun's protoplanetary disk and as similar type planets, it is plausible that they were also broadly similar in early planetary activity. The differences among Venus, Earth, and Mars today are the result of an ever-divergent evolution which belies an ancient commonality. In the beginning they were of a kind—a single category of emerging planet, and each a variation on a theme within the category.

Critically, while Venus has become a scorching and hellish cauldron, obliterating all traces of its early history, the same is not true of Mars. As we have seen, the Viking surveys (and indeed all subsequent space probe surveys) of the planet reveal widespread activity on Mars in its early history, but that seems to have ceased about three billion years ago, and although it left the planet in a mostly dormant state, a global record of all that had occurred up to that time was preserved. More poignantly, the Mariners and Vikings have shown that Mars' early activity was indeed broadly similar to our own, indicating material differentiation, a silicate-based crust, tectonic faults, and shield volcanoes not unlike those found on Earth. They have even provided tantalizing evidence that suggests a denser atmosphere and liquid water flowing through ancient rivers and in gigantic floods. As we will see in later chapters, more recent evidence, from Pathfinder in 1997 to the Mars Reconnaissance Orbiter of 2006, continue to affirm our instincts of a vibrant past—deep sedimentary layering, evidence of ancient lakes, seas, river deltas, and water-based chemistry—all revealing an ancient past involving tectonic and volcanic activity accompanied by biogenic and volatile materials, including water.

Mars' surface may now be covered in red dust and peppered by impact craters, but unlike Venus and Earth, subsequent activity has not been sufficient to obliterate all evidence of its early history. Silently waiting across its globe and under the surface is a vast record of its early history, including evidence of planetary processes and volatile activity perhaps similar to those occurring on Earth at the same time. Mars today represents the *calm after the storm* of planetwide activity perhaps relevant to our quest for origins and a cosmic context for life. As the Mariner and Viking space probes revealed, its surface is a vast and diverse domain, shrouded by dust but beneath which lies a four-billion-year legacy of rich and complex activity on a planetary scale and lasting hundreds of millions of years. Mars is a time capsule to a four-billion-year past, offering an unprecedented and completely unique opportunity to learn directly about planetary activity similar to our own and relevant to the emergence of life. The question is not whether Mars was Earth-like in its distant past; rather, the question becomes: "What *precisely* does Earth-like mean for young rocky planets, including Earth?" We are now confident that Mars holds important answers to everything that question encompasses.

Indeed, similarities past or present seen on a planetary scale must manifest themselves on smaller scales, all the way down to molecular chemical activity. Nature does not stop on a planetary scale. Mars' broad features and processes ultimately manifest themselves as a multitude of finer, more detailed processes yet to be explored. With an abundance of water and other volatiles interacting under conditions broadly similar to Earth, it is entirely plausible that prebiotic chemistry—that is, precursor chemical activity relevant to the emergence of life—could have occurred there. If so, its identification and analysis could provide major insight into similar chemistry occurring on Earth prior to the emergence of life. Indeed, we already have tentative evidence that prebiotic chemistry did occur on Mars. SNC meteorites (Martian meteorites found on Earth) show traces of organic matter, possibly formed at hydrothermal vents on ancient Mars, suggesting both hydrothermal activity and ancient organic chemistry. We therefore feel that the question of prebiotic chemistry on Mars is no longer as wide open. With early favorable conditions and even a hint of evidence from Martian meteorites, the search for complex and prebiotic chemistry on Mars is now of high priority.

What if such activity did occur? How far did it run? Might life have actually arisen on Mars? With broadly comparable conditions to Earth, the chances of life arising on Mars would have been about the same as on Earth. We cannot ignore those odds and they mandate a search for both prebiotic chemistry and the emergence of life there. Significant yet currently unknown chemical pathways were needed for complex organic chemistry to bring about the first microorganisms on Earth; and if basic life arose on Mars it may provide unique insight into those chemical pathways. Critically, we *can* carry out such a search because Mars has preserved evidence of its origins and ancient past. If life arose there, evidence of that event still resides on the planet.

If we consider life having arisen on Mars in its distant past, then the questions of its survival must also be considered. Could life there have survived the devastating transition that Mars apparently endured over three billion years ago? Might life have found ways of adapting and surviving to the present day? While traditionally we would reject such a possibility, recent discoveries regarding the diversity and robustness of life on Earth indicate that it is plausible that ancient life could have survived on Mars. As we have already seen, many environments on Earth, far more hostile than on Mars, are rich in microbial life. Our planet is dominated by life that has had enormous amounts of time to adapt to hostile conditions. None the less, the discovery of extremophiles in the harshest of Earth environments quenches the argument that Mars cannot harbor life because conditions there are too harsh. If life emerged there, it is possible that it found ways to survive. We must also remember that while Mars, in general, is more hostile than Earth from a human perspective, it is far from being utterly hostile to all life as we know it. Mars is a mostly dormant world, but there may well be environments capable of sustaining life—at subsurface aquifers, at high latitudes where near-surface water is more abundant, and at sites of volcanic activity, however localized or sporadic.

From general questions about the early Solar System to specific questions about the origin of life, and perhaps its survivability today, Mars has much to offer. It is a priceless laboratory. The certainty that it shared some degree of commonality with Earth in its early history, and preserves that past, offers a unique opportunity to tackle many of our great questions concerning the nature of life. Therefore, a third approach to exploring our origins presents itself—that of detailed study of Mars' early history and evolution; of tectonic, volcanic, water, and atmospheric activity; of complex and prebiotic chemistry; and of possible ancient and current life.

Whatever the answers, they will be significant. At the very least, we expect to learn much about the surface conditions and chemical activity prevailing on all the terrestrial planets in the beginning, including Earth. As virtually no record of Earth's earliest history has been preserved, Mars offers the only substantial record of activity occurring at the same time and broadly similar to our own. Even if no trace of life is found on Mars, a thorough investigation cannot but significantly enhance our perception of our own planet in its earliest epoch. Also, if it comes to pass that no prebiotic chemistry or traces of life is ever found, then that in itself will be valuable in constraining the conditions necessary for life on Earth.

Nevertheless, evidence of prebiotic chemistry, whether or not it ran through to completion—that is, gave rise to life—would provide fundamental insight into the origin of life. Should we discover evidence of an ancient biosphere, or even a trace of extinct life, the consequences will be profound. Such a discovery will reveal that life emerged on two separate worlds in one Solar System, and that life is not unique to Earth. It will for ever more place life within a universal setting and not purely an Earthly one. It will also constitute independent verification of the ancient context for life on Earth and an origin based in nature. There would, in addition, be other cosmological implications: providing new insights into the mechanisms and diversity of mechanisms through which life is possible, perhaps revealing universal processes at work that we simply cannot infer from Earth alone. Such a discovery would tell us that Earth and Mars are indeed members of a class of planet upon which life can emerge, and that even with their differences, life was possible on both. Life will have arisen on *a range* of terrestrial planets, adding weight to the idea that life may be common across similar solar systems and similar planets throughout our galaxy and beyond. And even if life on Mars was shown to be similar to life on Earth, implying a common origin, it would still be a discovery of immense value. Perhaps we could not glean as clear a view on the origin of life, but it would still affirm an ancient and natural context for our origins—a cosmic context for all life, and provide insight into the origin, robustness, adaptability, interplanetary proliferation, and universal nature of life.

Of course, if we were to discover living organisms on Mars today, the impact would be even more significant. Not only would we gain the insights mentioned above, but we would also have direct access to a second and potentially independent biology. The discovery of advanced and intelligent life-forms may be important to our human sensibilities, but the discovery of even one living microorganism on Mars will be enough to teach us about an entirely new biology. It would be a second "take" on how the Universe organizes itself at the most complex level known to us, forever changing our understanding and perception of what life is and how it works. We would gain a profoundly broader view of the working of life that could radically affect our management of Earth biology, from environmental issues to medical treatments. It could also change our view of the Earth. Never again could we look upon this planet as the sole bearer of life. We would know that there is life elsewhere, and that we share the Universe with other beings. Every time we looked into the sky at Mars we would know that we were looking at a world bearing life. It would become a beacon—a stepping stone—to a vast Universe beyond with unfathomable possibilities regarding the broadest context for life.

Mars is a cosmic stepping stone to a four-billion-year past and to an exciting future. It has lay silent for four billion years, accompanying us on our endless journeys around the Sun, awaiting our arrival, and now ready to reveal deep insight into our origins and broadest context. The detailed exploration of Mars is both timely and of our time. Through the tireless efforts of all who have come before, we are now able to formulate the right questions, we know where to look for answers, and we are able to pursue them effectively and efficiently. The scientists and engineers of the Mariner and Viking missions were correct after all. They had postulated an ancient chemical and perhaps biochemical parallel between Earth and Mars, and our very latest findings, set against our best ideas on the emergence of life, vindicate their stance. They had based their science, and their missions, on organic synthesis being dominated by atmospheric chemistry. That hypothesis, however, collapsed almost in sympathy with the arrival of the initial and deflating evidence from Mars provided by the Viking landers. Yet from the ashes of their hypotheses we find that only the specifics of their analyses were perhaps flawed or incomplete. Organic synthesis was vital to the emergence of life—though we now think that those events would have occurred at hydrothermal vents and within a carbon-dioxide atmosphere, which is precisely what the Viking and subsequent missions suggest Mars was like in its distant past.

And so our quest—to analyze the vast Cosmos at the molecular level at which life arises—has come into sharp focus on Mars. Our path forward becomes a specific set of questions followed by an exploration program of the Red Planet. We must understand its birth and early history, early tectonic and volcanic activity, and its original stocks of water, volatiles, and biogenics, the sort of surface environment that arose and whether hydrothermal and aqueous environments gave rise to complex chemistry. We must then search for evidence of prebiotic chemistry and life, past and present.

Significantly, we *can* go to Mars to conduct such intricate exploration and analyses because we now have the analytical techniques, instruments and space technology at our disposal to get the job done. This was not the case even a few decades ago. The Mariner and Viking missions, as pivotal as they were, could only do so much, but with the foundation they laid, combined with a developing and broader understanding of the nature of life and planetary sciences, we are now in a position to explore Mars in the hope of extracting deep-rooted answers about its past. This is a quest for our time because the efforts of all who came before have brought us to the point where we can take the next step. Whether to take that step is now a current choice, not some future aspiration.

While Mars is more hospitable than most other places in the Solar System, it will still be enormously challenging. The trade-off with Mars having preserved a record of its four-billion-year history is that we will have to dig deep—literally and metaphorically—to find the answers. Mars is an entire world with a complex history. To find answers we will have to visit dozens of sites from polar ice-caps to equatorial valleys, and from the highest volcanoes to the deepest impact basins. We'll have to drill into the surface and even return samples to the Earth. With our best efforts, Mars will push us to the limit before clear and meaningful results emerge; but this is not necessarily a disadvantage. Mars will push innovation in science and technology that we might otherwise be slow to develop, if at all; and we can be confident that our efforts, however difficult, will serve future generations well because our exploration of Mars will be neither arbitrary nor premature. Every future venture will build toward a robust and far-reaching picture about the Red Planet and Earth.

Even now, Mars guides us to a broader perspective of the nature of life and our place in the Cosmos. It is a catalyst for humanity to see itself differently. We are beginning to accept an idea that Mars itself partially triggered—that of broader context for life, of other worlds as other ordinary places, and of Earth as only one planet with life, among many. From the dawn of humanity, Mars has prompted questions about its nature that seem to transform into questions about our origin, nature, and destiny. It has always been one step beyond, but now we are catching up—stepping onto the real Mars, hopeful that it will finally reveal its ancient secrets and in so doing provide answers about our nature. Mars is about to serve once again in guiding us on our path through history.

Mars *is* now a real place. We are already there in earnest and there is no turning back. We are there because we relate to it. We can go there and kneel in the dirt, gaze upon a sunrise and feel some semblance of normality. Mars is far from hospitable to humans, but it is still the only other world we can discuss in such terms. Mars will ultimately become, literally, a stepping stone for humanity. If there is a future for humans in space, Mars will be planet number two. Humanity will have taken *the* substantive step into space via Mars, and no other world will even be quite as important to us. Humanity will have become a space-faring civilization through the exploration of Mars.

Mars beckons to us in the most pragmatic way: as a treasure trove of information concerning our origins and the questions of life in the Universe. It has virtually all of its secrets to reveal, and we are now on our way to uncovering them.

A Grand Plan

We need no longer purely speculate about our origins and the universal context for life. We now recognize the link between the origin of life and Earth's early environment, and the possibility of life emerging on similar planets throughout our galaxy. With broad parallels between Earth and Mars—especially in their early history—we are increasingly confident that Mars can play a substantial role in our quest to know our origins and the general nature of life. Our objective is therefore to learn all about Mars: what it can reveal about prevailing conditions in the early Solar System, about its own early planetary behavior and perhaps that of Earth, and about processes relevant to the emergence of life. And we must also search for any evidence of past and present life that may have arisen there.

Such a significant opportunity has prompted a new era of exploration. Beginning in 1997 with the launch of Mars Pathfinder—still in its infancy today—this international effort heralds a long-term program of exploration over the next three decades with Mars at the focus. It builds on the efforts of past Mars explorers—Brahe, Kepler, Galileo, Huygens, Schiaparelli, and Antoniadi—and most poignantly on the achievements of the first true planetary scientists, such as Lederberg, Calvin, Soffen, Vishniac, Sagan, and Murray, among others. With NASA, these scientists conducted the first ever reconnaissance of our Solar System and set the scene for an astronomical context for life and in finally revealing some semblance of truth about Mars. Our era of Mars exploration is now in the capable hands of a current generation of scientists who are already pursuing answers—Robert Zubrin, Jim Garvin, Chris McKay, David McKay, Mike Malin, Ken Edgett and Phil Christensen in the USA, and Gerhard Neukum and Colin Pillinger in Europe, to name but a few of the hundreds of people currently committing their energy and talent to the investigation of Mars.

While it remains a genuine possibility that we may stumble upon some spectacular discovery about life on Mars, we need no longer rely upon pure chance, because we now have the means to pursue a far-reaching program of scientific exploration where the results are sure to bring us closer to

K. Nolan, *Mars, A Cosmic Stepping Stone*,
DOI: 10.1007/978-0-387-49981-9_7,

understanding the connection between life and its planetary context. We are now organized: US teams working with NASA and (increasingly) European teams associated with ESA are now working in collaboration, and where they are pursuing separate agendas they are doing so in ways that minimize the unnecessary duplication of effort. All are clear on the important questions and on how to pursue them efficiently and effectively. Equally important, it is now realized that the best way to pursue our many and varied questions is in specific sequences from the global to the molecular scales, from geochemistry to biochemistry and from Mars' distant past to the present day. When we can answer more general questions about Mars we will then be better prepared to pursue more specific and complex ones. Pursuing questions in such an organized manner reveals how best to develop a program of exploration. There is no point, for example, in searching blindly for biomarkers if we have no idea of their specific nature or where they might be. It is now clear that the best approach is through a phased strategy for exploration, from orbital reconnaissance to robotic landers, sample return missions, and possibly human expeditions, with each phase feeding into the next, allowing for ever more informed lines of inquiry and the subsequent development of appropriate missions.

In this regard, ESA and NASA have already developed far-reaching exploration roadmaps for the coming decades, with many missions carefully planned in sequence. Indeed, these roadmaps have in part been realized through numerous missions currently on and around Mars. Pathfinder, Mars Global Surveyor, Odyssey, Express, Spirit, Opportunity, and Mars Reconnaissance Orbiter have all successfully traversed the interplanetary expanse between Earth and Mars in the last decade and today continue faithfully to probe and scrutinize the Red Planet in a highly orchestrated sequence, delivering unprecedented quantities and diversity of data.

Yet for all this effort, we have still barely begun. The task we have set ourselves is truly of mammoth proportions—to characterize an entire world that no one has ever set foot upon, down to its molecular level and over its four-and-a-half-billion-year history. Despite the historical legacy we are building upon, so ambitious is our agenda that it requires brand new methods and modes of exploration specific to this endeavor. We must devise methods of performing the types of science on Mars that we take for granted on Earth. We must learn how to undertake long-term exploration remotely yet efficiently and how to analyze and interpret the often unfamiliar findings. Over the coming decades we will also need technological innovation on an unprecedented scale—entry and landing systems, robotic rovers, smart landers, aerobots (airborne robots), and miniaturized yet sophisticated scientific payloads—all to be designed and built to work reliably for years or even decades. Also, unlike space observatories circling Earth—such as Hubble, Spitzer, XMM, and Integral—virtually all Mars probes must interact extensively with their environment whether in low orbit, gliding through the atmosphere or roaming the surface. Mars robots must be incredibly robust, reliable, adaptable, and as intelligent as we can make them. They must be capable of working without physical intervention by humans and be flexible enough in design to enable us to reprogram them to continue operating in an efficient manner, and to remedy them if they have malfunctions or enter a hostile environment.

The logistics of future Mars missions will also push the envelope. We must avail ourselves of the launch windows that open every 26 months or so, placing unprecedented demands on the development of new missions appropriate to each new phase of exploration. Not since the Apollo program has such a rate of space development been considered, let alone executed, and if we are to see this entire program through it will require at least several dozen robotic missions, with samples returned (and allowed to return) to Earth, and a massive investment for a human expedition. All of this will demand unprecedented and long-term commitment from the space community, governments, the international community, and society in general, and require great patience and steadfastness in the face of adversities that are certain to arise. Exploring Mars will push the limit of our endurance and tolerance.

Questions for Mars

Underpinning the scientific program for Mars is a new and extensive set of questions to be tackled in a very specific order. Our questions emerge from what we have learned about the origin of life on Earth and from the intriguing (yet incomplete) evidence of ancient activity on Mars, perhaps not dissimilar to Earth. Our questions are also prompted by the extreme difficulty in determining our origins from the scant evidence available on our planet alone and by the planetary context for life increasingly suggested from observations of the Cosmos in general. They are also motivated by virtue of the fact that Mars is now accessible.

First, we must determine the physical properties of the planet to an unprecedented accuracy—its size, geodesic shape, surface topology, and inner make-up. We must image and map the entire surface in exquisite detail to describe its morphology and geological history. We must determine the elemental and mineralogical make-up of the crust and determine the inventory of volatile and biogenic materials, especially water, that existed on

Mars in the past and are still resident today. We must then *run the clock forward* through Mars' history and attempt to unveil how it evolved—the rate of impacts of various sizes, its interior activity, whether it had an internal convective electric dynamo and hence planetary magnetic field, and the tectonic and volcanic activity that occurred. We must also determine the make-up and density of the early atmosphere, the climatic conditions that prevailed, and the extent and duration of surface and near-surface water.

Subsequent to a rigorous survey of Mars' planetary systems and characteristics, we must attempt to bring them together and understand the activity and interactivity that took place on the surface, within the crust, and across its planetary water reservoirs and atmosphere. We must also determine the extent of that activity, especially in ways relevant to the emergence of life—the complex and/or prebiotic chemistry that may have occurred, the energy mechanisms that drove such activity, and their duration. For this we must locate both ancient and subsurface aqueous sites, together with sedimentary basins and hydrothermal systems that could have brought about organic synthesis and provided mineralogical aid and stable conditions capable of driving molecular complexity and polymerization. We must also discover whether available energy systems could have acted as early forms of metabolic energy for primitive life. Finally, we must conduct a series of searches for evidence of ancient and current life on Mars. These searches will involve visiting many newly identified favorable sites of past and present water.

A Scientific Strategy for Mars Exploration

Our program for Mars requires a strategy that ensures controlled progress from general questions about the planet as a whole to detailed investigations of localized and complex behavior. Broadly speaking, we must first survey the entire planet from orbit in high resolution. Based on findings, we will then send robotic explorers to surface locations of chemical and mineralogical interest, especially where there is evidence of water. Finally, we will conduct extensive surface reconnaissance for evidence of prebiotic chemistry and life.

From orbit we will use optical cameras to image the surface at resolutions from several kilometers to better than 1 meter (and in stereo). Such a range of resolution allows local features and activity to be related to regional and planetary scale processes. Since much of Mars' surface is billions of years old, this strategy can reveal those geological features that are associated with impact, tectonic, volcanic, water, and wind activities, as well as their chronological context. Mars' early atmosphere can also in part be inferred from the extent of ancient volcanic activity as well as the number, size, and diversity of water features, because prolonged volcanism would have meant prolonged outgassing, while surface liquid water requires particular atmospheric conditions.

Global mapping of elemental and mineralogical abundances will provide details not only of the surface and internal planetary composition, but also of volatile and biogenic substances. Mineralogical analysis can also reveal details of early activity and conditions on Mars; while changes in mineralogy across differently aged surfaces and at different locations on Mars will allow us to track changes in planetary conditions over time, unveiling the planet's evolution. Overall, mineralogical analysis is extremely important in determining the extent of volcanism, water and atmospheric activity, hydrothermals, sedimentation, and many other processes.

Other surveys must also be carried out from orbit. Topological mapping of the entire globe will reveal the three-dimensional nature of surface features, greatly assisting our understanding of Martian geology and even revealing ancient water drainage patterns across the plant. Radio-sounding (a type of radar on board orbiting spacecraft) will provide maps of the Martian subsurface structure to depths of several kilometers, from which we can infer the existence of ancient geological features filled in on the surface and, most importantly, whether there are currently subsurface reservoirs of water and water-ice. Finally, to determine the make-up of the early atmosphere, we can analyze the atmosphere as it is today and from that infer its composition billions of years ago. Here, a technique called *isotopic analysis* reveals expected differences in the atmosphere, pointing to changes that have occurred through time and from which we can then work backwards to determine the make-up and density of the original atmosphere.

Individually, each orbital survey provides valuable new insight into Mars. When combined they represent a powerful resource for building an accurate picture of the make-up and early history of Mars from which we can plan subsequent missions. Indeed, as soon as orbital reconnaissance can adequately identify sites of chemical and/or biological interest, lander missions will be sent to the surface to perform more detailed analyses. While orbital analysis can reveal much, Mars is far from being another Earth. As it most likely never became a thriving, living world, any evidence of prebiotic chemistry and/or life will be extremely difficult to find, let alone understand. We expect that it will only be through detailed surface expeditions, scrutinizing the surface and subsurface at submeter, microscopic, and molecular levels, that such complexity may be identified.

We will therefore need to send landers to many locations of varying characteristics relevant to our search. Each lander will conduct photographic surveys of the local geology as well as microscopic imaging of rocks and soils to determine their morphology, texture, and chemical make-up, from which a context for their formation can be inferred. Miniature chemical and mineral detectors will reveal elemental and mineralogical composition to high precision, especially of water and water-based chemistry. Isotopic analyses of rocks will (as with atmospheric isotopic analysis) allow us to track the history of the materials within the rocks back through time, pointing to the extent of cycling of materials through the crust, atmosphere, and natural reservoirs of water, and revealing ancient activity on the surface of the planet. And, as with mineralogical variations across Mars' terrain, changes in various isotopes across sedimentary layers and at different locations will also reveal variations in the cycling of volatiles, from which changes in the climate over time can be inferred.

Through such means we will slowly build a picture of the history of each site, but it will only be subsequent to analyzing many such sites (or if we identify sites of special biological interest) that we can then consider conducting biological searches in earnest. We already think it likely, however, that the eventual search for biology on Mars will involve locations of significant past and/or present water activity, and given the current hypotheses on the origin life on Earth, we are particularly interested in finding sites of past and present hydrothermal activity and subaqueous sedimentation. Although searching for prebiotic chemistry and biology will be hugely challenging, we must remember that even the discovery of candidate sites will constitute success. Imagine, for example, having at our disposal evidence of a four-billion-year-old hydrothermal system showing evidence of rich chemical activity. We could only but gain invaluable insight into similar activity also occurring on Earth at that time. Such expeditions will be at the coal-face of exploration, providing long-sought-after and unprecedented insights into related processes on early Earth.

Our latest molecular analytical techniques may, however, go even further. Cutting-edge instruments to be sent to Mars in the next decade will be able to reveal the presence and structure of many complex and/or organic molecules. We may even be able to determine if such molecular species and functional groups were originally created inertly or by some type of biological activity. In ways that were not possible during the Viking missions, we may now be able to infer prebiotic and biological activity on Mars without having specific knowledge of the biology through which it occurred. This new and powerful science is now at our disposal. We can attempt to search for alien biology without knowing the specific character of that biology beforehand.

Of course, we must ultimately conduct direct searches for life. We think that the search for ancient life on Mars will mostly involve the search for fossilized microorganisms (called microfossils). Since life on Earth comprised only microorganisms for at least two billion years, we think it reasonable to infer that any life arising on ancient Mars would also have been microbial. If larger organisms did arise, our searches can, however, accommodate the detection of fossilized and other morphological evidence.

Finally, any search for current life on Mars requires the detection of liquid water now or in the planet's recent past. Should such sites be identified, we will then use a range of biological experiments, including improved versions of those used by Viking, to look for evidence of living organism morphologies, growth, metabolism, and by-products. If we find ourselves in a position to be carrying out such a search, it is hoped that we will by then understand enough of the planetary and local context to build adequately specialized biological experiments adapted to each locale and capable of revealing details of living versus non-living activity.

A Five-Phase Strategy for Exploration

The nature of the program—engaging the planet from the global to a molecular level, from past to present and from chemistry to biochemistry—coupled with experience gained from the Viking program, points to the need for a long-term and phased strategy of Mars exploration. Rather than trying to pursue too many lines of inquiry simultaneously, both NASA and ESA are now engaging their long-term exploration of Mars over five broad yet distinct phases.

Phase 1 involves orbital and aerial reconnaissance. Initially, orbiting space probes will survey the planet's morphology and geology, topology, chemistry and mineralogy, atmosphere and climate, magnetic properties, and interaction with the Sun's radiation fields. Since one spacecraft cannot accomplish all this, several will be needed, with each concentrating on a subset of the required investigations. For example, one probe may specialize in high-resolution optical imaging, while another concentrates on elemental and mineralogical studies. As will be seen in later chapters, this phase is already well under way, with Mars Global Surveyor beginning its analysis in 1998, Mars Odyssey in 2001, and Mars Express in 2004. These three orbiters represent our new-era Phase 1 Mars orbital reconnaissance; and so far they have scrutinized Mars in truly spectacular style, performing well beyond expectation. Today they continue to deliver unprecedented results that will take decades to analyze but have already brought about an even greater quantum leap in our understanding of Mars than any previous venture.

Subsequently, Phase 1 will also consist of a second generation of *follow-up* orbiters examining specific questions and evidence in even greater detail. The Mars Reconnaissance Orbiter (MRO), which arrived at Mars in March 2006, represents our first such follow-up mission. It is set to revolutionize our understanding of Mars yet again, perhaps even providing definitive answers about the planet's ancient past, as well as relaying images of sufficient resolution to plan human missions. Lastly, for Phase 1 reconnaissance, we hope to send a number of aerobots—extremely light glider aeroplane and balloon robots that can remain buoyant for hours and possibly days, and can travel great distances while providing exquisitely detailed surveys of huge tracts of the Martian surface.

Phase 2 involves robotic landers and rovers to sites of geological and geochemical interest discovered during global reconnaissance. Their task is to look for chemical and mineralogical evidence of past and present environments conducive to life-based activity. In particular, the first landers and rovers will search for water and water-based activity. With already existing reconnaissance data to about 1 meter resolution, landers will then create a seamless link of analysis to submeter and microscopic scales and provide *ground-truth* calibration for all orbiter data. Once again, this phase of exploration has already commenced through the Mars Pathfinder of 1997 and the Spirit and Opportunity Mars Exploration Rovers that landed on Mars in January 2004. While the Pathfinder mission was primarily a test of engineering capabilities, the science return from the mission was substantial as the first rover on Mars. Also, as of 2008, both Spirit and Opportunity are still operating far beyond their target mission life of just 90 days and have delivered unequivocal evidence of past surface water and water-based chemistry occurring in Mars' distant past. The hope is that, through sufficient orbital, aerial, and lander missions, specific details about Mars' past and current nature will emerge to allow identification of sites of potential biological interest.

In Phase 3, therefore, landers and rovers will be sent to candidate sites. These landers and rovers will be capable of looking for specific evidence of prebiotic chemistry, ancient fossils, and present life. They will also be highly specialized and sophisticated biological laboratories, capable of searching for organic matter and other complex molecules, metabolic waste, and by-products among many other plausible biomarkers. Mars Phoenix (2008) is the first Phase 3 lander, while two even more ambitious missions are already in the planning phase: NASA's Mars Science Laboratory and ESA's ExoMars, due to arrive on the planet around 2010 and 2013 respectively.

The reconnaissances during Phases 1, 2 and 3 will continue, often in tandem over the coming decades. There are, however, limitations to what can be done with robots alone. It is generally considered that the science we can do in laboratories on Earth is several decades ahead of what can be done with robots on Mars, and so many scientists therefore think that a sample return mission to Mars (Phase 4 of our 5-phase program) should be conducted at the earliest possible opportunity. Such a technical achievement and the ensuing science would no doubt have far-reaching consequences, allowing for extensive analyses to be conducted on Earth and over short timescales that might otherwise not be possible using robots.

Phase 5, the final phase, proposes human visits to Mars for both sociological and scientific reasons. While there is currently significant debate on the social, political, economic, and ethical issues surrounding such a mission, a substantial science case arguably does exist for sending people there. Simply stated, the increase in the amount and complexity of science that can be done by a human on Mars is perhaps comparable to the increase in sophistication between people and robots. For reasons we will explore later, there can be little doubt that the science return from a human mission to Mars will be significant; and to this end both NASA and ESA are gearing up for a human mission within the next 30 to 40 years. Irrespective of the eventual decision on such a mission however, there is no doubt that we are ready to explore Mars effectively and efficiently by robot, considering the multitude of significant scientific leads and the capability of our phased strategy. So let us now examine the scientific and program strategies in greater detail to see how we hope to uncover the detailed natural history of Mars and of any existing evidence of life.

8 The Search for Life: A Planetary Perspective

How we now manage the exploration of Mars is vitally important. Mars is an entire world whose surface is equal to the land area of Earth—a vast and unbroken landmass equivalent to all of Asia, Africa, Antarctica, Australia, North and South America, and Europe combined. Given how readily we have misjudged Mars in the past, we could easily do so again when exploring such a vast and unfamiliar landscape. This time we must pursue an appropriate path to explore the Red Planet—the phased strategy that is generally being pursued by American and other international space agencies.

We must first learn about the planet from a global perspective, determining its general dimensions, composition, and structure, as well as the nature and evolution of its internal and surface planetary systems. Only then can we properly progress to questions on the nature and behavior of the early atmosphere and climate, of volatiles such as water, of hydrothermals and of complex and life-related activity. Mars' bulk characteristics define its planetary type and bear heavily on questions of surface activity relevant to life.

Bulk Characteristics

Our starting point, therefore, becomes a thorough examination of the planet to unprecedented levels of accuracy: its size, mass, geodesic shape (that is, how spherical it is), and any irregularities in that shape. We must measure its bulk material composition, the masses of the various chemical constituents making up the planet, and their internal organization within the core, mantle, and lithosphere. We must also determine the nature and evolution of internal activity from its earliest history—by which processes and to what extent differentiation occurred, the nature and duration of internal convection, and if an electric-dynamo and resultant magnetic field originated. Such a magnetic field would have provided protection from

K. Nolan, *Mars, A Cosmic Stepping Stone*,
DOI: 10.1007/978-0-387-49981-9_8, © Praxis Publishing, Ltd. 2008

harmful cosmic and solar radiation while organic synthesis and prebiotic chemistry were developing, enhancing the possibility for the emergence and survival of life. Finally, we must also understand precisely how and when Mars' internal activity slowed and investigate the internal dynamic systems, if any, that persist today.

We can pursue our investigations from orbit and on the surface—for example:

- We can measure the shape of Mars (and the thickness of its crust) by monitoring minute variations in the gravitational pull of the planet on an orbiting probe.
- We can determine much about the material composition of the planet from orbital-based mineralogical and elemental abundance surveys.
- We can infer the internal structure of the planet by calculating its *moment of inertia*[1] from measurements of the gravitational influence of Mars upon an approaching spacecraft. With prior knowledge of Mars' shape, spin, mass, and material make-up, knowledge of its moment of inertia allows us to infer its internal structure, mass distribution, and even the depth of the various chemical discontinuities within the planet.

Orbital imaging and topological analysis[2] will also be hugely important to an understanding of the surface, giving an accurate impression of surface features resulting from ancient tectonic, volcanic fluvial, and other geological processes from which we can infer the nature and evolution of Mars' internal activity. Further, low-altitude orbiter and aerial reconnaissance craft can measure the planet's magnetic properties, revealing details of past and present magnetic activity. Determining the metallic content and the nature of convection currents within the planet will also confirm whether an internal electric-dynamo ever existed or the magnetic poles ever reversed.

On the surface we can carry out seismic experiments using a network of remotely controlled landers at various locations across the globe. In addition to monitoring current seismic activity, such experiments can tell a great deal about the internal structure of the planet's core, mantle, and lithosphere; and surface mineralogical and elemental analyses will provide vital details of the chemical make-up of the planet and hence the nature of past internal activity. Other instruments on landers could supply long-term meteorological data—such as wind speed, temperature, barometric pressure, and

[1] The *moment of inertia* of any spinning object tells how resistant it is to spinning and is determined from the object's shape; spin orientation, mass, and mass distribution.

[2] Mapping the three-dimensional relief of the surface.

humidity—that could be used in climate models to infer the past and present Martian atmosphere. Having examined the bulk characteristics of the planet we must then turn our attention to the nature, extent, and evolution of indigenous surface activity. Four major indigenous activities (tectonic, volcanic, atmospheric, and water-based) contributed to building and shaping Mars' early surface, and together with impacts on the surface during planetary accretion and heavy bombardment periods, it is of paramount importance to understand how all these features reacted, especially in the first billon years.

Tectonic and Volcanic History

Tectonic and volcanic activities are intrinsically linked to the bulk character and internal dynamics of a planet. They represent the reaction at the surface to the release of energy from and movement within the interior. As already discussed, Earth is still tectonically active in a way now quite well understood, i.e. crustal plate movement associated with the convective cycling of planetary material driven by heat from the core. Prolonged tectonic and volcanic activity has been central to creating and shaping the continents, oceans, and atmosphere and we now also suspect a deep-rooted connection with the origin of life. If we are to determine whether there was any potential for life on early Mars, we must first understand its tectonic and volcanic activity.

While the planetary mechanisms within Mars may eventually be found to be different to those on Earth, their manifestation on the surface may still have been quite similar from the perspective of life-related activity—providing geothermal energy, creating a crust of similar composition, and releasing similar volatile and biogenic materials to create an atmosphere and water reservoirs. Therefore, while it is imperative to understand the internal workings of Mars in order to understand the nature of its surface activity, it is not imperative for the interiors of both planets be the same. Whatever the composition of Mars' interior, we already know that it has manifested itself on the surface in ways sufficiently similar to Earth to warrant attention. Indeed, despite continuing uncertainty regarding the interior, evidence from orbiting spacecraft show that Mars' surface is abundant with tectonic and volcanic markers from its early history, bearing notably familiar characteristics. The Tharsis and Elysium bulges, Valles Marineris and many other rifts, the vast lava and sedimentary filled lowlands to the north, possible ancient hydrothermals at Syrtis Major, recently discovered hematite in Terra Meridiani (THEMIS), gypsum in the north polar sand sea, and the hundreds of regions showing deep sedimentary layering, are just some of the unequivocal evidence of a planet that was active for hundreds of millions of years or more in its early history. Also, with evidence of volcanic activity throughout Mars' history and even to the current geological era, we must consider the possibility of such activity giving rise to an atmosphere, water activity, and hydrothermal systems, and possibly even subsurface aquifers and hydrothermal systems persisting to the present day.

Tentative evidence of Mars' early internal dynamics and surface activity is already available, of course, though it is far from complete and often prompts as many questions as it answers. We can see, for example, that there are no tectonic plates on Mars today. None the less, magnetic *striping*[3] suggests that Mars once had convective cycling that produced an electric dynamo and magnetic field, as well as some sort of crustal movement similar to plate movement on Earth, lasting for millions of years. Irrespective of tectonic plates ever having arisen on Mars in the distant past, however, it seems that for much of Mars' later history the entire surface has acted as a single sealed cap similar to that found on the planet Venus. Such a *stagnant lid* must have radically affected the interior of the planet and given rise to tectonic, volcanic, and other surface processes that are currently unknown to us. On Venus, for example, the stagnant lid causes a buildup of temperature and pressure within the planet that *vents* once every 500 million years or so, melting and rejuvenating the surface on a global scale. And while Mars does not exhibit such extreme behavior and retains much of its original surface, there are many locations on the surface with unfamiliar features whose origin must be related to the nature of Mars' sealed cap.

We are still generally ignorant of the origin and extent of Mars' tectonic and volcanic activity and the sort of environment they initiated. This is largely because we do not yet know enough about the interior of the planet and, as already suggested, our first objective must be to determine Mars' internal structure and dynamics. In reference to tectonic and volcanic activity we need to determine (1) how heat from within the planet dissipates to the surface, (2) the nature of convective cycling within the mantle, (3) how long it lasted in Mars' early history, (4) how the original crust formed, and (5) how long it took to form. We must then determine the subsequent reaction of the crust to continuing internal activity—for example, the sort of crustal movement and rejuvenation cycles that may have occurred and the extent of volcanic outgassing. We must also determine, from Mars' earliest history to the present day, (1) the evolution of its crust, (2) how the stagnant

[3]Alternating strips of crust with magnetic properties pointing in opposite directions across the planet, only recently identified by Norm Sleep of Stanford University.

lid was formed, (3) how it has subsequently affected the evolution of the interior and of tectonic and volcanic activity, and (4) the surface environment that resulted. It is only with a clear understanding of the nature and full sequence of Mars' particular tectonic and volcanic history that we will finally be able to determine the sort of surface environment it created.

Coupled with investigations of Mars' planetary interior, we must also perform a geological analysis of its surface features and a geochemical analysis of its surface materials. A geological analysis involves determining Mars' tectonic and volcanic processes from the surface features we can observe. Numerous lava plains, crustal bulges, cracks and rifts in the surface, shield volcanoes, and sedimentary layering are everywhere evident, each retaining a small piece of the geological puzzle that is the history of activity upon the planet. Fortuitously, we *can* build a comprehensive picture of Mars' geological history simply because its development slowed down several billion years ago and the planet has retained such an extensive record of its past. We have a unique opportunity that is not available for Earth, for example, because virtually all traces of activity beyond several hundred million years ago have been erased by the tectonic activity and the physical and chemical weathering, etc., that has happened since. But with only a tenuous atmosphere and evidence of its ancient past clearly visible across the globe, much of Mars' geological history over four billion years is accessible to us. By constructing orbital maps of the globe at various resolutions from kilometers to single meter scales, together with some three-dimensional topological surveys, we can now begin comprehensively to characterize almost the entire surface geology.

We can also investigate Mars' tectonic and volcanic activity from a geochemical perspective, both from orbit and on the surface. An elemental analysis of the surface will reveal its composition, while a mineralogical analysis will tell *how* that surface was made and hence the tectonic and volcanic processes that were involved, because the crystalline structure and mineralogy of materials are a result of the processes that produced them in the first place. It is therefore possible, through different processes, to have two minerals of similar elemental composition but different mineralogy. Each mineral is unique to the particular volcanic, tectonic, temperature, and pressure conditions by which it originated, and an examination of Mars' mineralogy is therefore an important method of deciphering its past with respect to those processes. Mineralogy can even reveal the evolution of Mars' surface processes. By examining variations in mineralogy through sedimentary layers at various locations, we can also infer the surface processes at each location and at different moments in history.

Mineralogical analysis is one of the most powerful tools at our disposal as we attempt to determine the internal structure and tectonic and volcanic history of the planet. Further, because Earth and Mars have quite a similar composition, we can compare many of Mars' minerals with similar ones on Earth whose origin and formation are known. Taking the mineralogical information and tying it to pictures of surface features (i.e. from the Mars rovers) formed by various geologic processes (e.g. eolian, fluvial), we may even be able to infer the role of liquid water on the surface. Such *comparative planetology* can hugely enhance our understanding of the processes that gave rise to Mars' surface. There is no reciprocal argument, however; while we hope to learn about an early Earth by studying Mars, we are not trying to learn more about processes that are already known to us. But if we are eventually to uncover processes on Mars that are relevant to Earth's earliest history and the origin of life, we must first uncover the prevailing conditions at the time, and comparisons with Earth's mineralogy can play a significant part.

Impacts

While interior activity has radically shaped Mars' surface, so too have bombardments from space. During the earliest era of the Solar System all of the inner planets, including Mars, endured a pounding by fragments that were left over from formation of the planets themselves. Millions of rocky, metallic, and icy worldlets drifted throughout the Solar System, but as they attempted to achieve stable orbits around the Sun, their paths were disrupted and their fates were sealed by the gravitational influence of the larger and still emerging planets.

In an era that lasted for hundreds of millions of years and ended about 3.9 billion years ago, all the major worlds of the Solar System exerted their gravitational domination on those countless remaining fragments, mopping them up and suffering the consequences for doing so. In a final stage of dramatic activity called *the period of heavy bombardment*, the surfaces of all the rocky planets and moons were radically altered by a myriad of impacts.

While scooping up the largest fragments first, each planet suffered repeated poundings of truly gargantuan proportions. Imagine, for example, a worldlet several hundred kilometers in diameter colliding with the Earth at 30,000 kilometers per hour. If it occurred today, the energy from the impact would send shock waves and debris around the globe that would obliterate the entire surface. Much of the atmosphere would be blasted into space and fundamentally altered. All of Earth's oceans would vaporize, and the planet would become a sterilized, murky globe with an atmosphere of vaporized rock and toxic gases, showing no signs of ever having hosted life.

Incredibly, Earth probably suffered such a fate at least several times in its earliest history, with each impact resetting the entire planetary surface to one of utter desolation. But life still managed to emerge. Ironically, despite the devastation, many impactors brought fresh supplies of minerals, metals and volatiles. Although it is difficult to quantify, we suspect that those early impacts also brought a significant proportion of Earth's water and may have been a major source of the organic matter from which life emerged. There is little doubt that the bombardment period played a significant role in shaping the young planet. Eventually all the largest fragments were either absorbed or managed to achieve more stable orbits, becoming asteroids and icy comets. With increasingly smaller and fewer impactors, the period of mass bombardment eventually faded.

Mars endured a similar fate, and we can be confident that it was also radically affected, with each collision bringing similar devastation yet fresh supplies of volatiles and organic material. Significantly, while all traces of Earth's early bombardment have long since disappeared, Mars retains a substantial record of its impacts from that time. We have only to glance upon the gigantic southern hemisphere basin *Hellas* to realize just how devastating those impacts were. As originally revealed by the Mariner and Viking spacecraft, this impact, which created a basin 2,000 kilometers across and 9 kilometers deep, blasted off enough material to spread around the globe, alter the elevation profile of the planet and hence radically alter the planetary drainage pattern.

Hellas and the thousands of smaller collisions evident on the planet are testimony to the ferocity of that early period and the enormous effect of such impacts on the planet; and if we are to understand the true nature of early Mars we must also accurately chart the legacy of all of its impacts. We must build a case-by-case history of each impact site, determining its effect on indigenous processes such as tectonic and volcanic activity, the surface environment and climate, and the resources that were most probably brought to the planet. Even though there were no doubt differences in the bombardment histories of Earth and Mars—impact rates and material composition differences in their respective part of the Solar System, for example—an understanding of Mars' impact history will contribute significantly to an understanding of similar activity on Earth.

Mars' impact history is of fundamental importance in another respect—in determining the chronological sequence of the emergence of all of its surface features, from which we can then determine the evolution of the processes bringing them about. We can do this by mapping the number, density, and distribution of craters of various sizes across the planet. According to our best ideas on how planets form, a rocky planet slowly accumulates over millions of years from the merger of ever larger planetesimals, eventually leaving behind millions of fragments ranging in size from hundreds of kilometers across to boulders, rocks, pebbles, and dust. We surmise, therefore, that all the planets initially endured heavy bombardments from fragments large and small, and while the largest impacts were fewer in number, such were the overall number of impacts and their random nature that we would expect impacts of all sizes to have been evenly distributed across each planet.

On Earth, for example, there may have been several impacts of the size of Hellas on Mars, and those impacts should have appeared uniformly across the planet. But with indigenous planetary activity at that time, Earth would have quickly wiped away all traces of its initial impact basins and craters, leaving behind an unscathed surface. When most of the large and smaller fragments had eventually been absorbed, any new impacts were much smaller and fewer in number. Hence, any planet today showing high densities of large and small craters must still retain some of its oldest surfaces from the heavy bombardment era, while younger surfaces rejuvenated by indigenous activity will be devoid of large craters and basins and retain perhaps only a relatively lower density of smaller craters. Surfaces that are only millions of years old (as on Earth) are so new that they may retain no craters at all.

Therefore, by mapping the cratering pattern across any planetary surface, we can determine the relative ages of its various terrains from the size and distribution of its craters. This technique is particularly significant on Mars because its major indigenous activity seems to have occurred more or less in singular events. For example, in the mid-latitude highlands we see clear evidence of heavily cratered terrain most likely from the heavy bombardment period, while elsewhere we see surfaces characterized by younger tectonic, volcanic, aqueous, and eolian features. As these are all less cratered than the southern highlands, they must have occurred more recently. By mapping the number, sizes, density, and distribution of craters across all regions, we can build a picture of their relative ages through the Noachian, Hesperian and Amazonian periods and hence the chronological sequence of the activity that caused them. Although this technique is most appropriate to determining relative ages, comparisons with impact rates on our Moon and other planets and moons—combined with other analytical techniques such as ages derived from cosmogenic dating of meteorite samples—give us some constraints, perhaps eventually allowing us to attribute absolute times to those events and deduce a definitive history of Mars.

The Search for Life: Water and the Atmosphere

9

Life on Earth is totally dependent on water and has been since the beginning. Similarly, for life to have emerged on Mars, the availability of liquid water on or near the surface would have been a basic requirement. If water had been scarce, any chance for life would have been seriously constrained.

We know that liquid water resided on Mars long enough for river networks to form across the planet. This is crucial; it tells us that, when compared to scorching Mercury or the freezing outer Solar System, Mars was conducive to extensive water and volatile activity. It does not, however, tell us enough. There are currently unknowns regarding the length of time over which the river networks formed, how they formed and if they formed on the surface, beneath the ice, or some combination of both. Nor have we any knowledge of the general surface conditions on Mars during that period. The uncertainty is not about the existence of liquid water or the volatile activity occurred, rather, it's about the prevailing surface and climatic conditions, the extent of that activity, and whether it brought about complex chemical or biological systems.

The surrounding temperature and pressure are crucial to the liquifaction of water. If the temperature is too low, the water freezes; and even if the temperature is high, the pressure must also be high, otherwise the water vaporizes. On Mars today, water cannot reside on the surface as a liquid because even at times when the temperature rises above the melting point, the atmospheric pressure is too low and water sublimes, never achieving a liquid state. If liquid water existed on ancient Mars, the question then becomes: under what environmental conditions were the required temperature and pressure achieved, and for how long did such conditions persist?

As already mentioned in Chapter 5, one possibility is that Mars originally possessed a dense atmosphere that created a greenhouse effect capable of raising the temperature and providing adequate pressure. There are currently several concerns with such a hypothesis, however. Theoretical and computer models suggest that the Sun was about 30% dimmer in the

K. Nolan, *Mars, A Cosmic Stepping Stone*,
DOI: 10.1007/978-0-387-49981-9_9, © Praxis Publishing, Ltd. 2008

beginning, placing great demands on the atmosphere alone to create the required greenhouse effect. Studies also suggest that even if Mars' original atmosphere was five times as dense as Earth's current atmosphere (which is plausible) and composed of carbon dioxide and other greenhouse gases, it still probably would not have achieved temperatures above the melting point of water. Hence, while we cannot discard such a scenario, it is difficult to see how Mars could have achieved the required conditions via its atmosphere alone. Another possibility is of vast ice sheets covering much of the planet, beneath which rivers could form, heated by geothermal energy and sustained as a liquid by pressure from the ice above. In this scenario, a reasonably dense atmosphere would have still existed, but with surface ice sustaining liquid water beneath. A third possibility involves salts dissolved in water, allowing it to remain liquid at cooler temperatures, while a fourth possibility involves a combination of all, where a mild greenhouse effect coupled with geothermal energy could have supported the necessary temperature and pressure to sustain limited liquid water. It is clear, however, that if we are to discover the true nature of Mars' ancient river valleys and networks, we must first determine the nature of the planet's early atmosphere and climate.

There are many other important reasons to understand the nature of Mars' ancient atmosphere. First, the atmosphere was in itself a vital reservoir of volatile materials such as carbon dioxide, nitrogen, water, methane, and ammonia and whose original inventories would have critically affected the types and complexity of chemistry occurring on the planet. Second, the atmosphere and the ensuing climate would also have dictated the cycling and intermixing of volatiles and biogenics through the atmosphere, crust, and liquid water reservoirs. Finally, the atmosphere acted as a means of transport of material and energy through weathering, erosion, and deposition, vital to shaping both the surface and the prevailing environment. Currently, however, we know little of those issues, and our high priority is to determine the density, chemical composition, temperatures, and barometric pressures of Mars' original atmosphere, the resulting climate, and its duration. We must also determine the cycling that occurred between the atmosphere and crust, especially for water and carbon, and we need to know what happened to the atmosphere and the mechanisms that brought about its demise.

Mars Today: A Key to Its Ancient Atmosphere

On first consideration, it may seem incomprehensible that we could ever determine all of this. We must remind ourselves, however, that Mars preserves a substantial record of its ancient past and as our analytical techniques become more sophisticated we can gain better insight into that past. It will take time, but we are increasingly confident that we can eventually build an accurate picture of the atmosphere and climate from that period.

Today, Mars' atmosphere is extremely thin. While Earth's atmosphere has an average pressure of about 1,000 millibars, Mars' average atmospheric pressure is only 6 millibars—less than 1% as dense. It is composed of approximately 95% carbon dioxide, the remainder being nitrogen and trace levels of gases such as ozone and argon. Intriguingly, the fact that the atmospheric surface pressure is today remarkably close to 6 millibars, called the *triple point of water*,[1] suggests that the atmosphere is self-limiting, meaning that if the atmosphere were to become thick enough to facilitate liquid water, carbon dioxide would be removed as carbonate deposits, reducing the atmospheric pressure gradually back to 6 millibars. Mars is therefore a planet in dynamic equilibrium with regard to its current volatile materials. This is important because it tells us that the planet is neither static nor dormant. If environmental conditions significantly change, a corresponding change across its volatile reservoirs will also occur.

Why should we suspect that Mars' early atmosphere was different to that of today? First, volcanic and tectonic activity occurring on early Mars would have caused substantial outgassing, possibly leading to a dense atmosphere. Also, orbital images show that the walls and central peaks of many ancient craters are heavily eroded, suggesting a dense atmosphere in the past. Further, the widespread presence of river networks on Mars, whatever their nature, suggest an atmosphere that must previously have been denser.

Through sophisticated isotopic, geological and geochemical analyses, we are now in a position to attempt to determine the nature of that original atmosphere. As mentioned previously, isotopic analysis is a valuable means of tracking a natural system back through time. In the case of Mars' atmosphere, the key is to acquire a sample of its current atmosphere and examine its isotopic composition.

Isotopes are variations of a given element; for example, the element hydrogen, normally composed of one electron and one proton, can exist in other forms called isotopes of hydrogen. One of these, called deuterium, is heavier than normal hydrogen because it possesses an extra particle called a neutron in each atom. An atom of deuterium, while behaving chemically like hydrogen, is actually heavier because it contains the extra neutron.

[1] The triple point of water is the temperature and pressure at which water can exist as a solid, liquid, and gas in the same environment and below which liquid water is unstable.

Deuterium is very rare in nature—comprising less than 0.001 of 1% of any given sample of hydrogen. Most other elements also have isotopes. Carbon, for example, normally with six protons and six neutrons (and therefore called carbon-12), also comes as an isotope with six protons and eight neutrons, called carbon-14. Isotopes are useful because we generally know (under stable and natural conditions) the fractions of different isotopes to expect in given samples. If we examine a sample and find that those fractions have changed, the system must have undergone a process that removed one or more of the isotopes. In the case of Mars' atmosphere, samples of its present atmosphere examined from the SNC meteorites show less hydrogen than expected in its original atmosphere *and* an excess of five times more deuterium. This tells us that, over time, something happened to Mars' atmosphere that removed hydrogen *and* preferentially removed normal hydrogen over deuterium.

Estimates of the overall amount of original hydrogen suggest that Mars has lost more than 99% of its original atmosphere. Furthermore, the excess in deuterium suggests that a sizeable proportion of the atmosphere was lost through leakage into space. With no magnetic field currently protecting Mars from harmful radiation, cosmic and solar particles have relentlessly bombarded the upper atmosphere for billions of years, providing its constituent elements with sufficient energy to escape the planet. Since normal hydrogen is lighter than deuterium, it has been able to escape more easily, leading to an overall loss of hydrogen and the excess of deuterium that we find today.

While isotopic analysis from the SNC meteorites suggests a once dense atmosphere, it provides nothing like a sufficient picture of the actual density, composition, or evolution of the original atmosphere. The technique, however, is powerful and with further and extensive isotopic analyses on numerous atmospheric constituents (and especially inert gases like Argon), both from orbit and with robotic landers, we are confident of eventually building up a coherent picture of the original atmosphere.

There are other ways to determine Mars' ancient atmosphere. From a geological context and with high-resolution orbital imaging and topological mapping of the entire planet, we can characterize surface geological features with adequate precision to infer the atmosphere within which they must have formed. From water-based geological features, we can infer the extent of water activity and hence the atmosphere needed to sustain that activity. From dried river valleys and eroded craters and from sedimentary layering to the eolian forces that subsequently exposed them, they all required an atmosphere—and through extensive geological analysis we hope to eventually constrain the nature of that atmosphere.

Finally, we can also learn much about the atmosphere from a geochemical perspective. By analyzing the elemental and mineralogical composition of the surface materials, we can infer the methods by which those minerals were formed and hence the prevailing atmospheric conditions and initial stocks of biogenic and volatile materials. Mineralogical analyses from orbit and on the ground will point to the material nature and extent of cycling of atmospheric materials through the crust, the polar and glacial ices, and the standing bodies of water. An examination of variations in mineralogy across different layers of sedimentary deposition, and on differently aged surfaces, will enable us to track changes in the atmosphere and climate, revealing when and how it evolved.

Water on Mars

As mentioned at the beginning of the chapter, water is critical to life as we know it. The entire story of life on Earth—from its emergence to its evolution and within every cell of every organism that ever existed—has critically depended on liquid water for particular reasons.

First, water is an excellent *internal phase* for the living cell. Chemical reactants relevant to life can dissolve in water and move freely and uninhibited until they interact chemically, while many reactions that would otherwise never occur do so copiously within a body of liquid water. Furthermore, the water molecule is uniquely known as a *polar molecule* and is bestowed with properties relevant to life. A polar molecule has an imbalance of electrical charge across it. In the case of water (comprising two atoms of hydrogen and one of oxygen), each hydrogen atom suffers a loss of shared electrons making it slightly positively charged, while the oxygen retains a greater share of the electrons, making it more negatively charged. This polarity gives water intriguing properties. For example, the charge imbalance can *herd* and *guide* other molecules relevant to life from one location to another and into specific positions. As an example, water within living cells helps to support and position amino acids in the shaping of enzymes.

Polarity also bestows water with bulk properties that are important on a planetary scale. In particular, when compared to other compounds whose constituent molecules are small, water is incredibly robust in the face of changes to its surrounding environment. For example, it is because the freezing and boiling points of water are abnormally extreme that it exists as a liquid on Earth in the first place, while compounds such as carbon dioxide and ammonia can only exist as gases. Water also has very high specific and latent heat capacities, meaning that it is difficult to raise or lower its temperature. For example, it requires 10 times more heat to raise the temperature of water than that of iron. This makes water an excellent climate moderator, because it absorbs excess heat extremely well without wide swings in its temperature. Also, as it has a high latent heat, it is also very difficult to change the state of water from solid to liquid or from liquid to gas. While we witness such transitions readily in our everyday experience (within your ice box, for example!), in nature it is extremely difficult to alter the phase state of water, allowing for large swings in heat conditions on Earth to be regulated by its planetary water systems without severe alteration in their state. And even when water freezes, the ice produced becomes less dense than liquid water and remains at the surface. Hence, a river first freezes on the surface, protecting the liquid below and life within. When we consider that life on Earth existed almost exclusively within water for the first two billon years, the importance of this particular property becomes clear.

For all of these reasons, liquid water has proved to be one of the critical ingredients in the emergence, survival, and evolution of life as we know it; and we imagine that if life emerged on Mars, water must have been equally important. One of the primary goals for the current exploration of Mars is therefore to find evidence of water that can then reveal the conditions on the planet relevant to life, past and present.

While we are already aware that water played an important role in Mars' early history, we are far from understanding its full role. Therefore, if we are to understand the story of water on Mars, we must not only determine the original quantities, but also the amount that was indigenous and/or delivered through comet bombardment. We must determine the nature of the planet's hydrological cycle across its liquid reservoirs, the atmosphere, and polar regions and the sort of planetary environment they originated. We must determine how river networks and lakes were formed, how long they lasted and whether they were surface, subsurface, or subglacial in nature. We must also determine the evolution of the planetary water systems; for example: why are the river networks predominantly on the oldest terrain while flood channels are on younger terrain? This and similar questions must be pursued on an individual basis for every site indicating past water activity. We must then proceed to more specific questions about the cycling of water through the crust, the production of hydrated minerals, the nature of hydrothermal activity, aqueous erosion, deposition, and sedimentation. And we must investigate activity relevant to life—aqueous-based chemistry leading to organic synthesis, polymerization, metabolic energy production and even the emergence of life. Finally, we must also discover where on Mars water resides today and any water-based activity that persists under the surface, within the atmosphere, at the poles and even on the surface at niche sites, under specific weather conditions, or during periods of climate change.

As with the other planetary surveys, these question can be pursued first from orbit and then with follow-up missions to specific sites of interest, and once again we can approach our investigations both from geological and geochemical perspectives. Geologically, we begin with orbital reconnaissance by obtaining high-resolution images and topological maps of the entire planet, from which to assemble a comprehensive picture of geological features created and shaped by water. River valleys, flood channels and plains, lakes, seas and deltas, sedimentary layers, lobate craters, and mud flows—all were created on Mars billions of years ago and can be directly imaged from orbit. And while we have already accumulated overwhelming evidence of such features, it will only be after imaging thousands of individual sites to resolutions of 1 meter or better, followed by exhaustive analysis, that we will finally determine the full picture of past water activity on Mars.

Orbital mineralogical surveys are also immensely valuable in identifying sites of past and present water. Here, we can survey the planet for minerals that occur only in the presence of water. Not only that, but particular minerals will indicate the type of water environment within which they formed, pin-pointing lakes, seas, hydrothermal systems, and sedimentary basins. The precise nature, quantity, and geographical extent of each mineral should reveal the extent, duration, and variations in time of water at the site. One mineral of particular interest is called *hematite*—an iron oxide that forms when an iron-based mineral (such as those found in basalt rocks) comes into contact with water. It can form at numerous natural environments, including hydrothermal vents, volcanic settings, and within lakes and seas where the hematite precipitates onto the sea floor as sediment.

We can also look for chemical precipitates. These are salts that form in long-standing bodies of water and precipitate out as sediment on lake and sea floors. Metals such as magnesium found in mineral rocks when diluted in water, lose electrons to the water and become positively charged *cations*. Conversely, other materials such as carbonates, sulfates, and halides gain electrons when dissolved in water, becoming negatively charged *anions*. The seas of Earth are awash with a variety of cations and anions—metals from mineral rocks and carbonates produced when carbon dioxide in the atmosphere reacts with water. But when concentrations become too high, the dissolved cations and anions chemically combine to form salts that precipitate out of the water and settle on the sea floor. Table salt (sodium chloride) and limestone (calcium carbonate) are two common examples of precipitated salts from our seas.

With a surface composition similar to that on Earth, we expect that any past seas or lakes on Mars would also have produced precipitated salts that we could identify both from orbit and with landers. As we will discuss in detail in later chapters, NASA's Mars Global Surveyor, Odyssey, and Opportunity spacecraft have already worked in combination to discover hematite from orbit and precipitated salts at that same location on the ground. These far-reaching discoveries were made between 2001 and 2003 and represent unequivocal evidence of prolonged, surface water in a region called Meridiani Planum, verifying for the first time the existence of a shallow sea on Mars in its distant past.

There are numerous other ways of looking for water on Mars. Detailed topological maps will show local and global drainage patterns, revealing where water was likely to be, as well as allowing the volume capacities of ancient river and flood channels to be calculated. One of the most intriguing ways of searching for water on Mars today is via *Radio Sounding*. Here, a type of radar from orbit can probe the subsurface to depths of several kilometers and identify current underground reservoirs of liquid water or ice. An orbital scanning technique called *gamma-ray spectroscopy* will also reveal the presence of near-surface water, whether bonded to the soil as hydrated minerals or mixed as ice and slush.

With so many means at our disposal, we are increasingly confident that we can determine the prevailing atmosphere, water, and climate on Mars in its early history, from which we can then begin an earnest search for evidence of life-related activity.

The Search for Life: Past and Present

10

The search for the origins of life in the Universe may be fulfilled on Mars. If life emerged there and we find evidence of it, we will have verified a cosmic context for all life. Considering how we think life emerged on our planet, and its broad similarities to Mars, we cannot ignore the possibility of life having arisen there also. Such a possibility mandates a search.

While we can have little idea of the nature of any such life, we have good reasons to think that it would be based on carbon and, in particular, on complex organic molecules such as nucleic and amino acids. If life replicates, metabolizes, and evolves, then complexity in structure is needed and carbon is the most capable element of providing it. It may be that life based on different elemental building blocks can exist, but if life has arisen elsewhere it will more often be based on carbon. It is also worth noting that the diversity of life on Earth is due to its evolution and not specifically to its origin, and that carbon-based polymers—DNA, RNA, enzymes, etc.—are common to virtually all life on Earth. Therefore, while evolutionary forces will drive life on different planets in ways particular to each, and produce very different organisms, the functional building blocks making up all life may be quite similar everywhere.

We envisage three broad searches for life-related activity on Mars. First, we must look for evidence of prebiotic chemistry—complex, precursor activity such as organic synthesis, inorganic energy production used in primitive metabolic activity, as well a range of mineralogical activity and environments capable of providing stability and acting as catalysts for the emergence of life. The second search considers the emergence of microbial life on Mars during its early history, but became extinct when Mars slowed down over three billion years ago. Here we must look for evidence of ancient sites that might retain fossils of any such life. Third, we must consider the scenario of microbial life on Mars today, arising on ancient Mars and surviving to the present day in favorable locales and during periods supportive of life.

K. Nolan, *Mars, A Cosmic Stepping Stone*,
DOI: 10.1007/978-0-387-49981-9_10, © Praxis Publishing, Ltd. 2008

The Search for Prebiotic Chemistry

While it is more usual to ponder questions about the actual emergence and survival of life on Mars, our first scenario considers the question of prebiotic chemistry (precursor chemistry that, on Earth, eventually led to life) also arising on Mars—whether or not it led to the emergence of life there. With both planets broadly similar in the beginning, it is plausible that such activity arose on Mars. This idea is attractive to us. If prebiotic activity arose there, evidence may still exist. On Earth, subsequent organic life obliterated any evidence of the prebiotic conditions, rendering it virtually impossible to trace the chemical evolution that produced life initially. But with Mars retaining ample evidence of surface activity from over three billion years ago, indications of prebiotic chemistry might still exist. Investigations into that activity would, as a minimum, provide a new perspective on conditions prevailing on Mars and Earth during the emergence of life, and possibly point to specific environmental conditions and chemical pathways pertinent to the origin of life on Earth. In this respect, Mars is an unparalleled opportunity to explore our origins.

The search for prebiotic chemistry on Mars begins with the study of the bulk planet—its make-up, tectonic and volcanic activity, abundance and activity of water, and the inventory of volatiles and biogenics: hydrogen, carbon, nitrogen, oxygen, sulfur, and phosphorus among others. Only when these are characterized and quantified can we begin to tackle questions about the possible occurrence of complex chemistry at particular geographical and geological settings. We are particularly interested in searching for evidence of any activity leading toward self-reliant and self-replicating systems, involving organic molecules in inorganic settings with stable sources of inorganic energy that may have been precursors to metabolic energy.

Organic matter would have existed on Mars in the beginning. Indeed the SNC meteorites contain organic materials that probably originated on Mars. But if we are to understand the prebiotic chemistry that occurred on Mars, we must determine how organic matter originated, both from comet and asteroid bombardment as well as from indigenous organic synthesis on the planet.

The traditional view of organic synthesis taking place under an atmosphere filled with reducing gases such as methane and ammonia, and sparked by solar radiation and/or lightening, is no longer seen as the major contributor to organics on young planets. While such a mechanism would have produced some organic matter, it is now clear that methane and ammonia—so important in the process—cannot last long on planets that also have water vapor in their atmosphere, because water vapor reacts with those gases, producing nitrogen and carbon dioxide. Any organic synthesis via the atmosphere would therefore have been relatively short lived.

Instead, recent ideas point to organic synthesis occurring through water–rock interactions on and below the planetary surface and in the presence of carbon dioxide. A process called *reduction* fuses hydrogen (released through water–rock interactions) with carbon dioxide in the atmosphere, producing a range of simple organic molecular species. This would have been possible at sites on Mars where igneous rock and water interacted. Even if Mars was too cold to allow liquid water to exist on the surface, internal planetary activity would have created a geothermal gradient leading to water melting at some depth below the surface, which could then interact with the newly formed igneous rock. One plausible chemical pathway, for example, involves fayalite (a type of olivine found in basalt rocks) in contact with water, oxidizing to ferric hydroxide and ferric silicate, yielding hydrogen gas. The hydrogen would then reduce atmospheric carbon dioxide (i.e. combine with it), producing a variety of simple organic materials.

There is no reason to doubt that these, or similar mechanisms, were taking place on early Mars, and finding evidence of their existence is now a priority. Our search will focus in particular on finding evidence of past and present hydrothermal systems which occur on geologically active planets with liquid water. Hydrothermals form when molten rock solidifies, cools, contacts, and fractures. Water circulating though the region heats, expands, and circulates to colder regions and it is within such environments of igneous rock, rapidly circulating water, and elevated temperatures that hydrogen gas can be produced and released, subsequently bonding with carbon dioxide from the atmosphere to produce organic molecules.

We must also look for plausible sources of early metabolic energy from inorganic sources, in constant supply and within stable environments where organic synthesis was also taking place. Of course, many sources of energy are readily available in nature. Solar energy is an obvious candidate, but it is now considered not to be a primary source of energy in the emergence of life because the complex chemical pathways required to harness sunlight had not yet emerged. Instead, natural chemical and electrochemical energy already at the environments of early organic synthesis appear to be more likely and viable energy sources. Two plausible sources (among others) include energy from acid–alkali boundaries at subaqueous hydrothermal vents, and energy released through the weathering of igneous rocks by flowing water at hydrothermal and volcanic systems.

While we must consider (and look for) all possibilities of organic synthesis and energy production, searching for evidence of ancient hydrothermal systems is particularly important because of the possibility of both organic synthesis and energy production. As with other orbital surveys, we can search for geological and geochemical evidence of hydrothermals. Geologically, we will look at the most ancient landscapes on Mars for signs of ancient water activity close to volcanic sites. We can also look for ancient subsurface and subaqueous hydrothermals that have been exposed through weathering and erosion. Geochemical (mineralogical) surveys will be particularly revealing. Water–crust interactions happening under the particular temperature and pressure conditions of hydrothermal systems produce unique mineralogical signatures that are readily identifiable. For example, water-altered hydrous minerals—chlorates, smectites, and micas—as well as anhydrous feldspars and pyroxenes can all indicate sites of hydrothermal activity.

Once favorable sites have been identified from orbit, the next step is to send landing craft to the surface to characterize the site geomorphology, geology, elemental abundances, and mineralogy to determine the processes that occurred in the distant past, and whether volatiles and biogenics were involved and capable of bringing about organic synthesis and energy production. We can use isotopic analysis of surface materials to search for imbalances in the expected ratios—revealing any cycling of material through the atmosphere, liquid water reservoirs and crust, and suggesting hydrothermal activity. If variations in isotopic fractions are found on differently aged terrain, we may be able to determine the time of origin of varying organic species they contain, and whether they are indigenous to Mars or delivered from impacts. If and when carbon-based material is found, various techniques can be used to determine whether it is organic or inorganic, as well as identify the molecular shapes and types. For example, by determining the relative quantities of carbon, hydrogen, and oxygen as well as the atomic bond types, we can even identify classes of organic species such as fatty acids, amino acids, and nucleic acid bases.

While such tests are completely achievable by robotic explorers today, in reality it will only be after an exhaustive search at dozens of candidate sites that we may gain any insight into prebiotic chemistry on Mars. We may have to drill into the surface to extract pristine samples, or return samples to Earth for even more detailed analysis. However arduous, we are increasingly confident that we can determine whether prebiotic chemistry occurred on Mars and that our search will reveal hitherto unconsidered scenarios for processes relevant to the origin of life.

The Search for Extinct Life on Mars

For five-sixths of the age of the Earth, all life was microbial. Even today, the living cell is the cornerstone of all living morphologies. Entirely and from the beginning, membrane-encased water-filled cells have provided the protective environment within which the activity of life on Earth has functioned. Only after billions of years, when the biosphere of Earth had settled, did larger life forms emerge. For similar reasons we think that life originating on Mars would also have been microbial, remaining so for all of Mars' early and active period.

Let us imagine a situation where life originated elsewhere but took root Mars. Consider, for instance, life-bearing material from Earth blasted into space by asteroid and comet impacts during the heavy bombardment period. Some of that material could have found a trajectory toward Mars, eventually arriving and surviving entry to the surface and perhaps even finding conditions conducive to survival. Another scenario, called Panspermia, proposes that life, being an unlikely phenomenon, arose only once in all the Cosmos and is disseminated as microbes throughout the Universe and throughout the galaxies within nebulae, comets, and asteroids. As far-reaching as this sounds, theories by Fred Hoyle and Chandra Wickramasinghe reveal how plausible this may be. However, even if particulars of Panspermia are found to be incorrect or incomplete, there are strong reasons to suggest that microbial life can traverse and survive the distances between the stars and arrive on new planets, once there perhaps to take seed and evolve according to the local conditions. Whatever the considered scenario on how life emerged on Earth and may emerge on other rocky planets, all seem to point to microbial life being the first type to take root and remain for such long periods. A search for ancient life on Mars would therefore suggest, primarily, a search for microbial life.

Of course, if we have learned anything from past scientific endeavors, especially the Viking landers, it is that we must tread cautiously when considering scenarios for life elsewhere in the Universe. While we must strive to conceive of generalized biological principles that might apply everywhere, we are not yet privy to any other environment harboring life beyond Earth. So, while we think it valid to look for evidence of ancient microbial life on Mars, we must remain vigilant and objective, looking for all possible patterns and morphological structures that suggest the presence of life. It may be, for example, that Mars' particular make-up, size, and planetary history brought about life and evolutionary paths different to those on Earth; and we have no way of anticipating how these would impact on the nature of life on that planet. We are confident that a search for microbial life is valid, but we must also think, and look, beyond that scenario.

We are fortunate to be able to look for evidence of extinct life on Mars because its landscape and rocks preserve a record of its ancient history. There are probably numerous types of evidence available—morphological, sedimentary, chemical, and isotopic. Morphologically, we can search visually for fossils of individual microbes within rock samples. For fossilization to occur, rapid sedimentation and entombment are needed within impermeable minerals, and if we consider life on Mars originating and surviving in aqueous environments then we should look particularly for aqueous sedimentary rocks. On Earth, about 25% of all fine-grained siliciclastic rocks, 10% of all carbonates, and 50% of silicified carbonates contain microfossils, and a search for these minerals should also be a priority on Mars.

We can also look for fossilized evidence from larger features such as stromatolites—large fossils that show the life processes of microorganisms such as cyanobacteria. On Earth, these primitive cells live in huge masses that can form floating mats or even extensive reefs, calcium carbonate layers, and domes. On Mars we can look for similar large morphological structures, and any evidence identified visually would need to be followed up by detailed microscopic, chemical, and isotopic analyses to confirm a biological origin.

Chemical analyses of ancient strata would reveal biomarkers such as the presence of particular amino acids, lipids, and phosphates, pointing to a biological origin. Also, as with the SNC meteorites, we can look for various potential biomarkers—such as magnetite, iron sulfide, and low-temperature carbonates—that, if found, suggest a biological origin because they cannot remain in equilibrium when generated by inorganic means. An isotopic analysis of carbon-based compounds showing, for example, a bias of carbon-12 or contrasts in sedimentary organic carbon over inorganic carbonates, would also suggest past life activity.

There are many potential sites we can search. Ancient thermal springs and hydrothermal systems forming at volcanic sites, for example, might have acted literally as oases, providing water, reduced gases, and the chemical nutrients needed for life. Deposits from such systems—carbonates, organic-rich cherts, silicas, and iron oxides—would all preserve fossils in reasonable condition and could have survived to the present day. Ancient lakes and seas, perhaps seats of *life's last stand* on a dying planet, would also preserve microfossils within sedimentary layered carbonates, fine-grained siliciclastic sediments, and compacted cements. Sublacustrine springs occurring in association with volcanic caldera lakes can also produce deposits preserving fossilized evidence of microbial communities; while evaporate deposits such as sulfate and halide salts, deposited and crystallized on shrinking lake and sea floors, could retain an excellent fossil record of any salt-tolerant bacteria within. Even where the long-term preservation of such fossils may be questionable, brine inclusions within evaporates can provide excellent environments for preserving fossil microbes and biomolecules. Finally, we could look for cemented regolith, produced when water percolated through soils that were deposited further down as mineralized hard-pans. Such sedimentary cements would survive until today, preserving a fossil record of microorganisms trapped during the cementing process.

If microfossils exist on Mars, we are now hopeful that we can find them. The first step will be to search from orbit for geological and mineralogical evidence of ancient hydrothermals, volcanoes, lakes, seas, and sedimentary basins. We can also carry out direct searches of past life from orbit, where mineralogical analyses might reveal biomarkers such as phosphates, carbonates, evaporate salts, and silica-rich cherts. We will then follow up with lander missions capable of drilling into the soil, grinding rocks, gathering samples, and imaging to microscopic resolution to look for fossils within, as well as performing a range of mineralogical, isotopic, and chemical biomarker analyses.

Even with such leads, the search for evidence of ancient life will be enormously challenging. While we may have the means to perform analyses, the discovery of samples—even at the most favourable sites—could be a lengthy and arduous process. Mars is not, and probably never was, a thriving living planet and we are well aware that it will not reveal evidence of past life without exhaustive searches.

The Search for Extant Life

Extant life means biomass that is living and metabolizing today, or is alive and dormant, awaiting more favorable conditions for metabolism to recommence. If life arose on Mars in its distant past, the question of its survival to the present day must be considered. While analyses by orbiters and landers have revealed a barren planet incapable of supporting flora and fauna, we cannot conclude that Mars is currently incapable of supporting microbial life. Our broadening views on the nature of life, increased understanding of the robustness of microbial life on Earth and recent finding of quite diverse environments on Mars suggest that it might be capable of supporting basic life even today.

As mentioned previously, we are aware of bacteria on Earth that have survived conditions far harsher than many of the environments on Mars. It is now clear that microbial life in particular is capable of adapting to major and rapid environmental change, well beyond what would have been considered possible only decades ago. Microbial life arising on early Mars could have adapted to the changes it encountered.

We must also remember that despite the number of space probes currently on and around Mars, none is looking (or is capable of looking) directly for biology and so cannot tell us anything concrete about the presence of microbial life. The only search for microbial life was done by the Viking landers. While it is commonly accepted that no life was found and that the surface conditions at both landing sites were too hostile for life, we cannot draw any planetwide conclusions. Apart from the fact that some results were less than conclusive, even the accepted findings tell us little about *every* site on Mars, above and below ground. Also, the Vikings were limited in their observations—to search for evidence of life similar to that on Earth today. We now know that life on Mars need not metabolically resemble life on Earth. So diverse is organic life even on Earth and so limited were the Viking biological experiments that it is conceivable that they might not have conclusively found life even if it was present.

Note also that orbital surveys to date are incapable of resolving features smaller than a few meters across and can reveal little about local and small-scale activity occurring on the planet. Also, while recent lander missions have not revealed morphological structures indicative of life, this is not a surprise as those probes were not sent to sites that showed a preponderance to life. In summary, it is entirely plausible that if life arose on Mars, it could have evolved to survive through to the present at niche sites. None of our investigations to date can shed any light on the matter, and if we are to find life on Mars we must pursue specific searches aimed at revealing evidence of any form of microbial biology.

We can search for several types of evidence. First, we can look for direct evidence of current life—as in actual living organisms—as well as for signs of growth, metabolic activity, and chemical and biological by-products. Second, we can look for dormant life in a state of hibernation while temporally or geographically separated from conditions occasionally favorable to living activity. Finally, we can look globally for non-living remnants of recent life, as well as particular environments that recently, but no longer, supported life.

Central to the effort will be identifying locations of recent or current water activity on or near the surface, irrespective of how intermittent or apparently insignificant in terrestrial terms. Potential habitats include surface and underground hydrothermal and volcanic systems where geothermal energy could melt water-ice and whose geochemistry could yield, for example, hydrogen, carbon monoxide, and hydrogen sulfide—reactants in non-photosynthetic chemo-autotrophic metabolism. Since there is no significant water activity on the surface of Mars today, such a search must involve orbital techniques capable of identifying evidence of geologically recent water activity as well as currently active small vents and fumaroles.

Another plausible habitat might include deeper and possibly more extensive underground aquifer systems, where once again a geothermal gradient would sustain liquid water and possibly life today. Cryptoendolithic autotrophic bacteria (bacteria which on Earth live within rocks on the cold dry deserts of Antarctica) might also survive on Mars. While Mars appears to be substantially drier than even the driest desert on Earth, this need not be always the case, when, for example, periods of climate change might bring about moist conditions favorable to those organisms. However, even if such microbial communities can occasionally exist on Mars it is likely that, given present conditions on Mars, they would be in a dormant state and any search would be for dormant microbial biomass within rocks in regions suggestive of water activity in geologically recent times.

We can also look for extant life at saline environments where halophilic organisms could reside within evaporate crystals. On Earth, viable microorganisms have been isolated from salt crystals over 200 million years old and it is equally conceivable that halophilic organisms could reside in a dormant state within salt crystals on Mars until more favorable conditions arise for recovery and metabolism. We must also consider the permafrost regions of the high northern and southern latitudes. On Earth, microorganisms can survive for long periods within permafrost ice, and any search for life on Mars should also include consideration of possible microbiology in the Martian permafrost. Also in this category of potential habitats are the recently proposed near-surface frozen sea near Elysium and potential glacial activity on the flanks of volcanoes such as Olympus Mons. While the exact nature and extent of water and ice activity at such sites is ongoing, they suggest that periods of climate change and recent tectonic and/or volcanic activity—where ice flows, and even near-surface localized water flows—are possible away from the polar regions. This suggests a further range of potential habitats for life.

We can also routinely monitor the atmosphere for the presence and fluctuation in quantity of gases, such as methane, that do not remain in an atmosphere unless continuously replenished by volcanic, impact, or life activity. Depending on the rate and accuracy of monitoring, we can locate the region on the surface from which such gases are being generated and follow up with more detailed investigation.

Another suggested habitat has recently emerged—that of liquid water in limited amounts and for limited times on the surface of present-day Mars. Based on data collected from the Viking and Pathfinder missions, a diurnal water cycle on Mars suggests that water vapor in the air freezes by night and melts during the day. As the day progresses, the heat of the Sun causes this liquid water to evaporate back into the air. Viking landers actually photographed a thin layer of frost forming overnight, and now it is proposed (by Dr Gilbert Levin and his son Dr Ron Levin) that this frosty layer does not instantly revert into water vapor at sunrise. Instead, in the early hours of the Martian morning the atmosphere more than 1 meter above the Martian surface in too cold to contain water vapor and the water moisture remains on the ground. Data from the Mars Pathfinder support this theory, as the Pathfinder readings noted that the temperatures 1 meter above the surface were often dozens of degrees colder than the temperatures nearer the surface. This layer of cold air might provide a form of insulation, trapping water moisture below. Since the atmosphere is too cold to hold the water as a vapor and the ground is warm enough to melt the ice, the water melts into a liquid. This liquid water remains on the surface until the temperature of the atmosphere increases enough to allow the water to evaporate. In such a curious manner, the Martian soil may become briefly saturated with liquid water every day. Intriguingly, meteorological data might already confirm the presence of liquid water in the topsoil each morning, where black-and-white and color images show slick areas that may be moist patches. Whether or not this scenario turns out to be viable, it reminds us that there may be many hitherto unconsidered potential habitats. As was pointed out by Wesley Huntress Jr, "Where there is water and chemical energy, there is life. There is no exception."

Apart from looking for potential habitats supporting current or dormant life, we can also look for sites that are inhospitable to life at present, but contain non-living indicators of recent biological activity. For example, young and non-cratered terrain with fluvial features from relatively recent surface water flows might contain deposits indicative of extant life such as organic carbon and particular minerals. The polar ice caps, layered-deposition terrain, and high-latitude permafrost regions could also retain a record of biogenic gases, chemical and biological by-products, and even microorganisms brought to those regions from across the planet.

Strategically, we will carry out a search for extant life in a phased manner, with orbital surveys first searching for sites of potential biological interest, followed by chemical and biological lander missions to many of those sites. From orbit we will search for possible current and dormant habitats and sites possibly revealing non-living geochemical evidence of extant life. High-resolution orbital imaging, for example, can reveal evidence of frozen seas, geologically recent glaciation and glaciation deposits, surface and near-surface aquifers, and gullies. Orbital mineralogy mapping will reveal evidence of water alteration of soils, soils devoid of oxidizing dust and therefore possibly retaining by-products, minerals indicative of recent water activity as well as salts and other evaporate deposits that also point to recent water activity. Of particular interest will be orbital analysis of the polar layered-deposit regions and ice caps to identify sites of particular value and accessibility for follow-up lander missions.

Whereas similar searches concerning ancient life will involve many of the same orbital techniques, searching for current and recent activity will be especially challenging because of its scarcity. Instead we will need to look across the vast landscape of Mars for localized, meter-scale systems, which will require highly accurate and sensitive instruments on orbiters and aerobots operating over a long period of time.

Subsequent to orbital analysis we will send robotic landers and rovers to many sites for detailed chemical and biological analysis. These highly sophisticated robots will be specifically built to cope with their environment. They will acquire samples by burrowing and scooping soil, extracting ice cores from polar caps, and grinding and drilling into rock to take samples of wind-blown particle from the air. They will be equipped with a suite of portable instruments and laboratories—high-resolution cameras and microscopes to identify organisms and spores; spectroscopes that will not only detect carbon-based material and determine if it is organic, but will even characterize the molecular types; and environmental detectors capable of identifying biological by-products in the soil and atmosphere. Sealed chemical and biological laboratories, specific to each location, will conduct analyses for signs of growth, metabolism, or replication.

The search for present and extant life on Mars will represent the pinnacle of our exploration. If we find ourselves in a position to be designing and sending such specialized robotic landers, we will have made tremendous progress, and the results, when painstakingly acquired over many sites and with great patience, will finally begin to reveal answers with far-reaching implications.

A Global Strategy

Unsuspectingly, we have traveled a great distance along the path to Mars. We have sent space probes there for 40 years, and within another 40 we hope to send humans. All missions to date—successful probes like Mariners and Vikings and those that did not make it like Mars Climate Observer and Beagle 2—have laid the foundation for future missions. From bold aspirations and risks to meticulous mission designs, our efforts have brought both technical failures and triumphant successes, but all have contributed to radically altering our view of Mars and its value to us and have prepared us for our next step. Despite our incomplete picture of Mars, we have taken ownership of the planet through those missions. We see it as important, and increasingly we regard it as the next natural destination for humans to visit. Mars is no longer remote and has become a real place.

With our maturing view of the nature of life, we recognize Mars' central role in providing answers. We have drawn together many questions and have developed the technical means to pursue them. But it will not be easy. Mars is an entirely new world with a vast and ancient landscape. It will test our ability and endurance. If we are to succeed we must build a far-reaching strategy for exploration, spanning five distinct phases. This strategy emerges naturally from having formulated the right questions and the order in which to pursue them. Before we can investigate the details of water activity at given locales we must first survey the planet to find those locales, and before we look for biology, we must understand the planet's geology, mineralogy, and chemistry. It is now clear that a general but thorough planetary-scale investigation will lead naturally to follow-up orbital, aerial, and lander missions pursuing increasingly specific and complex questions about the possibility of life there.

There is also a consensus that we must pursue a program appropriate to our financial, scientific, technical, and sociological ability and desire at any given time. A rapid and aggressive program is not likely to succeed simply because it would demand too many resources at the same time. Such a strategy would also be scientifically ad hoc. Instead, a longer term and

K. Nolan, *Mars, A Cosmic Stepping Stone*,
DOI: 10.1007/978-0-387-49981-9_11, © Praxis Publishing, Ltd. 2008

phased approach will serve our exploration goals and better survive sways in financial, political, and social commitment. Furthermore, a phased program will allow for changes in our understanding of the planet through new and appropriately designed missions, with ever-improving technology providing a better capability to explore. For all of these reasons, a phased program over several decades is now seen as the approach with the best opportunity for success.

A Five-Phased Strategy for Mars

Phase 1: Orbital Reconnaissance

Phase 1 of our exploration program involves orbital and aerial reconnaissance with a view to characterizing the nature, extent, and evolution of internal, tectonic, volcanic, water, and atmospheric activity; and determining the original inventory of volatiles such as water, carbon dioxide, ammonia, and methane and biogenics such as hydrogen, carbon, nitrogen, oxygen, phosphorus, and sulfur. We must discover how they interacted, the planetary surface conditions that resulted, and their duration. We must then locate sites of prebiotic and potentially biological activity, especially at hydrothermals and other ancient waterways. Reconnaissance surveys will also characterize the planet as it is today: monitor current geological activity, unveil climate and weather patterns, such as the interaction of the atmosphere with solar and cosmic radiation, and identify water and other volatiles in their current planetary reservoirs. These surveys will add to a subsequent search for prebiotic chemistry and life past and present, as well as prepare the way for sample return and human missions. Numerous orbiters will be needed, each tailored to specific tasks. As we learn from one, another will follow, capable of higher precision and of different yet complementary surveys. Overall, global reconnaissance will involve a broad range of well-defined investigations, each requiring specialist missions and instruments.

Geological surveying will involve imaging the entire planet at various resolutions through kilometer, meter, and submeter scales. While a range of resolutions allows us to determine planetary and regional scale geology, higher resolutions to the decimeter scale will allow for characterization from orbit of individual site rocks and soils and for identifying the underlying processes of sedimentary layering, from which we can build a detailed picture of tectonic, volcanic, impact, water, and atmospheric contributions to the surface morphologies, unlocking the story of Mars' internal structure and geological history. Geological stratigraphic analysis will also allow for the determination of the ages of individual terrains on the surface and hence chart the geological evolution of the planet, as well as identify ancient sites of potential prebiotic chemistry and extinct life and younger surfaces with potential current habitats.

Orbital topological mapping will reveal height information of surface features. An instrument called a *laser altimeter* determines the elevation of the surface by firing a laser at the surface from orbit and timing the delay upon seeing a reflection. Laser light reflections from a low-lying basin will take longer to return to the spacecraft than those from a mountain top, for example. The altimeter can then convert the reflection timings into distances from the spacecraft, revealing the relative heights of surface features. By firing millions of pulses across the globe, a detailed elevation map of the surface emerges. Such a topological map has already been produced by the Mars Global Surveyor space probe. From over 600 million laser pulses fired at virtually the entire Martian surface, Global Surveyor produced a spectacular topological map of almost the entire globe with lateral resolution of about 100 meters and vertical resolution of just 1 meter.

Topological maps complement high-resolution imaging, revealing geological features in all three dimensions and hugely enhancing our ability to determine the underlying geological processes creating them. Topological maps can also reveal the water capacity of river networks and flood channels, allowing us to determine the quantities of water that once flowed through them. More generally, they allow us to see the *shape* of the planet to high accuracy and hence determine planetwide drainage patterns; pointing to where water might once have been and its final destiny.

Another major objective of orbital surveying is to precisely determine the mineralogy of the planet. With high-quality mineralogical maps we can finally begin to determine the material make-up of the crust as well as the processes by which it was formed. It is known that individual minerals form under individual planetary conditions, many of which we can infer through *comparative planetology* with mineral-forming processes on Earth. Given the broad similarity of the mineralogy on both planets, mapping the mineralogy of Mars will reveal the equivalent planetary processes taking place on that planet, such as the nature, extent, and evolution of tectonic and volcanic activity, ancient water systems, and planetwide erosion, deposition, and sedimentation. Mineralogy will also provide insight into the original quantities of various elements, volatiles, and biogenics making up the crust, including water. It will reveal interactions across the atmosphere, water systems, and crust—the cyclical processes that occurred and the climatic conditions that prevailed throughout Mars' history. It will allow for identification of hydrothermal systems and even evidence of prebiotic chemistry and possible life activity. As already mentioned, certain carbonates and phosphates are produced only by living systems on Earth, and identification of those minerals on the surface would strongly suggest past life activity. Finally, mineralogical analyses from orbit will also allow us to characterize Mars' activity today—the transfer of volatiles across the crust, atmosphere and poles, recent geological activity, and the effects of dust, for example.

Orbital mineralogical analysis involves mapping the surface using infrared light (heat) with thermal imaging cameras and spectrometers. Because each given mineral absorbs, re-emits and reflects heat in a unique way, we can determine the mineral content of the surface by observing heat absorption, emission, and reflectance patterns and comparing them with already known heat patterns for terrestrial minerals. This technique can also be used to determine whether a site is composed of bedrock, boulders, pebbles, or sand. Through what is called *thermal inertia*, different sized particles absorb heat during the day and re-emit it at night at different rates. If we monitor a given area during different times of the day, we can determine the thermal inertia of the various materials and infer the composition of the surface particles. Though thermal analysis is a powerful technique for determining surface mineralogy from orbit, difficulties can arise because the atmosphere and surface dust may affect the *truth* about surface mineralogy. While we can infer a great deal about the surface mineralogy from orbit, many of the trace-level minerals associated with sparse life activity may be difficult to locate.

In addition to mineralogical analysis, the elemental abundances of Mars' crust can be probed from orbit using instruments called *neutron* and *gamma-ray spectrometers*. These instruments "see" cosmic neutrons and gamma rays reflected off the surface, and from the reflection patterns infer the atomic structure of surface materials. A neutron detector can detect over 20 elements including carbon, magnesium, aluminum, silicon, and iron, while a gamma-ray spectrometer is particularly useful at detecting hydrogen that is most likely bound to water in the crust, making it excellent at detecting surface and near-surface water, hydrated minerals (minerals with water chemically bound to them), and soils that contain water-ice. While elemental analysis provides a powerful mechanism for estimating the original inventory of volatiles and biogenics, there are limitations to what can be done from orbit, particularly regarding spatial resolution. For example, a gamma-ray spectrometer is only capable of determining the presence of hydrogen to a lateral resolution equal to its height above the surface. If the orbiter is at an altitude of 300 kilometers, it can locate water on the surface with a lateral resolution of 300 kilometers. While this technique is useful for telling of the presence of surface water, it is insufficient to determine its location.

A new and quite revolutionary technique for detecting water on Mars, called *radio sounding*, will help us to detect even deeper subsurface reservoirs of water, and to higher spatial resolution. Here, a large radio transmitter and detector on board an orbiter emits radio waves toward Mars, and from the reflection pattern it can map subsurface structure to depths of 5 kilometers or more, enabling underground reservoirs of water and ice to be detected.

Yet another survey to be conducted from orbit is the determination of the isotopic fractions of atmospheric gases. As already discussed, measuring the *isotopic fractionation* of the various atmospheric components points to the density and composition of Mars' early atmosphere and how it has changed over time. This can be done from orbit using a device called a *mass spectrometer*, which samples gases from the uppermost part of the atmosphere and separates the various elements and their isotopes. Associated with this is monitoring of the interaction of the upper atmosphere with solar radiation. Using orbital *plasma detectors*, we can detect the interaction of the Sun's radiation with the upper atmosphere and any subsequent leaking away to space. This has already been witnessed and measured by the Mars Express orbiter, confirming that solar radiation has played an important role in thinning Mars' atmosphere for over the past several billion years.

Orbital surveys will allow us to monitor the presence and variation of trace gases associated with life, such as hydrogen, hydrogen sulfide, methane, ammonia, and various nitrogen oxides. By monitoring these over a prolonged period we can even begin to determine where on the planet they are emanating from for follow-up aerial and lander missions. Finally, numerous other experiments from orbit can monitor the current weather and seasonal changes to the polar caps, and can use techniques such as radio sounding and Doppler movement techniques to build a three-dimensional picture of Mars' current atmosphere.

Phase 1: Aerial Reconnaissance

Subsequent to orbital reconnaissance, we envisaged a follow-up survey with airborne explorers called aerobots—extremely light robotic aerogliders and balloons capable of surveying the surface in even greater detail.

Aerobots will survey vast areas of the surface in spectacular detail because they will be so close to the surface and may be airborne from periods of several hours using light aircraft to possibly days using balloons. They will carry similar instruments to those on board orbiters—stereo cameras, infrared mineralogical detectors, and neutron and gamma-ray spectrometers. But because they will glide hundreds of times closer to the surface, they will be proportionality more sensitive and accurate. Mineralogical imaging, for example, may be capable of detecting traces of carbonates (if present) in tiny amounts, or concealed at locations undetectable from orbit. Similarly, gamma-ray spectroscopy will provide spatial resolution in the order of kilometers or less, helping to pinpoint interesting sites for follow-up lander missions. Aerobots will also carry out atmospheric measurements that are impossible from orbit, such as directly sampling the atmosphere at numerous altitudes, providing unprecedented measurements on Mars' weather and climate.

Currently we can envisage two ways of pursuing aerobot reconnaissance: balloons and light aircraft. Recent tests with terrestrial high-altitude balloons have shown that Mars balloon missions are feasible. During the day, a balloon will heat and rise into the atmosphere where it will conduct a range of atmospheric analyses. During the night the balloon will cool and descend. A long tether attached to the underside of the balloon will balance the overall weight. While the tether would touch the ground the balloon would not, protecting it for the next assent. Further, the tether can be equipped with a host of mineralogical, chemical, and biological experiments directly sampling the surface soil on every landing and providing unparalleled surface reconnaissance across huge tracts of the Martian surface.

Alternatively, extremely light and solar-powered aircraft could remain airborne for several hours. Here we envisage the aircraft deployed from a descending lander mission, where the aircraft would be dropped into the atmosphere at high altitude and allowed to glide for perhaps several hours before landing. While it would not be able to carry out surface sampling missions (until it permanently landed!), an aircraft could potentially travel a great distance very quickly, providing unprecedented views of an entire region. What aircraft lack in surface analysis capability and longevity, they gain in their ability to survey even great areas of the surface quickly and with perhaps some degree of control. Aerobot missions to many locations and traversing great expanses of the surface would transform our perception of the landscape of Mars and provide a vital link between orbital reconnaissance and our most ambitious robotic and human lander missions.

Phase 2: Lander Reconnaissance

Phase 2 of our exploration involves sending numerous landers to sites of interest identified from orbit. This phase runs parallel with orbital reconnaissance but (initially) before aerial attempts. Although we are currently making great strides in the development and deployment of orbital and landing probes, aerobots will not be available for some decades.

Phase 2 landers are tasked with characterizing the geology, geochemistry, and mineralogy of sites relevant to subsequent biological investigation. In particular, the NASA mantra of *follow the water* especially applies because if we are to look for local and specific evidence of life, we must first find evidence of water activity. The data acquired from landers will also contribute to our understanding of Mars' broad history and evolution, complement orbital reconnaissance and provide *ground truth* calibration for instruments in orbit, increasing their accuracy, sensitivity, and effectiveness. They will also pave the way for sample-return and human missions—acting as technology testbeds, exploring sites for subsequent sample collection and return, and cataloguing resources and environmental conditions important to human missions.

Two categories of lander are available: stationary landers and rovers. Stationary landers can be physically larger than rovers simply because they do not need to move. They can therefore house a correspondingly large array of sophisticated mini-laboratories, cameras, and detectors. Stationary landers can channel most of their power into operating these systems, unlike rovers that must use substantial power simply to move. And, as with the Vikings and the ill-fated Mars Polar Lander for example, stationary landers can even harbor sizeable robotic arms, drills, and scoops capable of extracting samples from the surface and subsurface. Stationary landers are therefore more suited to sites worthy of prolonged and detailed study. Rovers, on the other hand, will generally (but not always) be smaller—to improve their mobility and reduce power consumption—and streamlined for roaming-type surveys and investigations. What they may lack in size is more than compensated by their maneuverabiilty.

As with orbital reconnaissance, Phase 2 reconnaissance will be implemented through a series of well-defined surveys. First, each site will have its geology characterized thoroughly. While on Earth a geologist can pick up a rock, crack it open and infer a great deal about the local and global conditions under which it formed, that luxury does not yet exist for Mars geologists. Instead they must determine the geology from remote evidence gathered by the lander and relayed back to Earth. Of particular value is imagining of the site at high resolution and across multiples scales. Landers will be equipped with numerous cameras, from panoramic stereo cameras capable of showing the site geomorphology from horizon to horizon and in three dimensions, to microscopes capable of resolving features down to micrometer scale. In this way we can seamlessly relate planetary and regional imaging from orbit to local and microscopic imaging at the lander site, connecting the landscape with small-scale rock and soil textures and composition and allowing a determination of the geology of the region in great detail.

Of paramount importance to this phase are detailed elemental and mineralogical analyses at each site and across all surface types—soils, sands, pebbles, rocks, boulders, bedrocks, sedimentary layers, evaporate deposits, and impact materials, among others. If we are to describe the history of a site and eventually search for prebiotic and biological activity, we must begin with detailed elemental and mineralogical analyses. Elemental analysis will reveal the chemical make-up of the surface materials, the presence of volatile and biogenic materials, and help with our understanding of the super-oxidizing nature of dust and soils. An instrument called an *Alpha-Proton-X-ray Spectrometer* (APXS) will become a standard bearer on virtually all landers because it is capable of firing alpha particles, protons, and x-rays at surface materials and infer their elemental composition from the pattern of back-scattered particles and x-rays. An APXS is capable of detecting the presence and abundance of many elements, including carbon, oxygen, sodium, magnesium, aluminum, silicon, and iron. An APXS attached to a rover is a truly powerful combination, allowing examination of the elemental composition of a range of surface types across many locations. And because of its high sensitivity, it will provide powerful insight into the make-up of individual component materials.

While mineralogical analysis from orbit is central to understanding the planet and locating interesting sites for landing expeditions, lander-based mineralogy will be of a different order, revealing new and potentially significant details directly relevant our searches. We can now virtually completely characterize the mineralogy of any given site, revealing the precise chemical activity occurring there in the past and the local context under which that activity occurred. Biological searches aside, lander-based mineralogical surveys represent a unique and significant program of investigation.

Site mineralogical analysis will also form the basis for future biological searches. We can, for example, determine the presence and extent of past aquifers from the presence of hydrated minerals, iron oxides such as hematite and magnetite and evaporated salts such as sulfates and halides. From an examination of water-altered basaltic rocks, carbonates, silicas, and other spring deposits, we can even detect evidence of ancient hydrothermal systems, as well as minerals involved with organic synthesis and energy production. We will be able to identify materials capable of harboring microfossils—sedimentary rocks and cements, siliciclastic cherts, and evaporate crystals, for example. In short, lander-based mineralogy is a critical step toward the exploration for biology on Mars.

Once again an array of instruments will be used—including the APXS, gamma-ray and neutron spectrometers. Thermal Emission Spectrometers, similar to those on orbiters and which infer the mineralogy from the heat signature of surface materials, will also be particularly effective, because they can examine individual rocks that have been cleaned and are free of the dust and atmospheric contamination that affect orbital-based mineralogical determination.

Phase 2 landers will also conduct isotopic analyses of the atmosphere and the crust. As already explained, measurements of isotopic fractionations allow us to *work backward through time* and infer Mars' original stocks of volatile and biogenic materials. Isotopic analysis on the surface, using a miniature mass-spectrometer capable of sorting and identifying isotopes, will far exceed orbital isotopic analysis in several ways. First, a lander can sample the atmosphere at its densest, rather than the extremely rarefied atmosphere available to an orbiter. Second, a lander can carry out an isotopic analysis of surface materials, revealing the cycling of materials that occurred through water reservoirs, the atmosphere, and the crust. For example, if we find that hydrogen in the soil contains an excess of deuterium similar to that found in the atmosphere, we can say that at one time atmospheric gases mixed with crustal material or water, from which deposition then occurred, or mixing at a hydrothermal system. By looking at isotopic fractionation across a range of elements and at different geological settings, we can begin to build a picture of the full range of cycling, of the size and make-up of the atmosphere, of initial quantities of volatiles and biogenics, the planetary climate, and the extent and duration of complex natural chemical systems. Furthermore, by measuring differences in isotopic fractionations within surfaces of different ages, we can even identify the evolution of the atmosphere and resulting planetary conditions over time.

Many other experiments will also be carried out: networks of landers capable of detecting seismic activity and mapping the planetary interior; atmospheric monitoring of wind, temperature, pressures, humidity, and dust content; magnetic properties of dust, water, and carbon dioxide frost formation—to name but a few. And while we can regard this phase of exploration as a step toward biological exploration and human missions, it will persist for many decades to come, helping us to construct a complete picture of Mars, past and present.

Phase 3: Robotic Biological Reconnaissance

All that we have learned from previous Mars missions, and from searching for our origins, tells us that we can be greatly mistaken in our investigations and conclusions about the nature of life. This is why we are now so cautious on returning to Mars. It is too easy to get it wrong—not only in drawing conclusions, but even in what questions to ask. Indeed, what Mars ultimately teaches us will be gleaned as much from our questions as from the answers. There is, literally speaking, nowhere else we can frame completely new questions about the origin of life or from where we can seek answers at present. Mars is not only forcing us to tread carefully, but is also forcing us to devise broader and deeper questions about the nature of life.

This is why we recognize a phased program to be such a viable path forward. If we are to have any chance of determining a biological context for Mars, we had better first understand it very well through orbital and lander surveys. Only then can we frame biological questions, identify sites worth visiting and devise missions and experiments appropriate to the task. It is perhaps naive to think that we could ever have found answers about biology on Mars without first learning of its geology, the story of water, its atmosphere, and biologically significant materials. Only with such information at our disposal can we then send biological landers unique to each location in search of specific answers about possible life-related activity there. Even individual categories of experiment—for example, one designed to look for biomarkers—will need to be *tuned* to the specific chemistry of the site it will investigate. We already know how to perform quite broad and sophisticated organic and biological tests, but the key is in knowing what particular tests to conduct at any given site.

Of course we are not pursuing questions of biology on Mars without *some* knowledge. While it is true that we do not yet know the chemical pathways leading to life on Earth, or the form of any potential life on Mars, our improved understanding of organic life on Earth, coupled with Mars reconnaissance to date, suggests that three categories of biological exploration are valid and worth while: the search for prebiotic chemistry, ancient life, and extant life.

In searching for prebiotic chemistry, we must visit locations on the oldest surfaces that also suggest ancient water and hydrothermal systems where organic synthesis could have occurred. There, we will first look for mineralogical evidence of prebiotic chemistry using thermal imaging and spectroscopy. For example, we can search directly for minerals that can bring about organic synthesis such as water-altered olivine, as well as hot spring deposits such as carbonates and silica. As with Phase 2 landers, neutron, gamma-ray, and APX spectrometers can determine elemental abundances, as well as volatiles and biogenics used in organic synthesis. Determining, through mass-spectroscopy, the isotopic fractions of the various biogenics will reveal the extent, evolution, and duration of chemical mixing, whether carbon-based materials are organic in nature and when they were created. Isotopic analysis may also reveal existing biases toward particular carbon isotopes (as with terrestrial organisms), again suggesting a biological origin.

The search for extinct life partially overlaps with the search for prebiotic chemistry—visiting sites of ancient water and hydrothermal activity, and there conducting mineralogical, elemental, and isotopic analyses for evidence of organic synthesis, nutrients, and inorganic sources of metabolic energy. The search for extinct life will also involve visiting sites with rock and mineral types capable of preserving fossil records—carbonates, cherts, evaporates, and phosphates found within aqueous sedimentary rocks and dried lakes. Although searches to date have not revealed carbonates beyond trace levels, we cannot conclude that they are not present at particular locations and buried by subsequent volcanic activity. At candidate sites, samples will be isolated and imaged using high-resolution cameras and microscopes, directly looking for morphological evidence of microfossils whether from large stromatolite type structures or individual microbe fossils within small samples. Determining the presence of organic carbon will require a stepwise approach. First, a quantitative analysis of organic carbon is required using, for example, a reactive *carrier-gas* and carbon-sensitive detector. Temperature-programmed experiments—monitoring the temperature of carbon combustion and energy consumed and released in the process—will also determine whether carbon-based materials are organic or inorganic. A second phase would involve attempting to characterize organic molecules, and while it would not be possible to perform typical terrestrial organic tests on Mars (because they require prior knowledge of the structure of the molecule), we can carry out an elemental and isotopic analysis using a miniaturized combustion/purification/mass-spectrometer experiment, and from this we can attempt an interpretation of the molecular characteristics. Chromatography and other separation techniques can also help to identify organic molecular species. Revolutionary new biomarker chips will also act as a means of comparing any organics found on Mars with thousands of known types of organic molecules.

As previously mentioned, the search for evidence of extant life itself involves three distinct searches: (a) living organisms and biomarkers of current life; (b) evidence of dormant life; and (c) non-living evidence of recent life. On searching for current life, we must identify and visit geologically young sites showing recent water activity, especially surface or exposed subsurface aquifers, for example, hot springs or fumaroles. Polar and high-latitude regions where ice may occasionally become liquid during periods of climate change, or sites of recent volcanic activity, could also retain evidence of former or current life. Once individual sites are identified, biological landers and rovers will visit them and carry out searches particular to each. They will first determine the suitability of local conditions for life, such as the distribution and chemical structure of oxidants, and will then identify terrain that is free of oxidants, as well as measuring soil acidity and oxidizing potential. The effects of moisture and wetting, soil solubility, availability of energy and nutrients, among others, will also be determined. Next they will look for evidence of microorganisms as well as chemical and biological markers indicating current metabolic activity and growth, using a range of biological experiments similar to those on board the Viking landers, but individually tailored to each site. For example, the Viking Labeled Release experiment, which looked for and detected metabolizing carbon compounds, could not identify them as being biological in origin because the soil chemistry was not known. On current landers, a more sophisticated version of this experiment could be used to specifically distinguish biological from non-biological reactions and produce a complete analysis of the site chemistry. Furthermore, while the Viking experiment incubated samples in conditions different to those of the surface, current landers will be able to simulate a range of incubation conditions more suited to the site. Another Viking experiment, the Pyrolysis GCMS experiment that searched for organic molecules, could also be improved to greater sensitivity and yield significantly better results.

The search for chemical and biological biomarkers would involve many of the techniques already mentioned for prebiotic chemistry and extinct life searches—looking for evidence of mineral and organic deposits and identifying their structure and origin. But this search, as well as for evidence of dormant life, will involve global searches at many locations and not just on the most ancient terrain (as for prebiotic chemistry and extinct life). A journey to the polar-layered regions and the polar caps will be of paramount importance. Above approximately 60 degrees latitude in each hemisphere, a near-surface permafrost could contain a record of any recent biological activity, and even dormant life awaiting more favorable environmental conditions. Finally, ventures based on the Viking lander experiments, but more sophisticated and tailored to each specific environment, could test for present life. Such experiments would attempt to amplify growth and initiate metabolic activity using an array of volatiles and nutrients specific to the environment and then look for radioactive tracers within released by-products.

Phase 4: Sample Return

We are increasingly confident that, through a program of robotic exploration, we can determine the precise character of the planet during the Noachian period, how it evolved into the planet we see today and whether life-related activity ever occurred there. Indeed it is only through such a phased, cost-effective program that we have found a viable path forward. Currently, we could do it no other way.

None the less, even if evidence of life-related activity actually exists on Mars, we cannot automatically assume that we will find it or fully characterize it with robots alone. Even after exhaustive efforts through multiple missions, it may come to pass that we do not locate any such evidence. The enormity of the task—to characterize an entire world through its history and to identify and understand life-related activity there—may simply be too demanding using remote robotic techniques alone. Even on Earth where we have expended great energy on determining the origin of life, we are far from definitive answers. And given that we are largely in the dark on virtually all aspects of the origin of life—whether on Earth or on Mars—the true scale of the job ahead becomes apparent.

Furthermore, even in the fortuitous situation where we actually find what we think to be fossilized microorganisms on Mars, we will have to go to great lengths to extract precise information from those findings. Once again, considering how difficult it has been to resolve the nature of organic compounds, biomarkers, and possible microfossils within meteorites such as ALH84001—upon which can be applied the full weight of cutting-edge analytical techniques here on Earth—we must be realistic in our expectations from robotic exploration alone. With limited life spans, restricted mobility and range, and a rigidly defined scientific payload that cannot be altered or extended when deployed, robotic landers and rovers can only do so much.

For all these reasons, there is a broad consensus among planetary science and exobiological communities that a vital step toward understanding Mars is to collect samples of Martian soil, rocks, and even its air and return them to Earth for analysis. With such materials at our disposal, our understanding of Mars would be greatly accelerated. We could perform analyses in weeks that would otherwise take years through robotic missions. We could conduct more sophisticated analyses. We could distribute the material across the world's leading research institutes and teams; and we could respond to initial findings with new experiments designed to extract yet further information from the samples, however elusive. Our capacity to do analytical science in a laboratory on Earth outperforms what will ever be possible with robotic landers, however sophisticated. For this reason, a sample return program is now seen as a key phase of Mars exploration.

Although there are many ways to implement such a program, the currently preferred option is for a multi-mission strategy. First, one or more missions will be sent to sites of particular exobiological interest. Each mission will comprise a robotic lander and/or rover as well as a Mars Ascent Vehicle (MAV). On arrival the rover will analyze numerous rock and soil samples, and place the most interesting in a sealed container attached to the MAV. Subsequently, a decision will be made about which site contains the most interesting samples, as there will be only a realistic option to return one to Earth. The chosen MAV will then launch into Mars orbit, deploying its sealed container. Another mission will launch from Earth, settle into Mars orbit, collect the sample container, and return it to Earth. Although there are many planetary protection and contamination issues to be resolved (discussed in detail in Chapter 19), it is currently envisaged, especially by ESA, that such a mission could be implemented by 2016. Such a feat would bring about a monumental step forward in our understanding of Mars and in how to undertake a sophisticated and precise exploration of that planet. It would also represent a milestone in space exploration—the first return mission from another planet.

Phase 5: Humans on Mars

As far as orbiters, landers, aerobots, and sample return missions can take us, there is also a current desire to send people to Mars. It is not unanimous, but it is an ancient desire, one that has already been recognized by previous generations by orbiting the Earth, climbing the highest mountains, trekking to the poles, and sending people to the Moon. The desire to explore is neither logical nor quenchable—it is innately human. And so, with an ever-increasing capability in space exploration, we will probably send people to Mars as soon as the opportunity arises. There will be heated debates over the cost, the risks and even the morality of such human ventures, but if we are able we will most surely go. That time is almost upon us.

Many of the issues on sending people to Mars are discussed in Chapters 18 through 20, but let us here consider one aspect to a human mission: the scientific case. Despite the controversy and debate surrounding the human dimension to such a mission, there also exists a valid scientific reason for sending people to Mars. Although there are many practical reasons why human exploration is superior to robotic exploration—better mobility, the ability to make important decisions on the spot, the capacity to do sophisticated analyses in the field and in field laboratories and to bring significantly better samples for return to Earth, and so on—the most significant scientific reason lies within the nature of the quest itself. Imagine, for example, a scenario in which we have determined that Mars was a clement world in its distant past and the emergence of life was quite probable. Imagine also that no trace of life-related activity has been found, even after exhaustive robotic and sample return missions. In such a scenario, will we be content to declare that life never existed on Mars? Are humans ever content with an open-ended question that they know can be answered, given the means to do so? Imagine another scenario in which we find widespread evidence of prebiotic activity from the Noachian period over an entire region on Mars. Will we be content to send a succession of robots to do the science? There are many conceivable outcomes from our robotic exploration of Mars, but few if any will infer that we have learned enough. There can be little doubt that the eventual outcome will point to a need for further exploration. There can also be little doubt, given the complexity and vastness of Mars, that a tangible demand will eventually arise for humans to travel there to conduct key scientific exploration. If it is technically and economically feasible, our phased robotic program will surely deliver a plethora of challenges that will require the sum total of humanity's ingenuity and talent—using robots as well as humans.

Conclusion

The five-phased scientific strategy for Mars represents a ground-breaking advancement in scientific exploration strategy. It is the optimal strategy that enables otherwise impossible feats in space exploration and ensures maximum returns. Astoundingly, how we implement this strategy has already been substantially decided, devised, and developed—especially by NASA and, increasingly, by ESA. Indeed the implementation of a phased program has been in operation for over a decade and is already delivering spectacular results.

PLATE 1a: The Pillars of Creation at the heart of the Eagle Nebula, lying at approximately 7,000 light years toward the centre of our Galaxy. The largest pillar is approximately one light year in length. [Credit: NASA, ESA, STScI, J. Hester and P. Scowen (Arizona State University)]

PLATE 1b: Seen more closely, the largest pillar reveals an outcrop or "finger" of dust and gas pointing directly upward. [Credit: NASA, ESA, STScI, J. Hester and P. Scowen (Arizona State University)]

PLATE 2a: A black rectangle, superimposed on the outcrop of dust and gas in the Pillars of Creation at the heart of the Eagle Nebula, represents the approximate size of our Solar System. [Credit: NASA, ESA, STScI, J. Hester and P. Scowen (Arizona State University)]

PLATE 2b: Dividing the tiny rectangle in PLATE 2a into a thousand pieces, Earth would appear as a tiny "Pale Blue Dot" (center right), providing some sense of the enormity of the Pillars by comparison. This actual image was acquired by the Voyager 1 spacecraft from a distance of more than 6.4 billion kilometers from Earth. Coincidentally, Earth lies near the center of one of the scattered light rays resulting from taking the image in the direction of the Sun. [Credit: NASA/JPL]

PLATE 3: The Orion nebula, where stars with proto-planetary disks, as well as massive quantities of water and various organic materials are being created. [Credit: NASA, ESA, M. Robberto (Space Telescope Science Institute/ESA) and the Hubble Space Telescope Orion Treasury Project Team]

PLATE 4: Blue Earth—a spectacular and vibrant world whose legacy has been defined by internal and celestial forces. [Credit: NASA Earth Observatory]

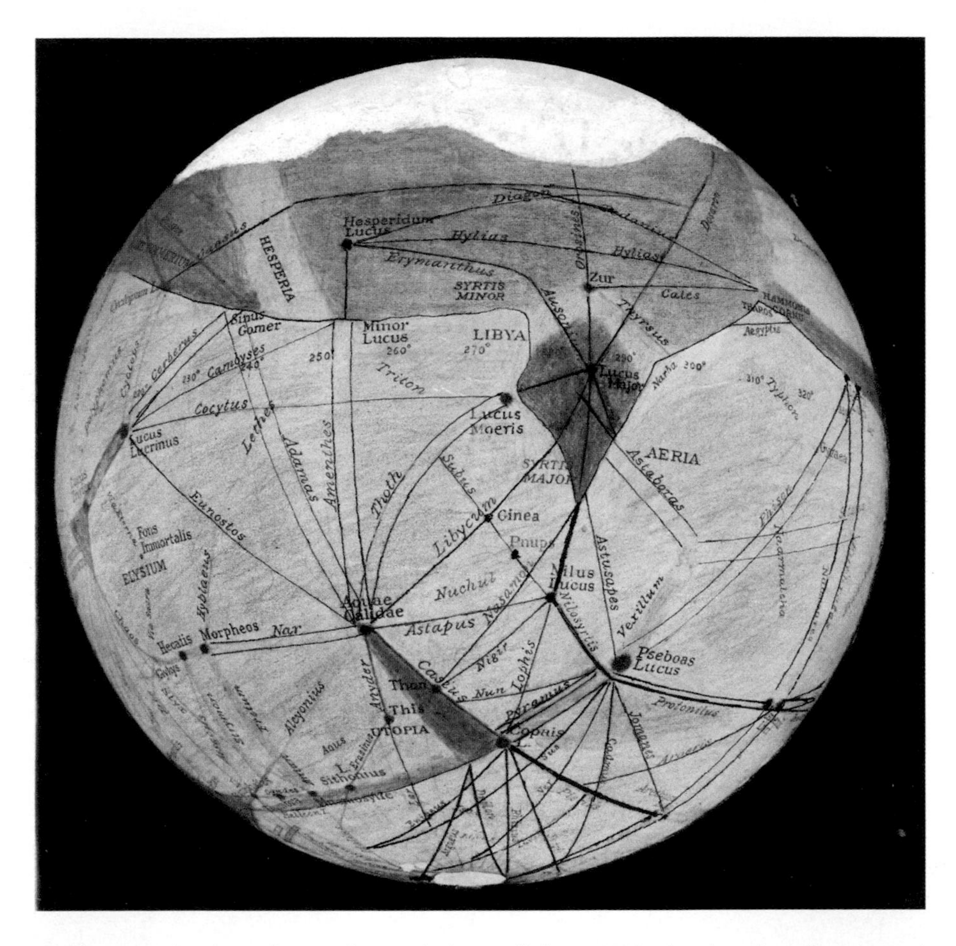

PLATE 5: A drawing of Mars by Percival Lowell from 1905 showing the alleged canal network. [Credit: Lowell Observatory Archives]

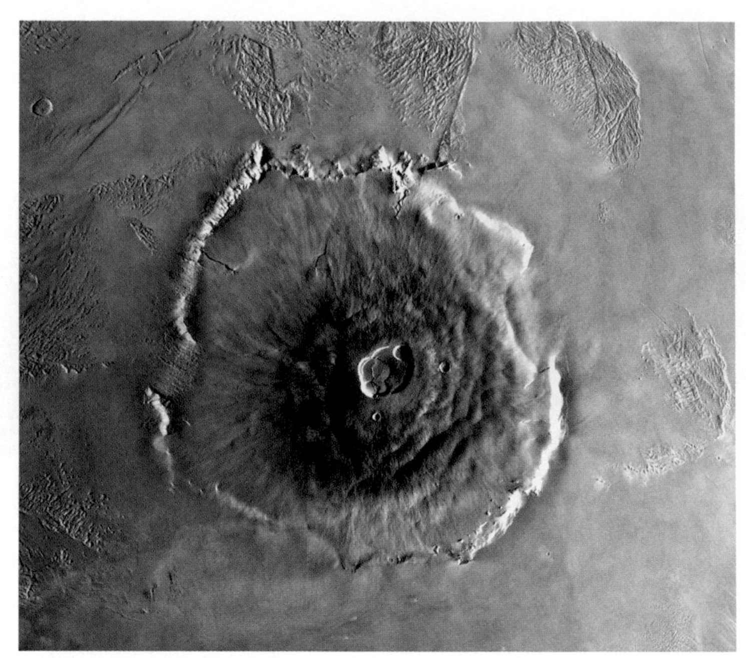

PLATE 6: Viking 1 orbiter photo-mosaic of the giant shield volcano Olympus Mons, which is 600 kilometers in diameter and 27 kilometers high. The entire volcano is flanked at its base by 6-kilometre-high cliffs. [Credit: NASA/JPL]

PLATE 7: Mars as seen by the Viking 1 orbiter. The center of the scene shows the entire Valles Marineris canyon system over 4,000 kilometers long and up to 8 kilometers deep, extending from Noctis Labyrinthus system of grabens in the west, to the chaotic terrain in the east. [Credit: NASA/USGS]

PLATE 8: A boulder-strewn plain of reddish rocks stretches to the horizon at Viking 2's northern hemisphere landing site in Utopia Planitia. Superficially it resembles the Viking 1 landing site in Chryse Planitia. The salmon-colored sky is caused by dust particles suspended in the atmosphere. [Credit: NASA/JPL]

PLATE 9: Viking 2 landing site at Utopia Planitia, showing a thin coating (less than 1/100 millimeter thick) of water-ice on the rocks and soil. Atmospheric dust picks up solid water and carbon dioxide, becoming heavy enough to sink to ground and form the visible layer. [Credit: NASA/JPL]

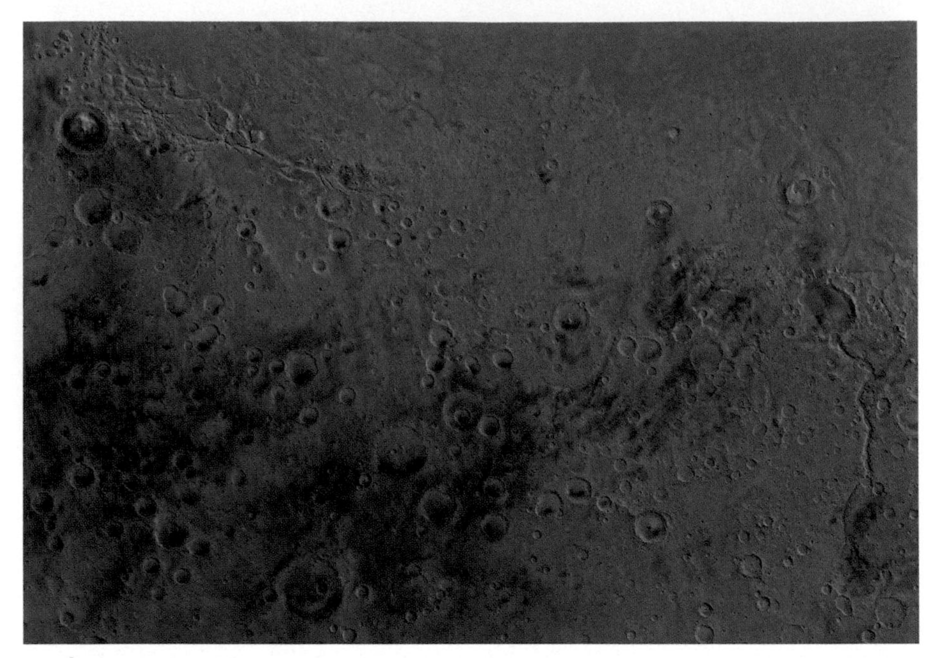

PLATE 10: The Aeolis region of Mars. The southern highlands are separated from the northern plains of Elysium Planitia by a highly dissected scarp. Ma''adim Vallis, an ancient channel, is seen running into Gusev crater on the far right. [Credit: NASA/JPL/USGS]

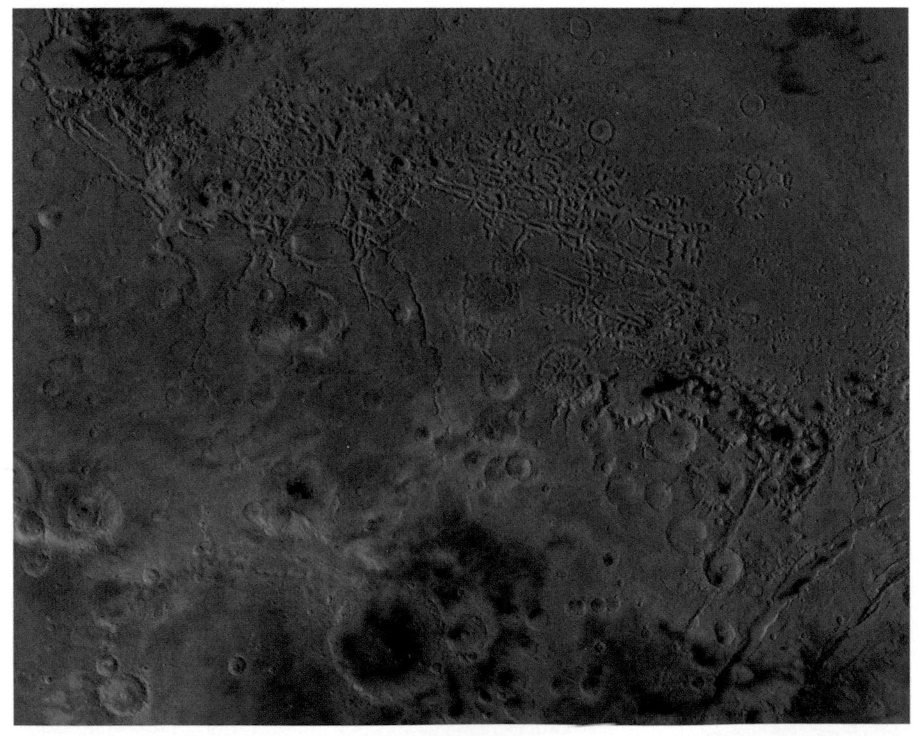

PLATE 11: The Nilosyrtis Mensae region of Mars showing heavily cratered highlands in the south separated from the northern lowland plains by a belt of flat-floored valleys, mesas, buttes, and the channels. [Credit: NASA/JPL/USGS]

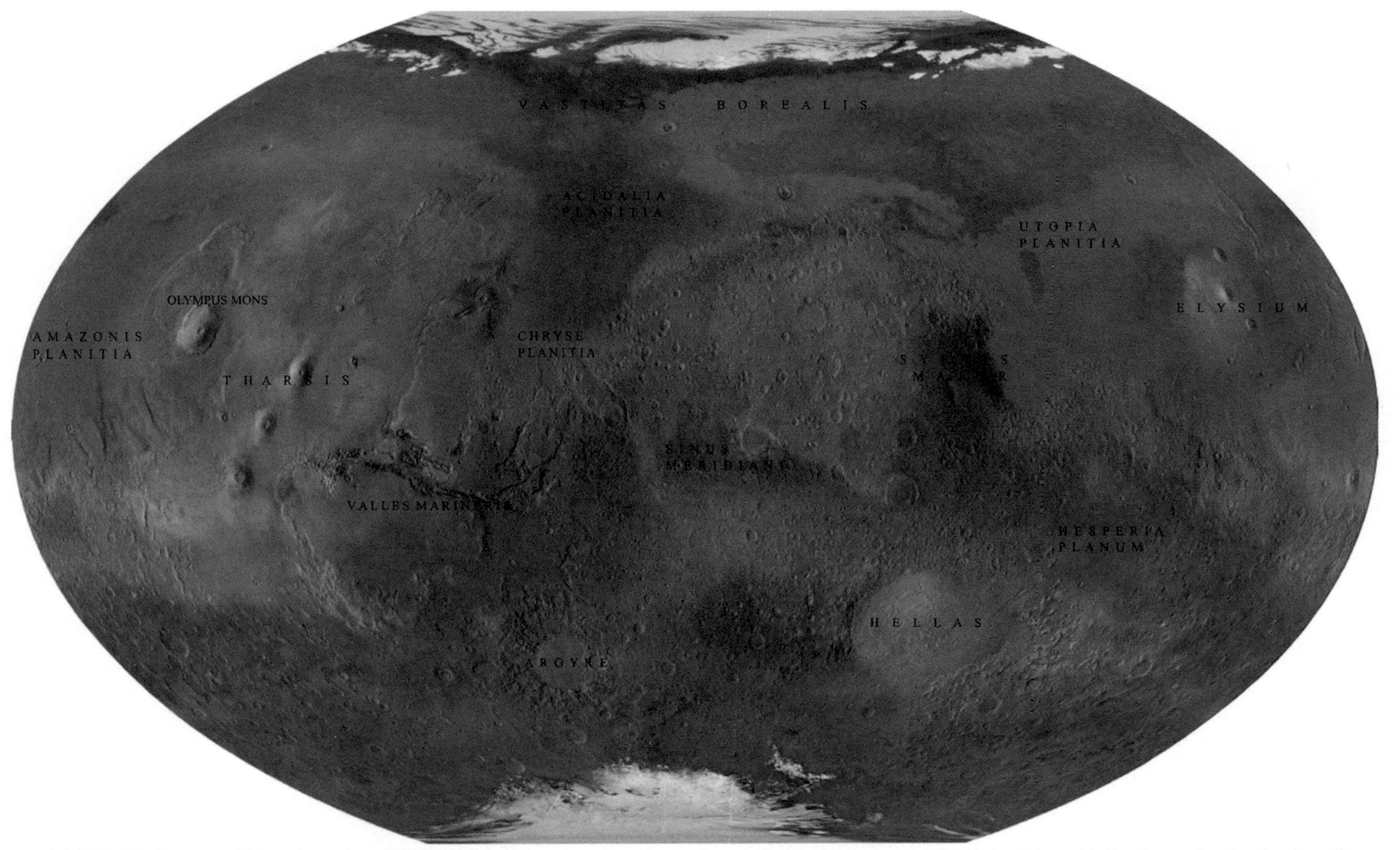

PLATE 12: A map of Mars based on MGS MOLA data, showing the major regions and features. North is at top. Notice in particular the heavily cratered southern hemisphere, and the smoother northern lowlands. [Credit: NASA/JPL/MGS MOLA Group]

PLATE 13: Part of Candor Chasma in Valles Marineris. The geomorphology is complex, having been shaped by tectonics, volcanism, and wind and perhaps by water. [Credit: NASA/JPL/USGS]

PLATE 14: Carl Sagan Memorial Station in Ares Vallis, where the Sojourner rover is seen examining a rock called "Yogi". Two hills called "Twin Peaks" are seen on the horizon. [Credit: NASA/JPL]

PLATE 15: A global image of Mars from Mars Global Surveyor revealing water clouds. Scenes such as this can only be captured through long-term reconnaissance of the planet. [Credit: NASA/JPL/MSSS]

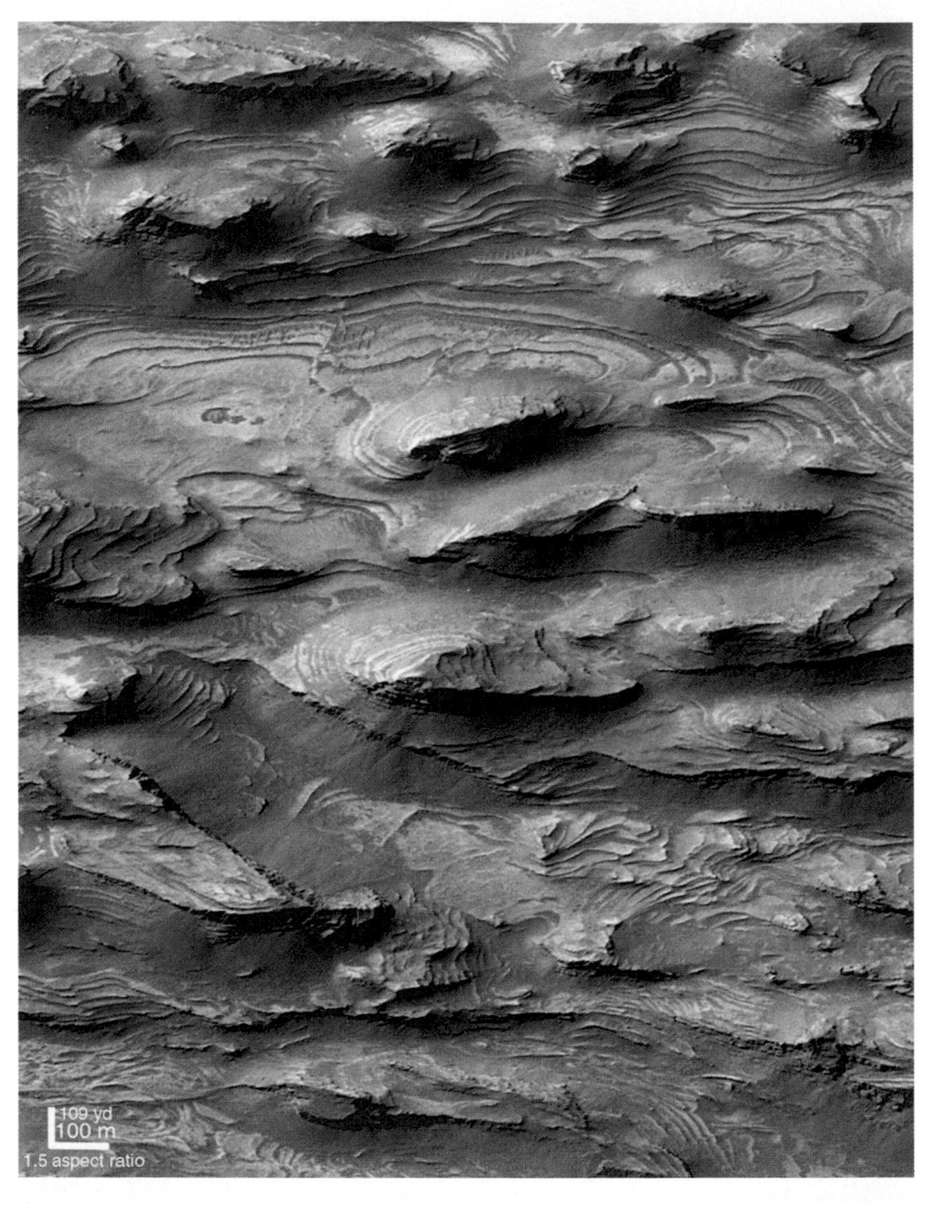

PLATE 16: Mars Global Surveyor's Mars Orbiter Camera (MOC) confirms the presence of layered outcrops within Valles Marineris. This high-resolution image reveals extensive layering on the floor of western Candor Chasma. [Credit: NASA/JPL/MSSS]

PLATE 17: MGS MOLA 3D Topological Map of Mars. Red indicates higher elevation and blue lower elevation terrain. Of note is the north—south planetary dichotomy, with the northern hemisphere both lower and smoother than the cratered plains of the south. [Credit: NASA/JPL/USGS]

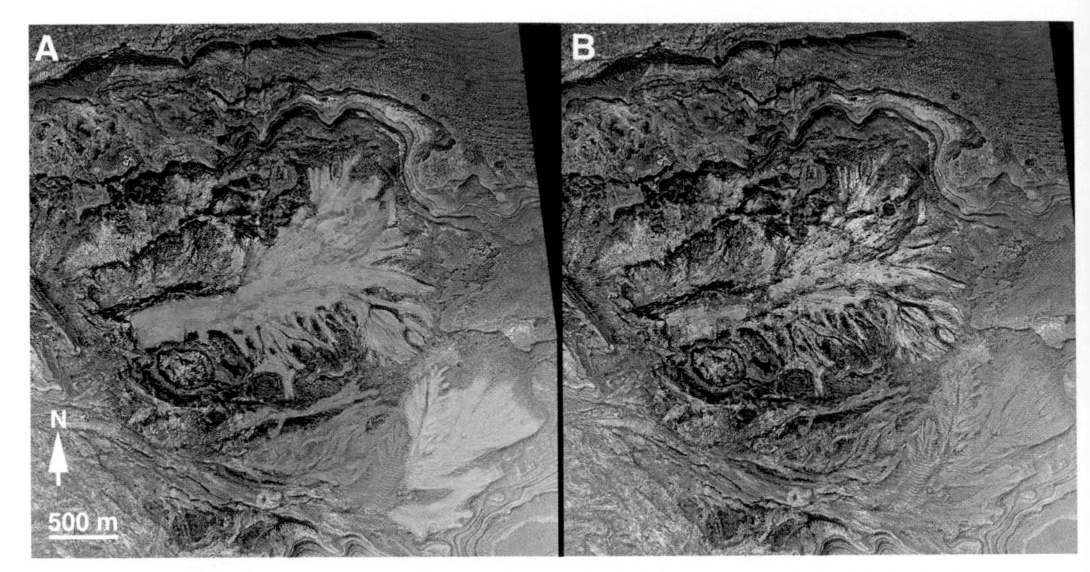

PLATE 18: MGS/MOC image of alluvial sedimentation in Melas Chasma, Valles Marineris. The water-based "fan-shaped" sedimentation is color highlighted in panel A (left). [Credit: NASA/JPL/MSSS]

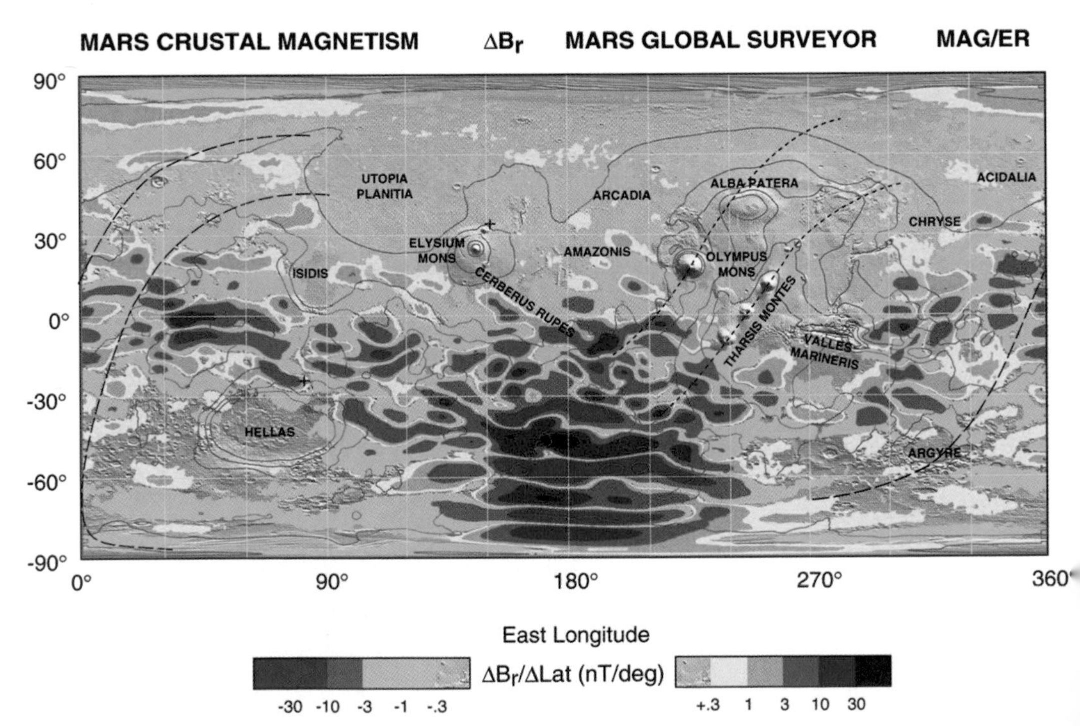

PLATE 19: MGS map of the magnetic fields retained within Mars's crust. The widespread "striping" pattern suggests plate tectonic activity on Mars in its distant past. [Credit: MGS Magnetometer Team led by Mario Acuna at the Goddard Space Flight Centre in Greenbelt, MD]

PLATE 20: From high above Valles Marineris, this 2001 Mars Odyssey derived view looks down upon a sight resembling parts of the desert west of the United States. Here the canyon is 150 kilometers wide, with the floor composed of rocks, sediments, and landslide debris. Within the canyon walls lie hundreds of layers revealing Mars's geologic history. [Credit: NASA/JPL/Arizona State University]

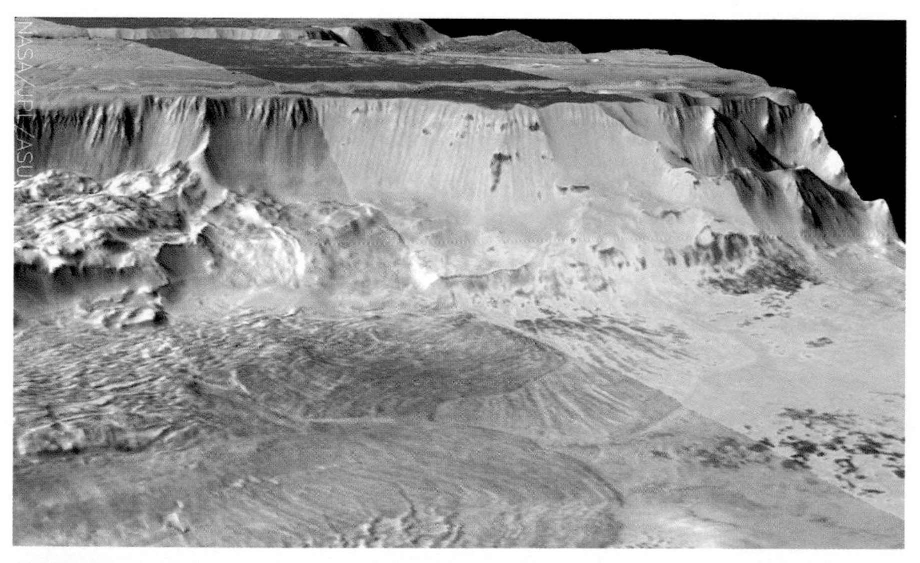

PLATE 21: Odyssey/THEMIS infrared image of Melas Chasma at night, revealing the abundance and distribution of surface materials. Rocks (red) retain their heat at night and stay warm, while dust and sand (blue) cool more rapidly. [Credit: NASA/JPL/Arizona State University]

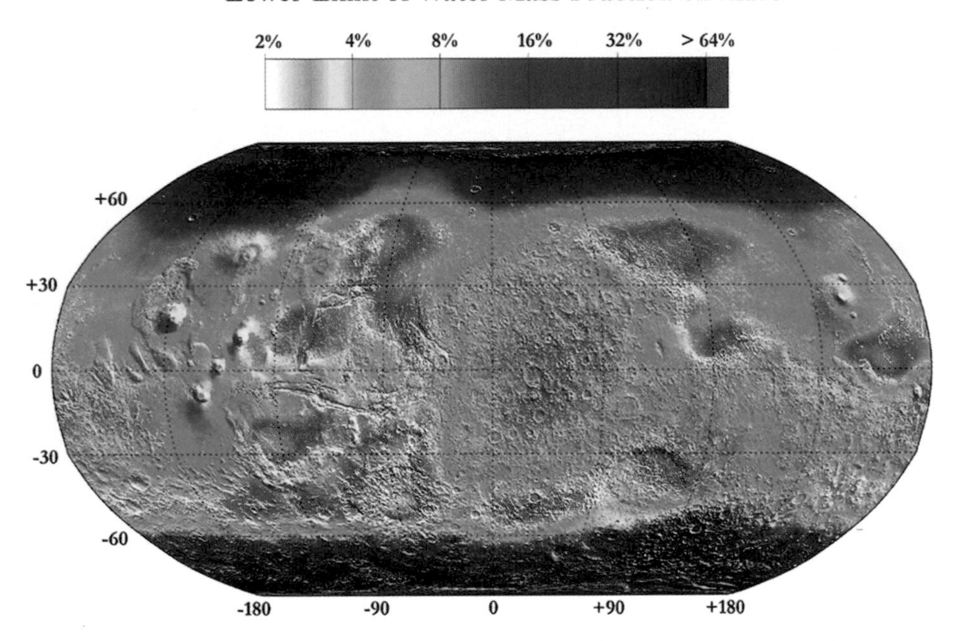

PLATE 22: Mars Odyssey map showing the lower limit of the water content of the upper meter of Martian soil. The highest water-mass fractions, exceeding 60% in places, are found in the high latitudes and polar regions. [Credit: NASA/JPL/Los Alamos National Laboratory]

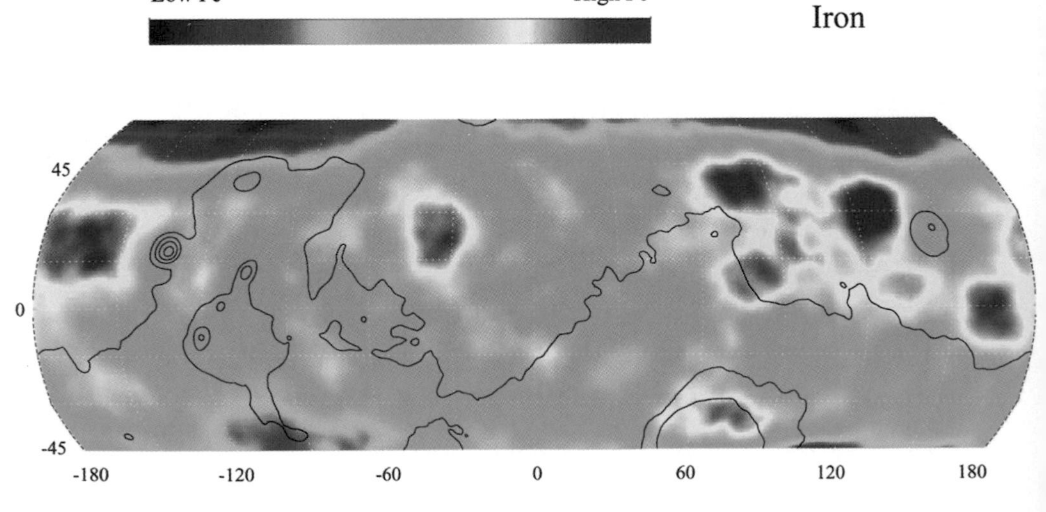

PLATE 23: A Mars Odyssey gamma-ray spectrometer map of the mid-latitude region of Mars for the element iron. Regions of lowest iron content are shown in blue while those with highest iron content are shown in red. Similar maps have been constructed for other elements. [Credit: NASA/JPL/Arizona State University]

PLATE 24: Ganges Chasma, part of the Valles Marineris, showing layering in the canyon walls to a depth of 5 kilometers. Outcrops of olivine, seen as dark red ribbons along the base of the canyon, suggest that the region has been dry for billons of years. [Credit: NASA/JPL/Arizona State University]

PLATE 25: 2001 Mars Odyssey THEMIS image of near-surface water-ice in a region on Mars approximately 70 degrees north. Blue shows ice at a depth of 5 centimeters while red shows an ice depth of more than 18 centimeters. [Credit: NASA/JPL/Arizona State University]

PLATE 26: False-color THEMIS image of a landslide in Juventae Chasma, part of the vast Noctis Labyrinthus (the Labyrinth of Night) in west Valles Marineris, perhaps created when tectonic faults opened and allowed subsurface water to escape, causing the ground to collapse. [Credit: NASA/JPL/Arizona State University]

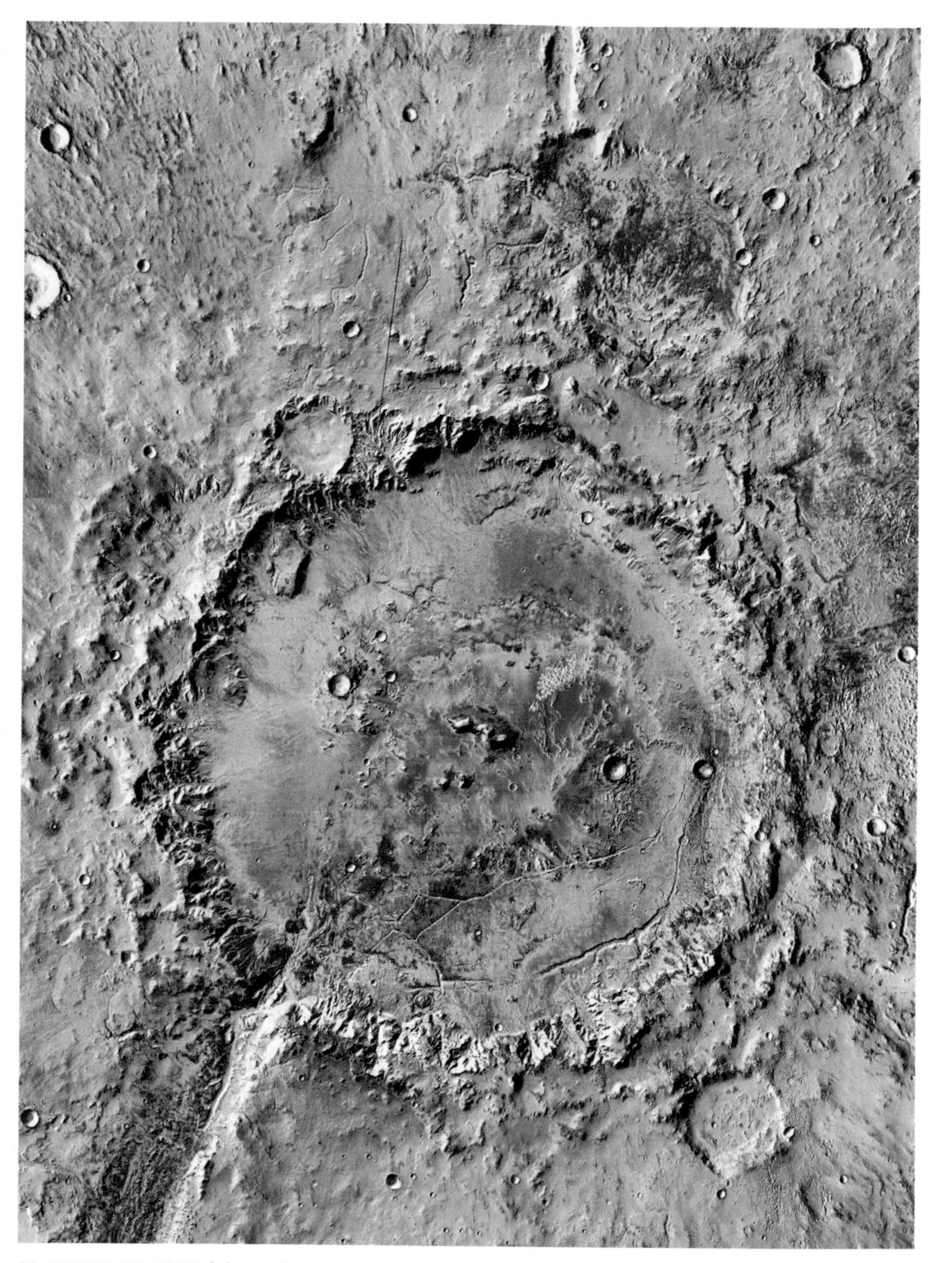

PLATE 27: THEMIS false-color image of Holden Crater. Rocks are shown as red, with sand and dust appearing blue. Holden Crater formed on what may be the longest watercourse on Mars—the Uzboi-Ladon-Margaritifer valleys—stretching from the Argyre Basin in the south to Chryse Planitia in the north. [Credit: NASA/JPL/Arizona State University]

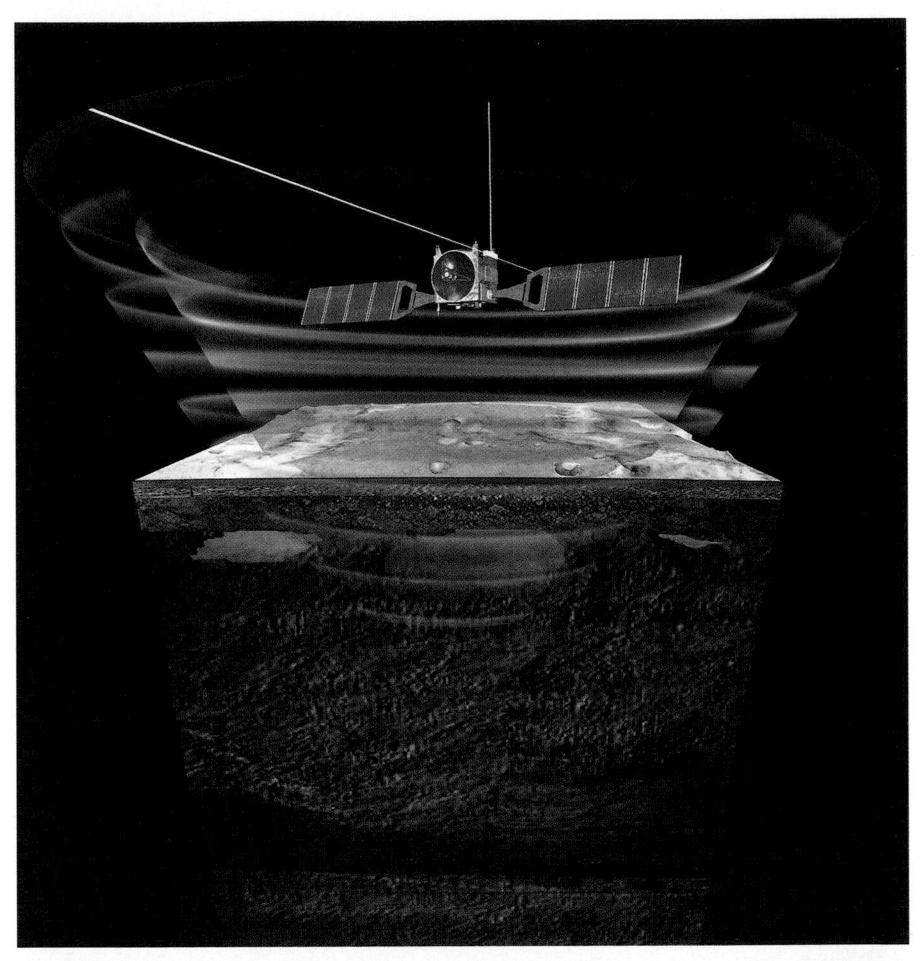

PLATE 28a: Image depicting Mars Express's MARSIS Radio Sounding instrument scanning for underground reservoirs of water and other buried geological features. [Credit: ESA]

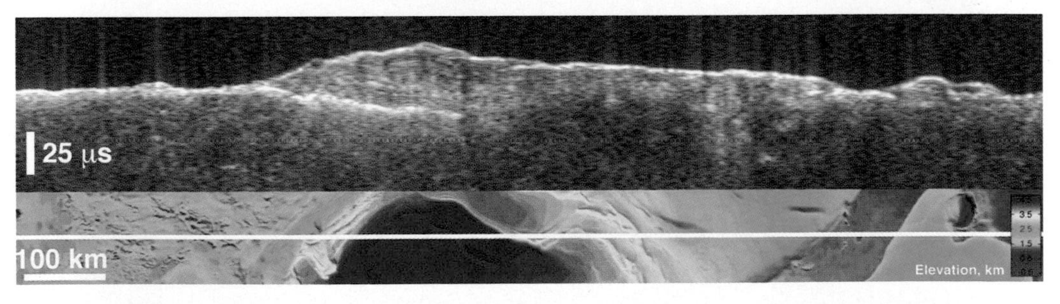

PLATE 28b: Mars Express/MARSIS subsurface map of the water-ice of the south polar region of Mars. The amount of water-ice in the region is equivalent to a layer 11 meters deep covering the entire planet. [Credit: NASA/JPL/ASI/ESA/Uni. of Rome/MOLA Science Team]

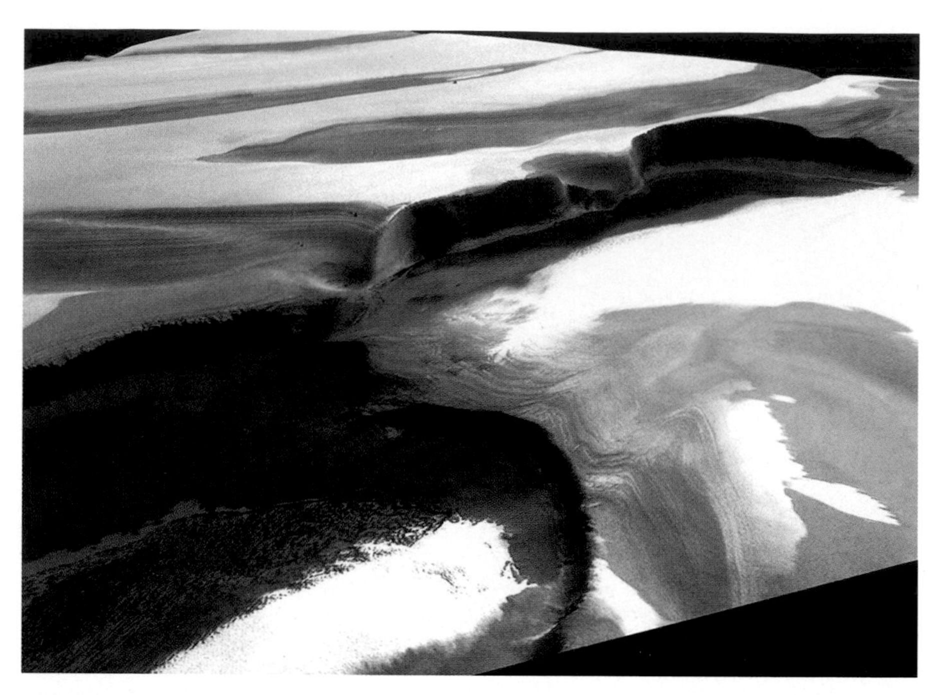

PLATE 29: Mars Express/HRSC 3D perspective view of the Martian north polar ice cap, showing layers of water-ice and dust. The cliffs are 2 kilometers high. [Credit: ESA/DLR/FU Berlin (G. Neukum)]

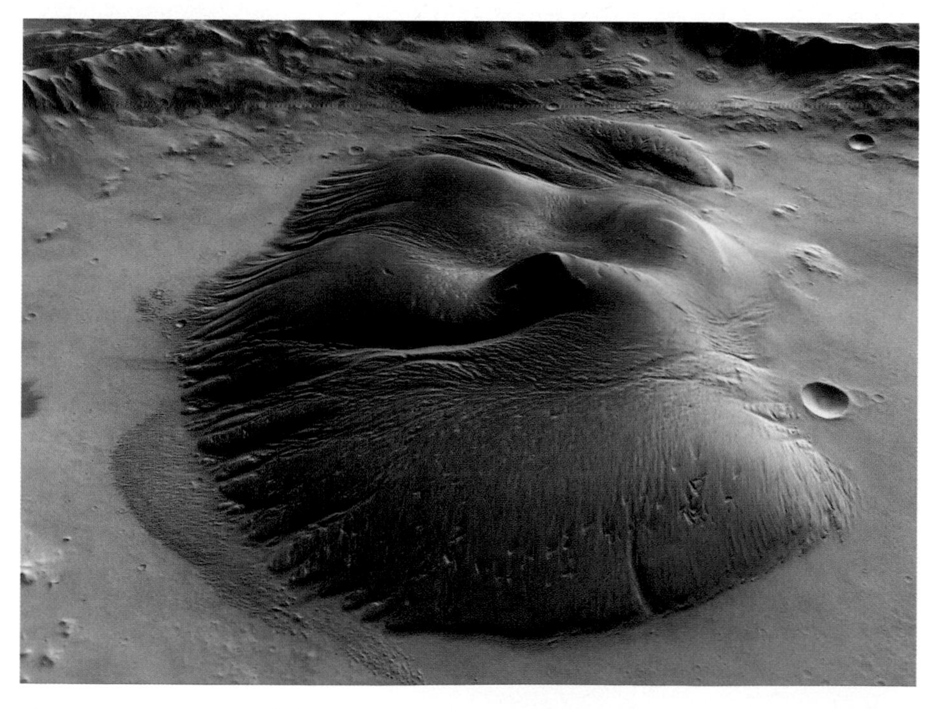

PLATE 30: Mars Express/HRSC 3D perspective view of Nicholson Crater located at Amazonis Planitia. This view shows the central part of the crater, measuring 100 kilometers across. [Credit: ESA/DLR/FU Berlin (G. Neukum)]

PLATE 31: Mars Express/HRSC 3D perspective view of Coprates Chasma and Catena in Valles Marineris, with a ground resolution of approximately 48 metres. The main trough, appearing in the north (top half) of this image, ranges from 60 kilometers to 100 kilometers wide and extends 9 kilometers below the surrounding plains. [Credit: ESA/DLR/FU Berlin (G. Neukum)]

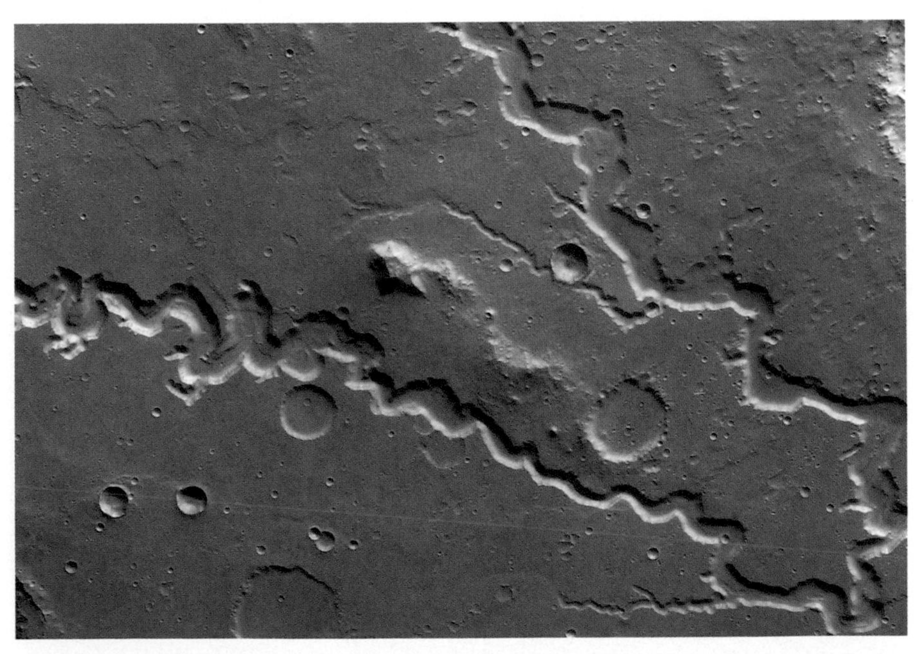

PLATE 32: Mars Express/HRSC image of Nanedi Valles valley, extending 800 kilometers diagonally across Xanthe Terra, southwest of Chryse Planitia. The valley exhibits meanders and a merging of two branches in the north. Erosion was perhaps caused by ground-water outflow, flow of liquid beneath an ice cover or collapse of the surface in association with liquid flow. [Credit: ESA/DLR/FU Berlin (G. Neukum)]

PLATE 33: Pack-ice on Mars? This image, taken by the HRSC, shows what appears to be a dust-covered frozen sea at Elysium Planitia near the Martian equator. This scene is tens of kilometers across, centered on latitude 5°N and longitude 150°E. [Credit: ESA/DLR/FU Berlin (G. Neukum)]

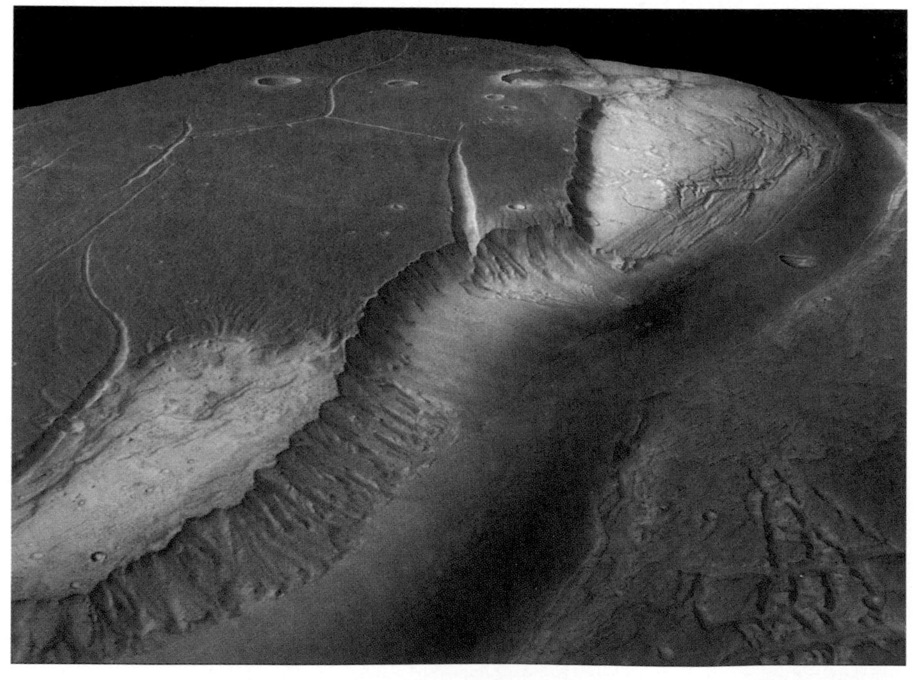

PLATE 34: Mars Express/HRSC 3D perspective of the Kasei Valles outflow channel—a gigantic channel 2,900 kilometers long and 500 kilometers wide that may have persisted on Mars for upwards of a billion years. [Credit: ESA/DLR/FU Berlin (G. Neukum)]

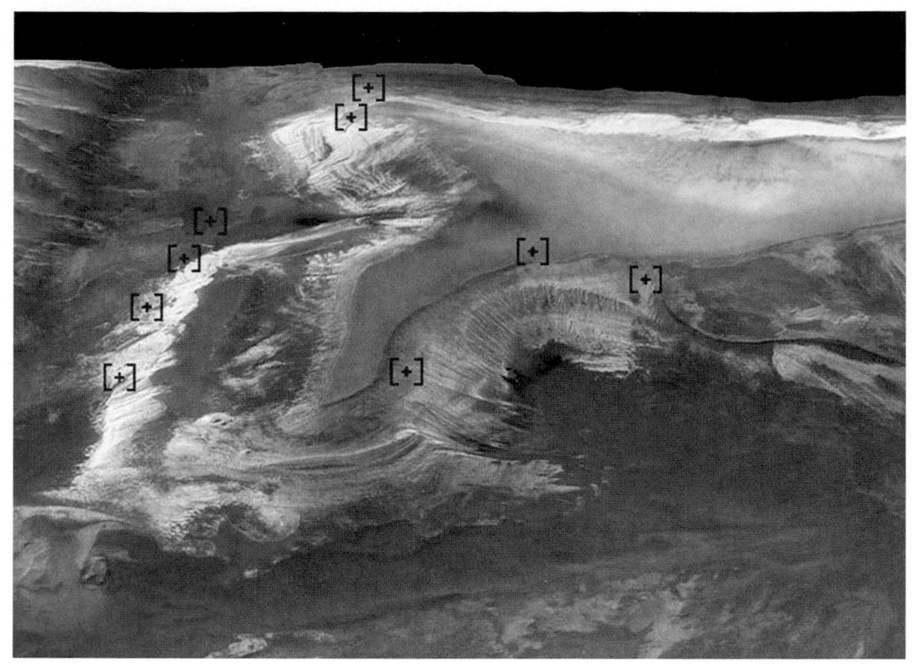

PLATE 35: Mars Express/HRSC 3D false-color perspective of Candor Chasma with a superimposed infrared image from OMEGA. This image shows bright and brown deposits (red markers) of the mineral kieserite (hydrated magnesium sulfate). [Credit: ESA/OMEGA/HRSC]

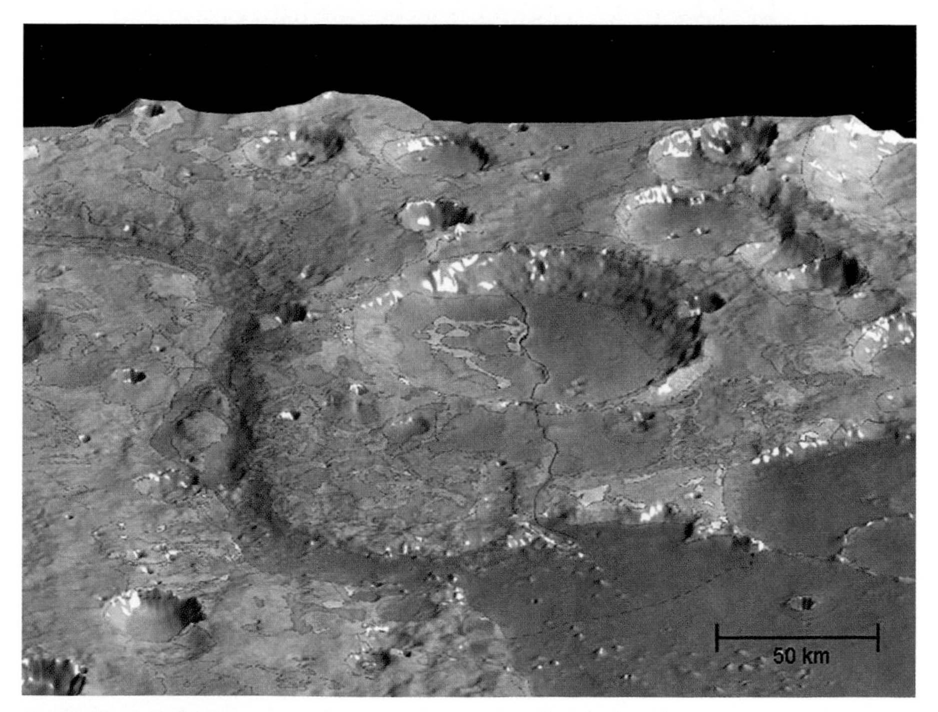

PLATE 36: A HRSC 3D perspective view of Marwth Vallis (gray), with a superimposed OMEGA map showing water-rich minerals (blue). While no hydrated minerals are detected in the channel, the outflow was so violent as to expose ancient hydrated minerals, revealing an earlier era when water was present. [Credit: ESA/OMEGA/HRSC]

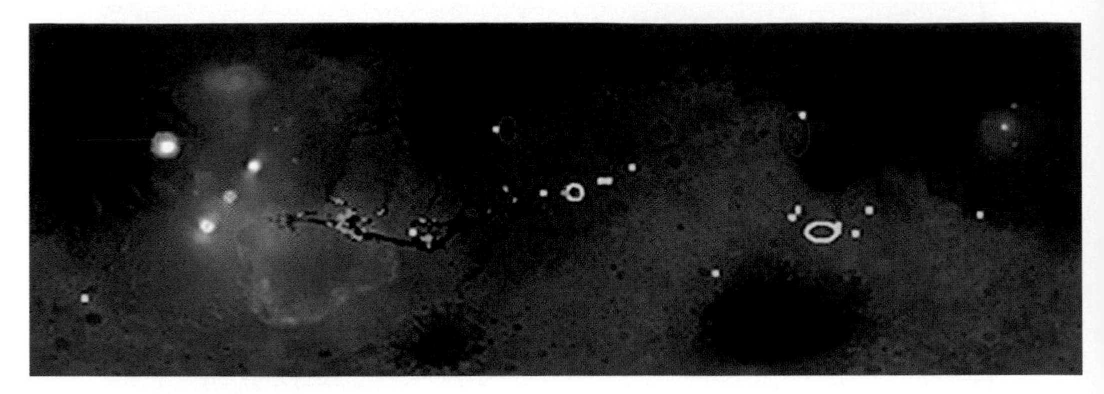

PLATE 37: The global distribution of hydrated (water-rich) minerals as discovered by Mars Express/OMEGA. The map is superimposed on an MGS/MOLA map. Red indicates phyllosilicates, blue indicates sulfates, and yellow other hydrated minerals. [Credit: IAS/OMEGA/ESA]

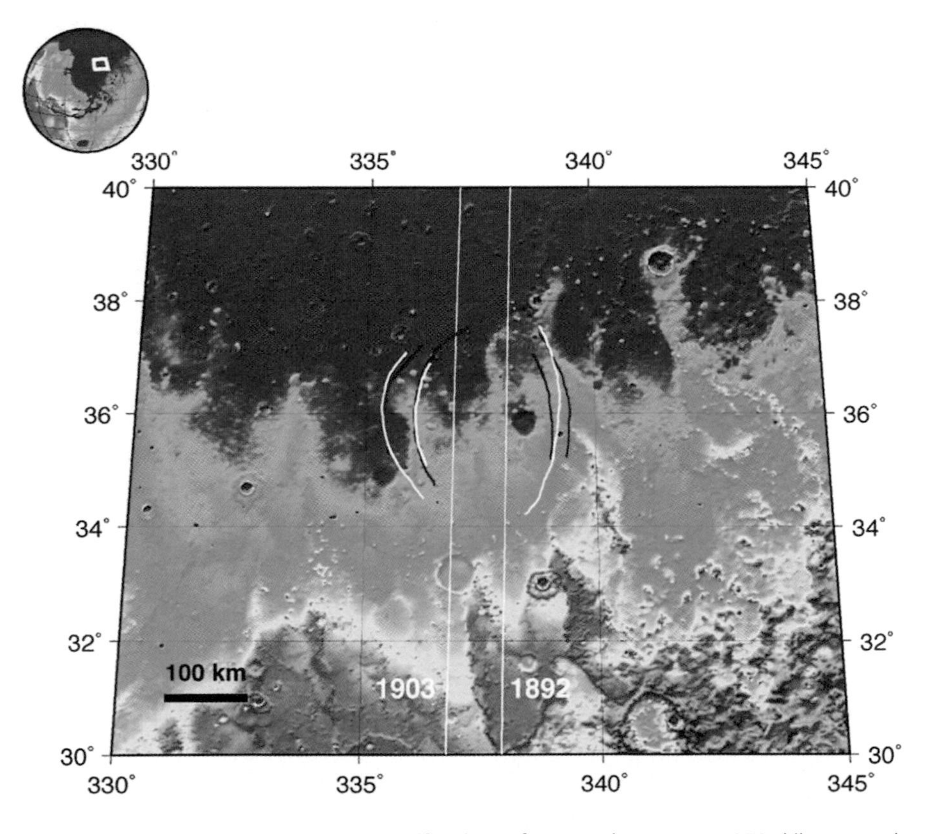

PLATE 38: Mars Express/MARSIS identification of an ancient crater 250 kilometers in diameter buried under Chryse Planitia, superimposed on a MGS/MOLA topology map. [Credit: NASA/JPL/ASI/ESA/Uni. of Rome/MOLA Science Team]

PLATE 39: Sedimentary deposits in Candor Chasma examined by MRO's HiRISE and CRISM reveal ancient fluid flows through cracks in the rocks on the sub-meter level, suggesting a possible ancient hydrothermal site as habitats for microbial life. [Credit: NASA/JPL-Caltech/Univ. of Arizona]

PLATE 40a: "Shoemaker's Patio," near Opportunity's landing site, showing finely layered sediments. Spherical wet sediment grains or concretions (nicknamed "blueberries") align with individual layers on the outcrop. [Credit: NASA/JPL/Cornell]

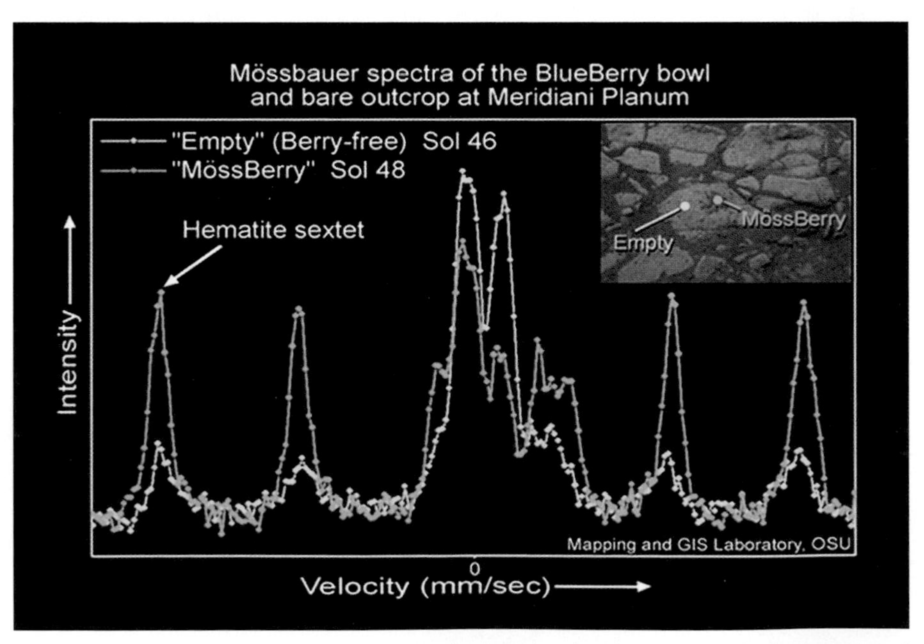

PLATE 40b: Spectra of an outcrop at Shoemaker's Patio. The blue line reveals the presence of hematite in an area called "Mossberry". The yellow line represents an area called "Empty" that is devoid of hematite. [Credit: NASA/JPL/Cornell/University of Mainz]

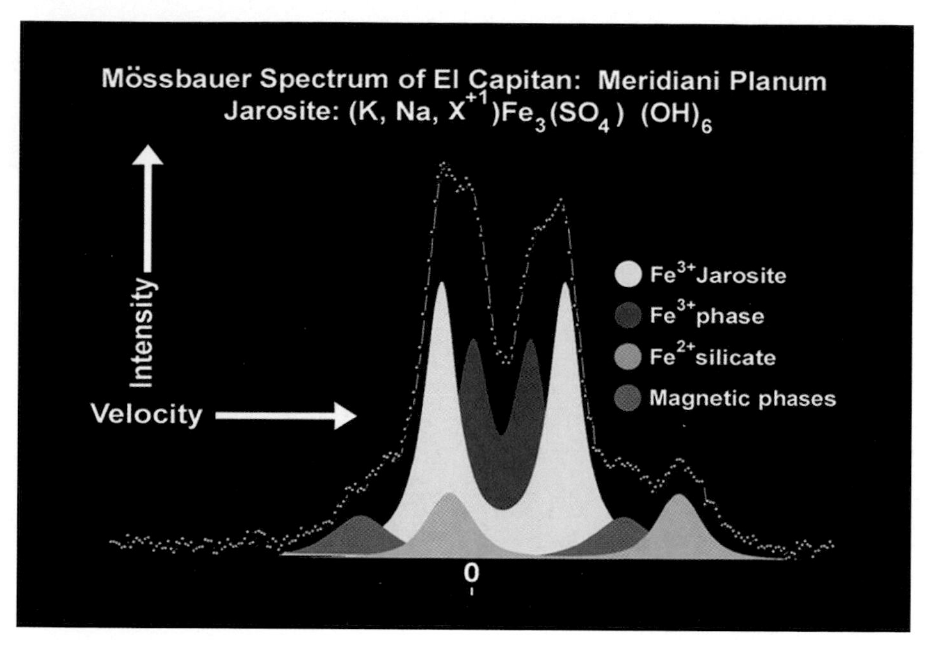

PLATE 41: Mössbauer spectrum reveals the iron-bearing mineral jarosite at "El Capitan" in Eagle Crater, suggesting water-driven processes in the region at one time. [Credit: NASA/JPL/Cornell/University of Mainz]

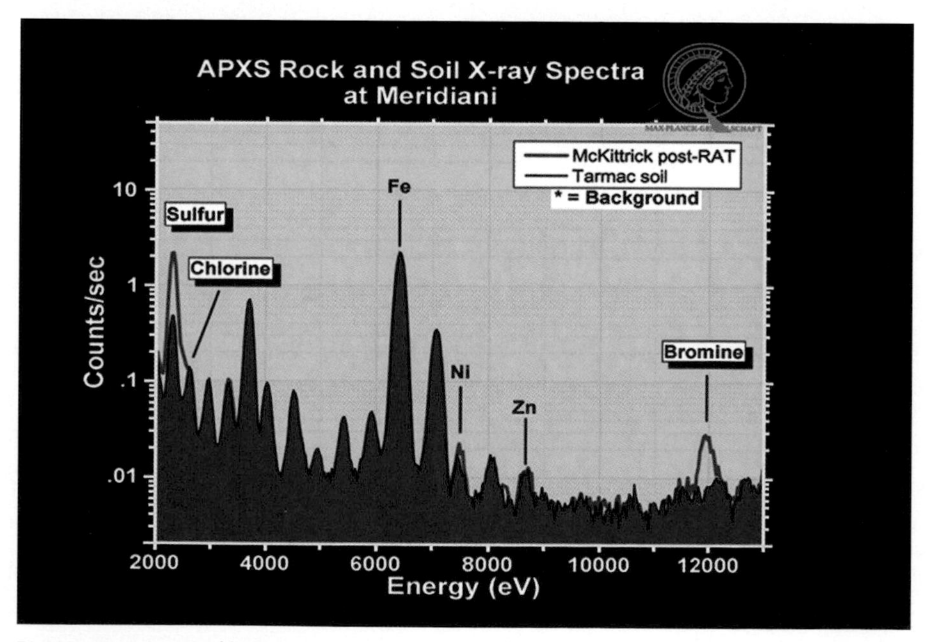

PLATE 42: Spectra of high concentrations of sulphur and bromine, arising when watery brine evaporates, in a rock dubbed "McKittrick" near Opportunity's landing site. [Credit: NASA/JPL/Cornell/Max Planck Institute]

PLATE 43a: Color mosaic taken by MER Spirit having left Bonneville Crater, approximately 600 meters from the base of Columbia Hills. [Credit: NASA/JPL/Cornell]

PLATE 43b: MER Spirit panorama of the impact feature called East Basin to the northeast of Husband Hill, taken on Sol 653 (November 3, 2005).The rim of "Thira" crater is visible on the horizon, 15 kilometers away. [Credit: NASA/JPL-Caltech/Cornell]

PLATE 43c: Endurance Crater and the surrounding plains of Meridiani Planum as seen by MER Opportunity's panoramic camera. [Credit: NASA/JPL/Cornell]

PLATE 43d: MER Opportunity panorama of the Payson outcrop on the edge of Erebus Crater on Sol 744 (February 26, 2006). Layered rocks are observed in the crater wall, while to the left a thin layer of spherule-rich soils overlies outcrop materials. [Credit: NASA/JPL-Caltech/USGS/Cornell]

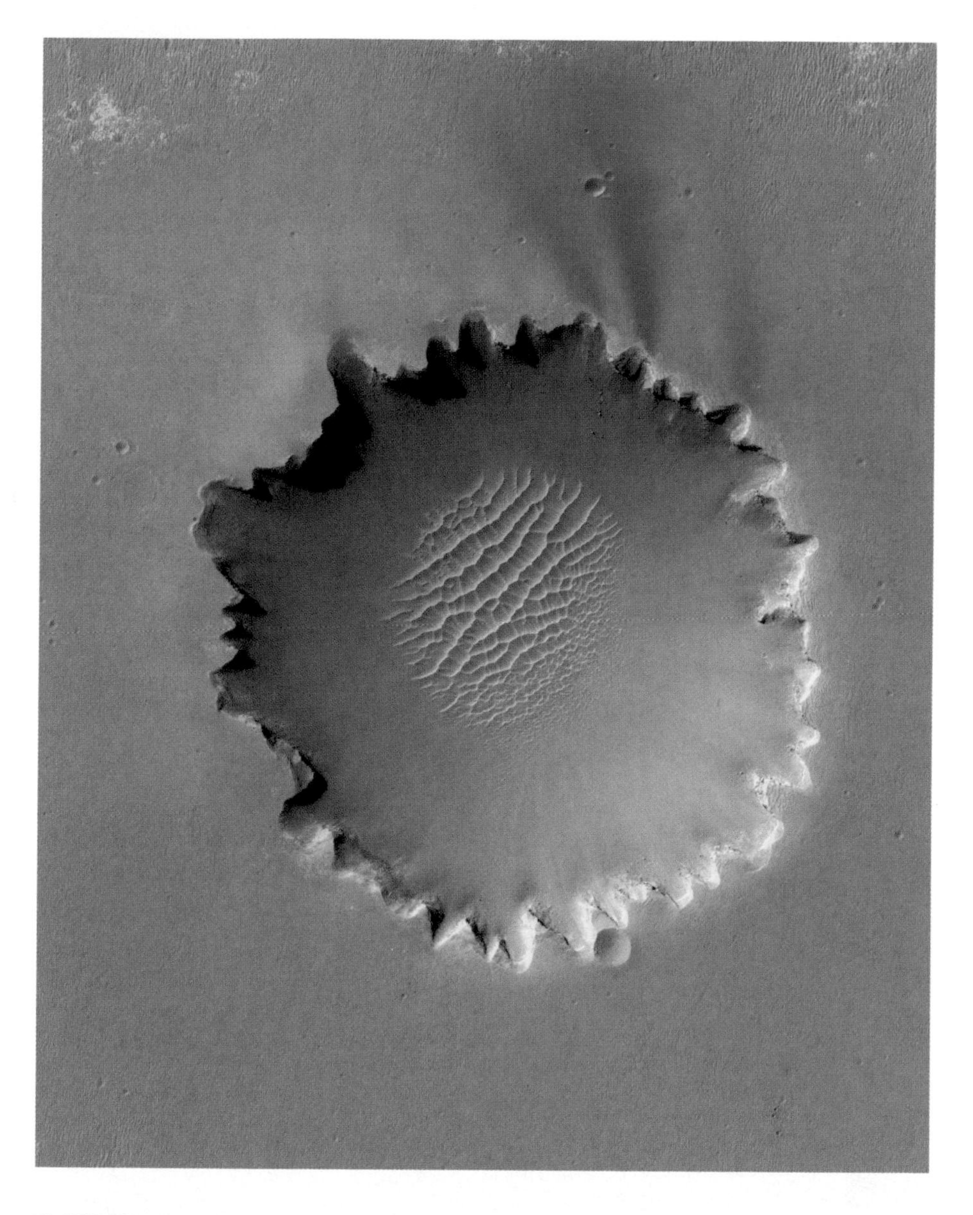

PLATE 44: MRO/HiRISE image of Victoria Crater, taken as Opportunity arrived. Such images bridge the gap between orbital and lander reconnaissance. [Credit: NASA/JPL-Caltech/Univ. of Arizona]

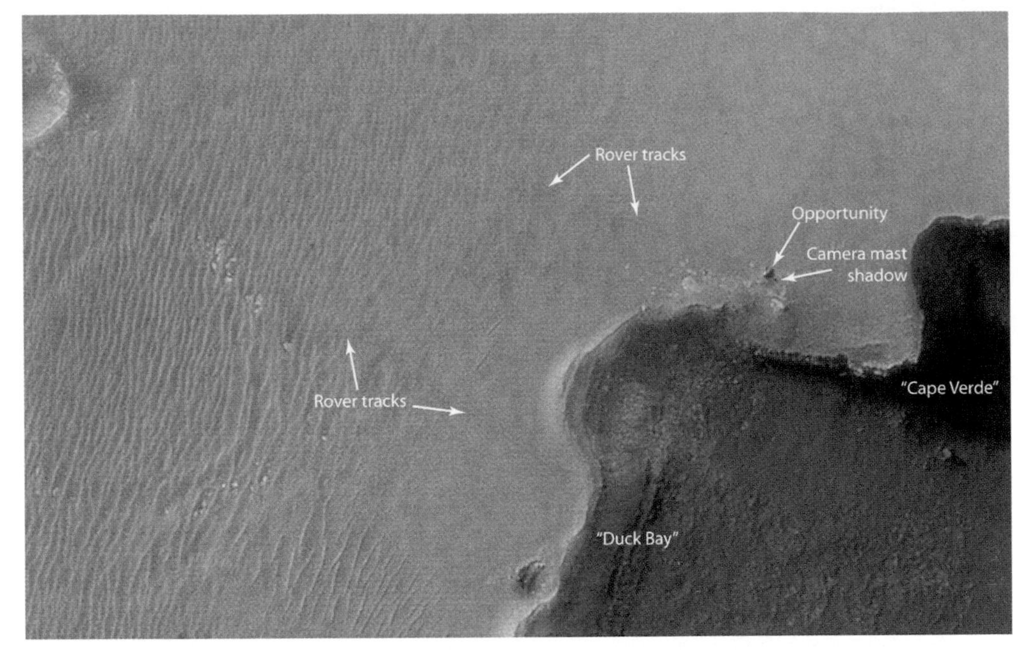

PLATE 45a: MRO/HiRISE image of Victoria Crater, showing the arrival of MER-Opportunity. [Credit: NASA/JPL-Caltech/Univ. of Arizona]

PLATE 45b: MER Opportunity image taken at the rim of Victoria Crater, with a simulated image of the rover superimposed. The walls of Victoria Crater reveal approximately 15 meters of exposed sedimentary layers for Opportunity to investigate. [Credit: NASA/JPL-Caltech/Univ. of Arizona]

Global Programs 12

On October 4, 1957, the Soviet Union launched a small spherical probe called Sputnik 1 into Earth orbit and with that the space age was born. At breakneck pace during the late 1950s and early 1960s, the USSR and the USA competed aggressively in a new era of politically motivated space exploration. During those early years every facet of each mission had to be conceived, invented, tried, and tested from scratch. There was no precedent, no wisdom of old or well-established field of expertise guiding their efforts. It was all new and risky, with as many failures (some involving loss of life) as there were successes.

Within just a few years both nations were achieving quite reliable space flight, with the USA successfully sending a space probe to Mars. At the time, planetary scientists were still grappling with ideas on the nature of planets (and especially Mars) that were at times closer to nineteenth-century thinking than anything we are familiar with today.

The new generation of planetary scientists working with NASA became the world's first space scientists. They quickly realized the unparalleled opportunity that space technology afforded—the ability to actually go to the planets and make a close study of their true character. They set their sights firmly on tackling the great questions of the Solar System, planning missions to virtually all the planets. Through a maturing perspective they guided a reluctant world into a new realization of an ancient connection between life on Earth and the grand Cosmos into which we were, at last, taking our first steps. Unsuspectingly, space science was playing a critical role in bringing about a new view of our world in a broader context.

The original motivation for space activity was not, however, the pursuit of science; it was Cold War competition that ensured massive investment into space activity by both sides. Despite the emergence of a civil space program in the USA, the first robotic expeditions to the planets essentially piggybacked on the vast budgets, technical sophistication, and engineering know-how that developed for Cold War ambitions. Between the two superpowers, dozens of robotic explorers were sent to all the planets of the Solar System.

K. Nolan, *Mars, A Cosmic Stepping Stone*,
DOI: 10.1007/978-0-387-49981-9_12, © Praxis Publishing, Ltd. 2008

By the late 1980s, culminating with the twin enigmatic Voyager missions to the outer planets, the first planetary reconnaissance had been completed, radically changing our understanding of the entire Solar System and of our place within it. We had ventured far, yet we also realized that we had barely begun a journey of exploration that was increasingly regarded as important by both the scientific and the wider community.

By the end of the 1980s the Cold War had ended and with it went the primary purpose for the USA and Soviet space programs. The USSR, in financial difficulties and close to collapse, struggled to sustain its space program. NASA also found itself at a crossroad. To survive it would have to define a new, worthwhile role. In the early 1990s the then director to NASA, Dan Goldin, took the bold step of seeking the views of the broad US science community on how best to proceed. The argument put forward by leading scientists was clear: if NASA was to be worthwhile, it should pursue science programs of importance to society by tackling fundamental questions on the origin and nature of life, Earth, and the Universe itself. Dan Goldin accepted that advice, from which the NASA *Origins* program emerged in the mid-1990s.

Through Origins, NASA would pursue answers to fundamental questions on the origin of the Universe, galaxies, stars, the Sun, and the Earth, and attempt to determine the origin of life on Earth and on other worlds. With government ratification, NASA successfully redefined its post-Cold War role with fundamental scientific exploration as one of its priorities. Space Science had finally emerged as an independent and important pursuit on its own merits.

Similar programs emerged across other great science organizations of the world. The European Space Agency (ESA), though smaller in size and scope, had actually led the way since 1984 with its *Horizon 2000* program, through which it also aimed to tackle fundamental science through space exploration. Funded by individual nations across Europe, ESA was by then also impressively demonstrating how space exploration can be pursued through international partnerships. Fuelled by the emergence of computer and global network technologies, many other global cooperatives formed throughout the 1990s across astronomy, cosmology, fundamental physics, Earth Sciences, and Life Sciences, thus heralding a new era of science made possible by sophisticated technology that we still enjoy today.

Of emerging interest was the new field of astrobiology: the question of biology from a universal perspective. Mergers in astronomy, space science, life sciences, geology, and chemistry were beginning to reveal details on the origin and nature of life on Earth hitherto unconsidered. Evidence of other solar systems, widespread and complex organic chemistry in space, and new ideas on origins on Earth set the whole question of life in a broader, cosmological context. Increasingly, the questions of life on Earth and of life in the Universe were regarded as the same question approached from different perspectives, and one that could now be tackled and perhaps answered through science. For the first time, national and international science organizations ratified astrobiology and exobiology programs. Upon the crest of activity, the search for our origins and for life in the Universe emerged as a valid scientific pursuit.

Once again, Mars was seen as important. In 1995, 19 leading US scientists produced a pivotal document titled *An Exobiological Strategy for Mars Exploration*, in which, based on nearly 20 years of analysis of Mariner and Viking data, they summarized the current state of knowledge of Mars, examined the plausibility of prebiotic chemistry and past and present life, and devised an optimal five-phase strategy for Mars exploration. This document underpinned NASA's far-reaching 1996 *Mars Surveyor Program* and 1997 *Space Science Strategic Plan* (SSSP). The strategic plan in particular brought into focus many of the fundamental questions posed by both Origins and the Mars Exobiological Strategy, from which NASA developed roadmaps, enterprise goals, science goals, science objectives, programs, and missions.

Through the SSSP, the big picture of fundamental space science was married to the methodologies of modern and mainstream business and industry—visionary science pursued on a pragmatic, world-class level. Priorities were set, exploration roadmaps were drafted and project plans and programs were set along a 20-year time line. Central to the SSSP was transparency, efficiency, and accountability. Programs and missions set on tackling fundamental questions would require and enable the development of revolutionary new technologies, and strategic ties with education and public outreach efforts would maximize the benefit to society and the younger generation. Metrics for measuring success across scientific, educational, technological, and financial goals were put in place, and it was expected that the new NASA philosophy of *faster, better, cheaper* would maximize quality, minimize cost, and allow for more missions more quickly. ESA followed suit. As with NASA's Space Science Strategic Plan, ESA's enhanced *Horizon 2000+* program updated and extended the original Horizon 2000 program for the twenty-first century. Astrobiology was also set as a priority, as embodied in ESA's 1998 document *Exobiology in the Solar System, and the Search for Life on Mars.*

By the late 1990s the way forward was becoming clear. Both NASA and ESA had set out a multitude of roadmaps and programs, to 2020 and beyond, set on tackling cosmological origins and the question of life in the Universe with Mars as a prime destination. NASA in particular quickly laid out an aggressive and ambitious roadmap for Mars exploration, largely based on its 1995 Mars Exobiological Strategy and broadly following the suggested five-phased mission strategy. First, orbital reconnaissance missions would characterize the entire planet—its geology, mineralogy, and topology and with particular emphasis on detecting water. Next, a series of lander and rover missions to sites of exobiological interest would conduct detailed chemical and mineralogical analyses, as well as pursue NASA's new mantra of *follow the water*, because where there is water there is an increased chance of finding evidence of life as we know it. This would lay the foundation for the third phase—exobiological landers capable of sophisticated biological analyses, set on answering questions about prebiotic chemistry and life, past and present. Sample return missions would then follow, returning rock, soil and air samples to Earth, which was justified because our current analytical techniques are decades ahead of anything we could do on Mars. Finally, human missions to the Red Planet by 2030 would bring scientific and human exploration to new heights, setting us on a bold and permanent path into space. From the mid-1990s on ward, one or more space probes would be sent to Mars every 26 months for the foreseeable future. Along with ESA, UK, French, Italian, and Japanese missions and valuable input from independent groups such as The Planetary Society, The Mars Society, and the international Mars Architectural Team, the world was finally ready to return to Mars by the late 1990s, and this time for an entire planetary survey.

The successful return to Mars commenced in 1997 with the NASA Pathfinder mission. Pathfinder was a two-part lander comprising a stationary landing-deck craft and a small rover. The mission was a resounding success—far outperforming all of its goals, which included testing new technologies for landing, roving, and conducting science. The icing on the cake was verification of the Pathfinder landing site as an ancient flood plain, adding to the argument that Mars had supported surface liquid water in its distant past.

No sooner had the Pathfinder mission completed when the next mission, Mars Global Surveyor (MGS), arrived in September 1997. MGS was the first successful reconnaissance orbiter since the Viking missions of the 1970s. Operating for four times its planned mission life but finally failing in November 2006, MGS carried out high-resolution mapping of the entire planet, a complete global mineralogical survey, and three-dimensional topological maps to exquisite resolution. Single handed, MGS had transformed our understanding of Mars once again and we will continue to use its data for decades to come.

While MGS's findings have been hugely important, technical problems on arrival at the planet were an ominous sign of difficulties ahead for the still emerging Mars program. MGS had been developed as one of the first missions of the *faster, better, cheaper* program—costing just 250 million dollars—less than one-tenth the cost of the Viking program. As a fuel-saving strategy, MGS was designed to enter Mars orbit using a revolutionary new method called *aerobraking*. Instead of using expensive fuel and thrusters to achieve orbit, MGS would extend its reinforced solar panels and drag against Mars' upper atmosphere. Over hundreds of orbital passes, it would eventually slow down and settle into its planned orbit. Unfortunately, a fault in one of the solar panels meant that MGS could not proceed as planned. It would have to execute aerobraking maneuvers over a longer period and settle into a less effective orbit, which would drastically curtail its science program, or it could extend the aerobraking maneuvers for another full year (with less pressure on the solar panels on each pass) and settle into its planned orbit one year later, but then achieve the vital orbit necessary to execute its full science program. The latter course of action was chosen and, with a one-year delay, MGS finally settled into orbit and started its scientific mission in late 1998.

The technical difficulties experienced by MGS were a warning and a concern, and questions were asked about the *faster, better, cheaper* approach to space exploration. While it was universally agreed that the scientific objectives and phased strategy for Mars exploration were robust, there was less confidence in NASA's ability to actually carry out the program—and what was to immediately follow greatly intensified this concern.

True to its aggressive exploration time line, NASA sent another two probes to Mars in 1999—Mars Climate Observer (MCO) and Mars Polar Lander (MPL)—arriving at Mars in September and December 1999 respectively. Devastatingly, both missions failed on arrival at the planet. Within a time frame of just four months, NASA's Mars program had been dealt a very serious blow. As one TV News broadcaster proclaimed: "NASA would otherwise had to have paid billions of dollars for such negative publicity." The loss of both missions was not only a huge blow to NASA and the hundreds of engineers and scientists directly involved in the missions, but it cast serious doubts over its entire strategy, with even the US Congress demanding an explanation.

Several internal and congressional inquiries were conducted, uncovering many reasons for the failures. In particular, serious flaws were revealed in NASA's new methodologies. In a nutshell, NASA had gone too far with its cost-cutting and efficiency measures—from cutting too many corners during design, testing and implementation to poor interorganization communications across NASA, JPL, and independent contractors. Even the space probes were shown to be poorly designed with inadequate onboard redundancy or backup systems on both of the failed probes. For example, even though MGS came in dangerously cheap at just 250 million dollars, MCO and MPL *combined* cost just 190 million dollars. It was clear to all that they were built too cheaply and developed too quickly under inadequate and unacceptable conditions, which consequently culminated in extremely high-risk missions with a near-certainty to fail.

NASA received an embarrassing bloody nose from the failures and the resulting inquiries, but it was just a bloody nose. At least clear reasons for the failures were identified. NASA could also demonstrate through other successes that its broad strategy had many strengths and was enabling space exploration in a way that was not previously possible. By learning from theses failure it vowed to do better—by slowing down the pace of development and the Mars program time line, by improving standards and communications, and by allocating the resources needed to deliver missions with a high probability of success. Subsequently, all inquiries recommended that NASA continue with the Mars Surveyor Program, but with the agreed less aggressive time lines and with improved resources, care and attention to subsequent missions. NASA had pushed too far, to a point of unacceptable high risk and diminishing return, and had paid the price. From the failures, however, it learned valuable lessons that have served subsequent missions well. Indeed, following almost immediately, the 2001 Mars Odyssey orbiter incorporated many of those requirements and has been a resounding success, producing mineralogy maps hundreds of times more accurate than even those from MGS and images of the entire planet at a vital resolution between those of Viking and MGS. With enormous relief, Odyssey heralded a more mature approach to exploring Mars. The right balance had now been found and NASA was on its way once again.

Subsequently, the historically close approach of Mars in 2003 brought with it a bounty of Mars missions. In January 2004 NASA sent two identical rovers, Spirit and Opportunity, which have been unprecedented successes. ESA also sent two probes—the ESA *Mars Express* reconnaissance orbiter and the UK *Beagle 2* exobiological lander—both arriving in December 2003. While Mars Express has also become a resounding success, Beagle 2 was tragically lost on arrival. It was a hugely sophisticated probe, scientifically ahead of its time, but an ESA inquiry into its failure revealed similar flaws in design and implementation to MPL and MCO, from which the UK and ESA have also learned a tough lesson.

While the successes of MGS, Odyssey, Express, Spirit, and Opportunity are undeniable, the losses of MCO, MPL, and Beagle 2 are sharp reminders of the difficulties of pursuing major science on other planets. Along with other past failures and the inability during that time for other planned missions—such as the French CNES *Net Lander* Mars robot network—to be fulfilled it became clear on both sides of the Atlantic around the turn of the millennium that although our space science goals were by then extremely robust, our technical ability to pursue them was perhaps less so.

In response, both the USA and the European/ESA nation states, independently, have recently recognized that far-reaching science needs equally far-reaching space exploration technologies and programs appropriate to the twenty-first century. Our science objectives demand better programs and missions and better technology than has recently been available, and both have now openly declared that revolutionary new ways of pursuing space exploration must be a priority if we are to establish a long-term, successful, and sustainable future in space.

To this end, the USA announced its bold *Vision for Space Exploration* in February 2004, while ESA announced its *Cosmic Vision* and *Aurora* programs, with both heralding new approaches to space exploration. Gone is the ethos of pursuing space exploration on a mission-by-mission basis. Instead, both NASA and ESA, independently (and increasingly in cooperation) will pursue integrated strategies across all of their programs. Generations of new *enabling* technologies will be invented to match our far-reaching scientific and human exploration requirements for decades to come. Grand roadmaps, phased programs, and cutting-edge business and industry models will underpin all efforts, positioning both the USA and Europe to successfully tackle the great science questions laid down in the late twentieth century. Emerging from the tragedy of Space Shuttle losses and the embarrassment of Mars probe failures is a singular and increasingly global strategy for space exploration.

A Vision for Space Exploration

While such events as the Exobiological Strategy for Mars and Space Science Strategic Plan laid firm foundations for ground-breaking new science, NASA has until recently pursued those goals through essentially tried and tested methodologies dating back to the 1960s and 1970s—all individual missions proposed, implemented, and managed by individual teams. While this has served well, a shortcoming to this approach has been a lack of integration on the various programs, whether robotic or human, and on different class probes to a given destination or similar class probes sent to different destinations. Of course, transfer of know-how from mission to mission has been largely achieved, but not in a truly integrated sense. Such a shortcoming, coupled with the tragic losses of Challenger and Columbia and the failures of various robotic explorers such as MCO and MPL, brought about a realization in the USA that NASA was unlikely to achieve its lofty goals without radically updating its traditional methodologies.

To address this, a grand new *Vision for Space Exploration* (VSE) for the USA was announced by President George W. Bush in February 2004. Ambitious space programs, and new ways of implementing them, are proposed for the next 30 years, pursuing fundamental science questions while also heralding a new era of human space exploration. VSE retains many of the questions proposed by the Origins and SSSP initiatives, but now also proposes an ambitious program of human exploration to the Moon, Mars and even to other destinations in the Solar System such as the near Earth objects (NEOs), the asteroids, and the moons of Jupiter.

What is radical about the VSE, however, is how its goals are to be pursued and achieved. Instead of a multitude of individual missions, the US space program will be pursued in a closely integrated way. Each mission—robotic or human and irrespective of its destination—will be developed with all others in mind, and new and enabling technologies and know-how will be applied across all missions, improving quality, safety, and overall capability. New NASA agencies will bind the US government and NASA more closely, enhancing the communications paths required for strategic planning. It is only through such integration and the adoption of new working methodologies and practices that NASA can achieve the lofty goals set out by the Vision for Space Exploration.

NASA, in response, has already reorganized internally, taking upon itself to implement as much of the Vision for Space Exploration as is possible within a new culture of safety and excellence. New *agency level* planning and strategy will oversee all activities, ensuring that programs and missions happen in an optimal way, with guaranteed consistency in standards, capabilities, and shared information. NASA's new strategic context and approach will fully adopt well-tried and tested modern business models—corporate focus, prioritized requirements, spiral transformation, and managerial rigor. It will develop a *systems-of-systems* capability for sustained exploration across multiple programs and adopt a *spiral* (phased) approach to all exploration, where one spiral is followed by, and enables, the next.

Critical to the success of the VSE is the identification of numerous *key technical challenges*: redundancy and margins, reusability, modularity, autonomy, humans in deep space, in-space assembly, reconfigurability, robot networks, prepositioning, energy-rich missions, and space resource utilization—all of which will be pursued via 15 *capability roadmaps*, ensuring the emergence of new enabling technologies for successful robotic and human missions to Mars and beyond.

As for how science and exploration will be pursued, 13 *strategic roadmaps* encompassing 18 distinct objectives have now been identified. They include:

- The exploration of low Earth orbit via the Space Shuttle and the International Space Station (ISS)
- The use of the ISS as a testbed for prolonged human mission in space and on the Moon
- Robotic missions to the Moon early in the next decade
- Human missions to the Moon no later than 2020
- The establishment of a permanently occupied Moon outpost
- Robotic missions to the asteroids and moons of Jupiter
- The search for habitable planets around other stars
- A long-term program for Mars to determine its history and in particular the story of water and life there
- Improved aerodynamic entry, Mars orbital and docking capability, precision landing, resources extraction and utilization, communications, and critical data for human missions
- Human missions to Mars
- Human missions to Near Earth Objects, asteroids, the moons of Jupiter, and beyond.

While roadmaps are not funded programs, they set out science and exploration goals, implementation approaches, milestones, dependencies, and required capabilities from which NASA can then make key decisions before initiating its programs.

Since the last manned mission to the Moon, the US space program has often struggled to see a sustainable path forward for space exploration. From the mid-1970s to the present day, even such giant programs as the Space Shuttle and International Space Station have been pursued without clarity of purpose. However, through the Vision for Space Exploration this is no longer the case. With clear and long-term goals across science and human exploration, and an identified sustainable strategy for their pursuit, the USA has once again identified a way forward and the means to become a long-term space-faring nation. There will, of course, be problems, delays and setbacks; but the space program and NASA in particular are already committed—ideologically, structurally, and financially—to pursuing the broad aspirations of the Vision for Space Exploration with a view to comprehensively studying Mars via robotic exploration in the coming decades, and to land people there some time after 2030. Everything now being instigated by NASA is toward those aims.

Aurora

ESA has also realized that if great science and exploration are to be tackled, a mission-by-mission approach will no longer suffice. Instead, far-reaching exploration of the Solar System—and in particular of Mars—requires a long-term, integrated strategy enabled by revolutionary technologies capable of the task. To this end, ESA have developed two initiatives: *Cosmic Vision* is a core strategy for tackling fundamental questions through space exploration; while the *Aurora* program outlines ESA's strategy for both robotic and human exploration of the Moon, Mars, and the Solar System.

Through Aurora in particular, Europe recognizes and endorses an integrated way forward for robotic and human space exploration, in a spirit of international, peaceful cooperation. Europe alone will not send people to the Moon or Mars, but it hopes to fully participate with other nations on such ventures, contributing critical technologies and resources. In this way, Europe can play a key role in what is arguably becoming a global program culminating in human missions to Mars and even beyond. The technical and sociological benefits, unprecedented international cooperatives, and the scientific returns form the hallmark of Aurora.

ESA has different operational challenges to NASA, however. Since Europe is a collection of individual nations, ESA must be managed differently. It depends on individual national budget commitments, which can vary from nation to nation and over time. ESA must therefore operate in a less binding environment, yet strive to provide maximum return to contributing states. Thus, while Aurora must address Europe's long-term space program goals in an integrated manner, it must also navigate the uncertainties brought about by the multination environment within which it exists.

Hence Aurora has been developed as a stepwise, non-binding program of building technical capability to enable Europe to be integrally involved in future major space exploration, including an international human mission to Mars. Countries currently committed to Aurora—which are not all ESA nations but include Canada—can become involved at various levels of commitment. For example, any individual nation can become involved with only robotic exploration, or human exploration, or both. Furthermore, planned missions will be grouped into one of two classes and individual states can decide whether to contribute to one, or both. *Flagship* class missions are large-scale missions critical to the success of each phase, while *Arrow* class missions act as smaller testbeds in support of later Flagship missions. As individual countries are allowed to opt for one or more five-year phases over the 30 years of Aurora, a flexible system of commitment has been devised where no one nation is burdened with unrealistic commitments that are binding throughout the entire program. One nation, for example, may decide to become involved in just a part of one Arrow class mission over a single five-year period, while another can commit to both classes of mission and over multiple periods.

As with VSE, Aurora has identified the need for revolutionary and enabling technologies for propulsion, life-support, resource extraction and utilization, robotics and navigation, among others. It is recognized that far-reaching exploration requires an integrated approach and that those enabling technologies must be applicable across all missions. While Aurora does not require universal support across Europe, is has already been ratified by the European Space Agency Council at Ministerial Level as well as ESA's internal Exploration Program Advisory Committee (EPAC). Currently 14 participating states of ESA are committed, including the UK, France, Germany, Italy, and Canada. Ireland—among the fastest growing economies and richest states per-capita in Europe—has not.

Intriguingly, Aurora will follow a broadly similar time line to NASA's Vision for Space Exploration. Over the next 15 years, Aurora is set on fully exploring Mars via orbital reconnaissance, robotic landers, and sample return. It will also execute technology testbed missions to Earth orbit and the Moon and lay the foundations for human missions to the Moon by 2020 and, hopefully, to Mars by 2033. A decision to send people to Mars will not be made until well into the next decade when Europe's technological readiness and that of other global partners can be more properly assessed.

Roadmap to Mars

For the first time, there is political endorsement on both sides of the Atlantic for a long-term program for Mars culminating with a human mission. This is unprecedented; prior to 2003 no government or space agency would commit to any plan that included a human mission to Mars, yet today US and European governments as well as NASA and ESA are actively engaged in pursuing that goal. This commitment is real and not some pipe-dream or optimistic proclamation by special interest groups. Governments and space agencies have committed budgets, expertise, and a substantive part of their future to such a venture and, from now on, the further we proceed the more difficult it will be to turn back. The drive toward fully exploring Mars and culminating with a human mission is no longer aspirational; it is earnest, big space exploration, and it will be a brave politician or space agency administrator who reverses direction. A future where humans land on Mars is neither guaranteed nor yet committed to; but everything now being pursued is with that goal in mind.

The survey of Mars began in 1997 with the success of Pathfinder, and continued with the orbital reconnaissance of MGS, Odyssey, Mars Express, Mars Reconnaissance Orbiter, and the twin rovers Spirit and Opportunity. While recent missions were originally devised as part of NASA's and ESA's traditional modes of operation, they led seamlessly into VSE and Aurora. Many new robotic missions are now in development or are nearing readiness for launch during available windows well into the next decade. They are set on greatly improving our understanding of Mars, initiating exobiological searches and extending our technical know-how.

The first of the new missions is the enigmatic Mars Reconnaissance Orbiter (MRO) which arrived at Mars in March 2006 and took orbital reconnaissance to a new level. Following, in 2008, is the NASA exobiological *scout class* mission called Phoenix, the first exobiological lander of the current program. NASA will then launch the Mars Science Laboratory (MSL), a long-range exobiological rover, in 2009. Also planned for around 2011 is a second scout class mission, followed in 2013 by a hybrid science and communications orbiter—the first space probe sent to another planet specifically to provide communications capability. ESA are also following suit. Two Flagship missions are now planned: ExoMars, an exobiological rover to launch in 2011, and a Mars Sample Return (MSR) mission planned for 2016. Two Arrow class missions will also test new technologies required for the MSR mission—an Earth return capsule and a friction-assisted Mars orbital entry craft.

The jewel in the crown for the coming decade would be a sample return mission to travel to Mars, collect samples from the surface, and return them to Earth. Here ESA is the major driver, though it is probable that such a significant feat will require a joint effort between ESA and NASA, and perhaps space agencies from other nations. Although NASA has recently reduced its development efforts for a sample return mission within the next decade, it is highly plausible that it will engage in such a mission before 2020. Spurred on by its historical legacy in space and recent successes by NASA and ESA, Russia's Federal Space Agency has also recently revitalized its space activities. Through its new "Program for Deep Space Exploration" (2006–2016), it will conduct a sample return mission to Phobos, one of Mars' two moons, in 2009. Called Phobos-Grunt, this enigmatic mission will send a robotic space probe to Phobos, land on the moon's surface, collect samples, and return them to Earth by July 2012. The mission will help to reveal the origin and nature of Phobos and provide new insights into both the early Solar System and the formation of Mars. It will also provide vital know-how and experience on how to carry out sample return missions, complementing American and European ambitions and perhaps even prompting a more ambitious multinational sample return mission to the surface of Mars itself. Indeed, in a spirit of international cooperation, The Planetary Society, the largest space interest groups in the world, is currently exploring the possibility of sending an experiment containing living organisms on board the three-year return mission to Phobos. Called LIFE (Living Interplanetary Flight Experiment), the experiment would help scientists to better understand the nature and robustness of life, and its ability to traverse the distance between the planets.

Subsequent to Phobos-Grunt, the first aerobot missions may take place in and around 2016; while deep drill missions, set on penetrating the surface of Mars to depths of dozens of meters in search of evidence of life, may take place as early as 2018. Through all of these missions, the coming decade will be critical to maturing our understanding of Mars, in learning how to carry out successful and complex missions, and in affecting a decision on whether to send humans there.

In conjunction with the robotic program, plans for long-term human exploration are now also underway. Having safely returned to operation in 2005, the Space Shuttle can complete construction of the International Space Station (ISS) by 2010. Once complete, the Shuttle will be retired from service, while the ISS will act as a platform for long-term human missions in space, in preparation for missions to the Moon and Mars.

Several robotic missions to the Moon are also planned as precursors to human missions there. The first of these, called Lunar Reconnaissance Orbiter (LRO), to be launched in 2008, mimics recent Mars orbiters—carrying out detailed imaging, high-resolution 3D global topographical mapping, searching for water-ice near the Moon's South Pole and mapping the surface mineralogy. The data gathered will lay the foundation for all future missions to the Moon, help with site selection and resource utilization (which may include, for example, creating breathable oxygen from lunar surface materials). Other NASA lunar robots are also in the planning stage and may include a Small Lunar Lander to test precision landing systems and a Small Circular Orbiter for testing orbital maneuvers and maintenance.

Subsequently, NASA and ESA intend to send people back to the Moon. While ESA regards this as feasible (with an international partner) by about 2024, NASA is already committed, having announced several major initiatives: the development of a new rocket capable of carrying a 125-tonne load into low Earth orbit (LEO); the development of a new Crew Exploration Vehicle to replace the Space Shuttle by 2012; and the development of a range of lunar landing systems, including lunar outpost habitats, lunar regolith moving equipment, and *in-situ* resource utilization technology. When complete, NASA will be able to send up to six people or over 20 tonnes of cargo to the Moon on any one mission. As outlined in NASA's 2007 budget and with technical details provided in its 2005 document *Exploration Systems Architecture Study* (ESAS), a significant proportion of funds available to the agency will be redirected toward the first human mission to the Moon since 1972, now earmarked to take place in 2020. Initially, each lunar landing will last about seven days, with each mission contributing to the establishment of a permanent lunar outpost. Eventually, lunar missions will allow up to six people to travel to the lunar surface on any given mission, for weeks or months at a time.

The myriad of robot and human missions to the Moon will provide a stepping stone and vital testbed for a human mission to Mars. The keys to a successful human mission to Mars are to know Mars thoroughly and develop technical, infrastructural, and human readiness through lunar exploration. It is hoped that by 2020 we will have a much clearer picture of life on Mars—a goal of profound importance in its own right, but also a critical requirement when considering a human mission to the planet. Should all go according to plan, the human lunar program will also contribute significantly toward implementing a human mission to Mars. The rockets needed to send people to the Moon are also capable of being assembled as human-rated Mars spacecraft in LEO. The Crew Exploration Vehicle that is used to take people to the Moon, will also be used to transfer people to their Mars spacecraft, while orbital docking and assembly technology, lunar outpost habitats, lunar life-support systems, lunar regolith moving vehicles, and *in-situ* resource utilization technology are all critical to a successful Mars mission and will by then have been designed, built, and field tested. Finally, with extended stays on the Moon, many health and safety issues, so vital to a long mission to Mars, will also have been researched.

Subsequent to a successful return to the Moon, NASA plans to pursue a dual strategy of sustaining its lunar program while commencing on a human Mars program. At the same time, ESA also plans to implement its strategy toward a human mission to Mars. Although a great number of critical issues toward a human mission will by then have been addressed through robotic Mars exploration, a sample return mission, and the new human lunar program, a human mission to Mars will be an enormously challenging proposition to both NASA and ESA. Furthermore, scant research and development will have taken place on a range of other critical aspects to a human mission: human-rated Mars aero-capture; Entry, Descent and Landing (EDL) systems; nuclear thermal rockets; Mars In Situ Resource

Utilization (ISRU); reliability of technology over a three-year life span; landing site mini-nuclear reactors; and microgravity and radiation countermeasures. Each of these is of fundamental importance, yet each is an unprecedented technical challenge to be overcome before a human mission to Mars is possible.

Furthermore, even with the myriad of robotic missions proposed for Mars up to 2020, we will still not have gained anything like a complete understanding of the planet. Thousands of years have been spent exploring our own world, yet a great many unknowns about its nature and evolution persist. Although Mars is less active, it is still an entire world with an ancient legacy and as far-reaching as the missions up to 2020 seem, they will not be sufficient. Thus, from 2020 onwards, further reconnaissance, lander, and test mission will be needed and these will be accompanied by a series of automated human-scale missions, bringing supplies, and testing technologies and infrastructural technology for a pending human mission. Very optimistically, NASA and ESA both envisage a two-and-a-half-year human mission to Mars, launching on April 8, 2033 (the best launch opportunity until 2048) and returning on November 25, 2035.

Although this time frame has been mooted by both agencies, given the technical and human requirements to be overcome from 2020 onwards, it is more likely that such a mission will not happen before 2040. When it happens, however, it will be the culmination of an integrated and global effort that will have stretched science, technology, and indeed all of human capability to its limit. It will be a feat of such enormous proportions that the world will have had to change to make it so.

Stepping Stone to Mars

The Vision for Space Exploration and Aurora are hugely impressive because they place one of the most central of human pursuits—the scientific quest for our origins—high on the space agenda. Governments and space agencies now fully endorse the search for our origins, for life in the Universe, and for life on Mars. Such an unprecedented commitment constitutes one of the biggest quests humanity has ever undertaken. The debate is over, and we are now on our way to discovery, with Mars as a prime target.

Despite the problems, contentions, and dilemmas likely to be faced in the future, these programs can fulfill our hopes and expectations. The US and European governments can now feel more comfortable with proclaiming their intent because not only are both programs built on important scientific and sociological grounds, but their phased and non-binding nature means that governments on both sides of the Atlantic can commit to long-term projects without committing to long-term and fixed budgets. VSE and Aurora enable governments to endorse long-term space exploration free of the constraints normally set by shorter term political infrastructures. VSE and Aurora are our best attempts yet at becoming a permanent space-faring people.

Whether by design or by coincidence, it is fortuitous that both VSE and Aurora have emerged at the same time, with similar goals and following similar roadmaps and time lines. Such serendipity may be pivotal, because, as ambitious as VSE and Aurora may seem, either program alone may not be capable of fulfilling its goals to completion. It is highly plausible that, as the time draws closer for a human mission to Mars, the USA and Europe (among other nations) may need to share the load in order to succeed. We must remember that, currently, we are not yet capable of sending deep-drilling robots to Mars, or of implementing a sample return mission, let alone a human mission. Virtually all the work to achieve such lofty goals is still to be done and it is not clear that it can be achieved in the given time frame, even under the VSE and Aurora frameworks. If we are to send humans to Mars, the minimum requirement will be the full and successful implementation of both VSE and Aurora combined, and in this way they may represent a path toward a single, global human mission to the Red Planet.

To this end, ESA, NASA, and other space-faring nations are, as of 2006, engaged in a process of discussion regarding commonality in aspirations and plans for space exploration, with a view to international cooperation where commonality exists. In particular, the convergence of NASA's and ESA's ambitions for the Moon, Mars, and Solar System exploration in general mandates cooperation wherever possible. In the words of Daniel Sacotte, Director of Space Exploration at ESA and organizer of an international meeting in 2006: "Activities linked to space exploration will return economic benefits and will provide answers about the origin and distribution of life in our Universe." With such a shared vision of the benefits of long-term space exploration, from economic to philosophical, all space agencies engaging in discussions now envisage a structured international cooperation mechanism arising in the coming years. Called a *Global Reference Architecture*, it will provide internationally agreed and ratified mechanisms that ensure the long-term sustainability of space exploration *plans*, allowing for their aspiration to be realized. From the success of the Mars phased robotic strategy, the VSE and Aurora phased strategies for all upcoming exploration have (in part) emerged; and from the parallels within both, a new and enabling global strategy for long-term space exploration is now emerging, with Mars firmly in our sights.

The path to Mars has been long and very winding. We have even taken the wrong turns on many occasions—in antiquity, in the nineteenth century, and in the latter half of the twentieth century. Despite this, we have been unflinching in our fascination and sense of connection with the Red Planet, and now we are reaping the benefits in so many ways. We have been able, resting upon great science questions about Mars, to build a far-reaching strategy that might just work. By devising the Vision for Space Exploration and Aurora, we have already demonstrated a maturity in pursuing far-reaching science and exploration which builds upon our legacy with Mars and inspires to enlighten future generations.

First Steps: Pathfinder 13

Despite the superlative technical success of the Viking program, it was to be 21 years before we would successfully return to Mars. The reasons for the long delay, with hindsight, are quite understandable. Certainly there was no lack of will on the part of many scientists to continue with Mars exploration after Viking. Although some of the original Viking team had become jaded and disappointed by the sterile nature of the surface, others recognized that Mars remained barely understood and that it was revealing itself to be a genuinely fascinating place. It was not simply some sterile ball in space; rather, it was an entire world with a vibrant past of which we knew virtually nothing. And as the 1980s brought a deeper appreciation of the complexity of the planet and of a new possibility of life there, an impetus to return to Mars arose once more. The jitters of Viking were fading and we were ready to tackle the Red Planet again.

There were several attempts to return to Mars in the intervening years, though all were unsuccessful. Phobos 1 and Phobos 2, each of which was an orbiter/lander mission built and launched by the Soviet Union in 1988, were both unsuccessful. Phobos 1 was lost *en route* to Mars, while Phobos 2 achieved orbit and before it failed transmitted 38 images of Mars and its moons, Phobos and Demos, back to Earth. Also unsuccessful was the 1993 US Mars Observer—a significant advancement on the Viking orbiters—carrying a sophisticated science payload including high-resolution imaging and mineralogical analysis instruments. So capable was Mars Observer that many of its instruments were later incorporated into the Mars Global Surveyor and Odyssey orbiters.

Each failed mission was a massive blow to planetary and Mars science, representing an abrupt halt to our aspirations for, and our ability to carry out, required new science on Mars. And so with the emergence in the 1990s of NASA's *Origins* program, *Space Science Strategy* and its *Exobiological Strategy for Mars Exploration*, the time had come for a new, long-term strategy for Mars exploration. Through an optimized, long-term phased program the risks of planetary exploration could be better managed, while

K. Nolan, *Mars, A Cosmic Stepping Stone*,
DOI: 10.1007/978-0-387-49981-9_13, © Praxis Publishing, Ltd. 2008

enabling far-reaching exploration set on gaining a full understanding of Mars.

Pathfinder

The first successful mission of the current era of Mars exploration began on July 4, 1997, not by the arrival of another orbiter but by the successful landing and deployment of a surface rover, the first ever to operate successfully on Mars. Called *Pathfinder*, the combined lander and rover mission set down on Ares Vallis where, it was suspected, there might remain evidence of a gigantic flood in Mars' distant past.

Pathfinder was actually a test mission, one of the first of NASA's new faster, better, cheaper *Discovery* class program. Its main objective was to test new planet-entry and rover technology, as well as sophisticated scientific instruments including an Alpha-Proton-X-ray Spectrometer (APXS) and a thermal-imaging camera, capable of determining the mineralogy and some of the chemistry of the landing site. From the outset, Pathfinder was a spectacular success. Upon entering the general region of the planet, Pathfinder was able to determine the planet's gravitational influence upon it, allowing for a factor of 3 improvement on a determination of the planet's axial tilt, a better determination of the size of the planet's core, an improvement on how Mars precesses and a *sharpening* of the position of every known feature on the planet's surface. All of this was accomplished by applying Newtonian mechanics on a payload of a few hundred kilograms heading toward Mars. From here on, there would be no waste in our efforts to understand the planet.

Entry Descent and Landing

The first major new test for Pathfinder, however, was a revolutionary new way of landing on a planet's surface. Because Mars has an atmosphere, all spacecraft landing there require a heat shield to absorb the heat generated while slowing down from tens of thousands of kilometers per hour to just hundreds of kilometers per hour within the upper atmosphere. Even within a tenuous atmosphere, friction forces are great enough to raise the temperature to over 1,200°C—more than enough to destroy any unprotected craft. Once a heat shield has absorbed the friction-generated heat, it is jettisoned at an altitude of about 10 kilometers above the Martian surface, whereupon a parachute opens to allow the craft to descend more slowly. Because the atmosphere of Mars is so tenuous, however, further mechanisms are needed to slow the descending craft even further to assist in safe landing.

In the case of the Viking landers (as with all lunar landings), retro rockets attached to each lander would fire when close to the ground, providing an upward trust that would slow the craft for safe landing. But in the case of Pathfinder, an entirely different and new descent and landing system was to be tried. Instead of using costly retro rockets, the lander was encased in airbags which, when close to the ground would inflate and absorb the impact on the surface. Pathfinder was encased in four independent airbags, with each airbag consisting of six intersecting spheres made from a woven polymer fiber called Vectran. Similar to Kevlar, Vectran is used in bulletproof vests and could survive the initial surface impact, followed by numerous rough and tumble bounces across even a rocky surface.

In a spectacularly choreographed sequence, Pathfinder entered Mars' atmosphere on July 4, 1997, where its heat shield absorbed the intense heat of the entry. Subsequently a large parachute opened and slowed the craft. At an altitude of about 9 kilometers the protective heat shield was jettisoned, whereupon the lander, still in mid-air, descended about 20 meters from the parachute along a metal tape to give room for the airbag inflation sequence. A radar altimeter detected the surface just 32 seconds before impact, causing the airbags to inflate at an altitude of just 1,500 meters. After another eight seconds, at an altitude of 300 meters, the airbags were fully inflated, whereupon three small retro rockets attached to the entry back-shell fired, bringing the lander to a vertical halt just 12 meters above the ground. The tether was then cut, dropping the airbag-encased lander to the ground which, on impact, bounced 15 times and came to a stop 1 kilometer downrange. Pathfinder had landed!

This new and quite revolutionary mechanism had worked incredibly well. As was later realized, Pathfinder landed in some of the rockiest terrain known on Mars and yet the airbags survived without a single puncture. Pathfinder's first objective—a new and cost-effective method of delivering robots to Mars—had been demonstrated and tested successfully. Three hours after landing, the airbags deflated to reveal a three-petal lander within which was the tiny rover. Another innovative feature of this setup was that no matter which way the lander had settled on the surface, it could self-correct its orientation by opening only the require petal(s) to do so.

On July 5, when NASA verified that the Pathfinder lander petals opened successfully and the tiny rover was undamaged, it was announced to the world that this newest of Mars robotic outposts would be named *Carl Sagan Memorial Station*, after the pioneering astronomer, exobiologist, and communicator who had transformed the world's perception of the Cosmos, and who had sadly passed away just seven months before Pathfinder landed (Figure 39).

Figure 39: Carl Sagan Memorial Station in Ares Vallis, where the Sojourner rover is seen examining a rock called "Yogi." Two hills called "Twin Peaks" are seen on the horizon. See Plate 14 in the color section. [Credit: NASA/JPL]

The Carl Sagan Memorial Station was not just a repository for the enclosed lander; it was a sophisticated lander with a spectacular stereo panoramic camera that delivered unprecedented images of the surface of Mars, many of which were in three dimensions, revealing for the first time the relief of the planet's surface. The station could also measure wind speed and direction as well as atmospheric temperature and barometric pressure, thus acting as a weather station. Further, the station provided two separate sets of ramps from which the rover could drive onto the surface; it also acted as a critical two-way communications link between the rover and Earth.

Sojourner

With two ramps available, the little 10-kilogram rover should have had a choice of routes to the surface. Unfortunately the front ramp remained aloft, not touching the ground because part of the airbag assembly had not retracted fully. None the less, late on July 5 or Sol 2 (the second Martian day of the Pathfinder mission), the rover lifted itself up on all six wheels and drove capably down the rear deployment ramp to began its 90-sol excursion on the Martian surface. The rover had been given the name *Sojourner* in honour of an African-American reformist called Sojourner Truth (actual name Isabella Van Wagener) who had made it her mission to "travel up and down the land" advocating rights for all—including women—during the US Civil War. Her name was chosen by a 12-year-old girl called Valerie Ambroise as part of a competition run jointly by The Planetary Society (an outreach and advocacy volunteer organization set up by Carl Sagan, Bruce Murray, and Louis Friedman) and JPL.

With a successful deployment of the rover to the surface, the Pathfinder mission was already a resounding success. With safe deployment, Sojourner could now begin a serious scientific exploration of the locale and a new era in the scientific exploration of Mars had begun. While the stationary lander could survey the region in spectacular detail from horizon to horizon, Sojourner could be directed to specific locations of interest for close examination with its APXS and be capable of determining a great deal about the elemental abundances and mineralogy of the surface soils and rocks. From these measurements we could then infer the geological history of the region, as well as calibrate future orbiters with this newly acquired *ground truth* information about Mars' mineralogy.

Another revolutionary aspect of this mission was its operational logistics. With a time lag in communications between Mars and Earth of about 11 minutes, real-time operation of Sojourner from Earth was not possible. Instead a system had to be developed where the rover was instructed to carry out a particular task, whereupon it would do it unaided and relay the results back to Earth. At the beginning of each sol, an image would be taken and relayed back to mission control. They would then quickly determine a location or *way point* to which Sojourner should drive, and this would then be relayed to Sojourner via the lander's high-gain antenna. Sojourner would then navigate its own way to that point via its onboard navigation system, assisted by a Laser Proximity Detection System that was designed to help Sojourner to avoid hitting obstacles *en route*. Through this ingenious system, NASA/JPL learned how to manage remote robotic exploration on Mars that not only served Sojourner but also lay the foundation for all future rover expeditions to Mars.

Rock Garden

While it was suspected that Ares Vallis may be a flood plain, some were skeptical of direct evidence of an actual ancient flood being identified on the surface. But from the first lander images it became immediately clear that the proposition of a flood plain had been correct. Strewn about the landing site, and from horizon to horizon, lay an enormous vista of loose rocks, exactly as would be expected if deposited by a gigantic flood. In a clever move, NASA, realizing that there were enough rocks in the immediate vicinity of the landing site to conduct a thorough analysis, decided to call their arena of investigation the *rock garden*, as well as to name individual prominent rocks and boulders with names such as Barnacle Bill, Couch, Yogi, and so on, according to their resemblances to Earthly objects and popular characters spotted by members of the investigation team. What began as a practical solution to the identification of individual features instantly turned into a fantastic public outreach exercise. With identifiable and humorous names, the general public could also begin to identify with the site and become familiar even with individual rocks. For the first time, the public at large became interested in locales on Mars along with the expert scientists carrying out the mission.

Over the next 92 sols, Sojourner traversed the Martian rock garden, taking images and many APXS measurements. Along with the lander, over 17,000 images, 16 complete rock and soil analyses, and over 8 million temperature, pressure, and wind measurements were transmitted back to Earth. Indeed the lander camera was so capable that it provided both close-up textural images as well as infrared mineralogy images of many rocks. On analyses back on Earth, the findings provided critical clues to Mars' distant past. First, analysis of the distribution of rocks across the surface verified that they were most likely deposited by a massive flood in Mars' distant past. Further, rocks were found to be of two types: basalt and andesite. Basalt rocks are found on all the inner rocky planets, including our Moon, indicating that Mars' surface was created by similar processes to the rest of the inner planets. Andesite, however, is associated with plate tectonics on Earth, suggesting some sort of crustal recycling on Mars in its early history. But with no evidence of plate tectonics on Mars, the discovery of andesite pointed to hitherto unknown crustal processes that would need further investigation. Rounded pebbles and conglomerate rocks also strongly suggested copious amounts of water in the region at one time; yet it also became clear that the entire region had been completely dry for at least two billion years. Magnetized magnetite dust (iron dissolved in water and then freeze dried) found at the site also indicated the presence of water in the past, and with a myriad of weather measurements, it was also possible to substantially characterize the climate of the region, revealing details about prevailing conditions on Mars today.

An Unqualified Success

With a 30-day primary mission, Pathfinder lasted three times longer than expected, but, with failing batteries, the lander ceased communications with Earth on September 27, 1997, and the Sojourner rover was also lost. From every perspective, however, the mission was a great success. Apart from laying the foundations for the next 30 or more years' exploration of the Red Planet, and verifying that water had flowed on the surface in its past, Pathfinder also connected humanity with Mars Exploration in a way that Viking perhaps never had. With the emergence of the Internet in the 1990s,

NASA and JPL had the vision to place the latest images from Pathfinder onto their website, and with the help of commercial companies such as Sun Microsystems and Silicon Graphics, over 560 million web hits were made on the Pathfinder site—the most for any event on the Internet to that date. Anyone with an Internet connection could now explore Mars almost in real-time with the tiny Sojourner rover; and with a microchip attached to the lander containing several million names submitted through The Planetary Society, humanity was now seriously on Mars *en masse*. Pathfinder had truly lived up to its name, literally laying the path for our return to Mars technically, scientifically, and sociologically. It had reminded us, as Viking had shown 20 years previously, that Mars is a real place that we can genuinely relate to. More people had witnessed the live exploration of another planet than ever before in history. They came to know Mars through identifiable, named features in a personal and surprisingly meaningful way. The scene was now set for a return to Mars.

Mars Reconnaissance 14

The success of Pathfinder trail blazed a new era in Mars exploration. As Sojourner worked upon the surface of Mars, a series of reconnaissance orbiters, set on analysing the entire planet in unprecedented detail, was being developed. The first, called Mars Global Surveyor (MGS), arrived in September 1997 just as the Pathfinder mission was coming to an end. In 1999 the Mars Climate Observer (MCO) was lost on its approach to Mars, but MGS was joined by NASA's Odyssey orbiter in 2001, by ESA's Mars Express in 2003 and by NASA's Mars Reconnaissance Orbiter (MRO) in 2006—giving a total of four robotic reconnaissance spacecraft orbiting Mars, three of which are still delivering images, geological, topological, mineralogical, and other data. Together they represent the first comprehensive survey of another planet and are set on addressing the great unknowns about Mars' long and complex history.

Mars Global Surveyor

As the first of the new generation of Mars orbiters, Mars Global Surveyor was a worthy successor to Viking (Figure 40). Its objectives were ambitious: to image the entire planet and characterize its surface morphology at unprecedented resolutions; to determine the global composition and distribution of minerals; to create a three-dimensional topological map of the entire planet; to monitor surface–atmospheric interactions; and to act as a communications satellite for future lander missions, among other functions (Figure 41).

To achieve all this, MGS was equipped with a sophisticated array of instruments. First is the Mars Orbital Camera (MOC)—a spectacular optical camera designed by Mike Malin and Ken Edgett of Malin Space Science Systems (MSSS) with two resolution modes: a wide-angle mode with a 280-meter resolution providing regional context views of the planet, and a narrow-angle mode with maximum pixel resolution of just 1.5 meters (and

K. Nolan, *Mars, A Cosmic Stepping Stone*,
DOI: 10.1007/978-0-387-49981-9_14, © Praxis Publishing, Ltd. 2008

Figure 40: Mars Global Surveyor (MGS), which in 1997 heralded a new era of Mars exploration based on a long-term, phased strategy. [Credit: NASA/JPL/Corby Waste]

with a recent adaptation providing 0.5-meter pixel resolution, allowing surface objects as small as 2 meters to be identified). Using both modes, it has been possible to examine both regional-context geology as well as individual features with sufficient clarity to infer their underlying process of formation. Equally important on MGS is its Thermal Emission Spectrometer (TES) designed by Phil Christensen. This camera is capable of looking at the surface across many thermal (infrared) frequency bands, from which the surface mineralogy can be determined. With a resolution of 3×6 kilometers, TES was specifically designed to produce a mineralogical map of the entire

Figure 41: A global image of Mars from Mars Global Surveyor revealing water clouds. Scenes such as this can only be captured through long-term reconnaissance of the planet. See Plate 15 in the color section. [Credit: NASA/JPL/MSSS]

planet. A third instrument, arguably the most enigmatic on board, is the Mars Orbital Laser Altimeter, or MOLA. Using an onboard laser, MOLA fired laser-light pulses at the surface and observes their reflection. By measuring the time delay of each reflection from a given point on the surface, MOLA could determine the distance to that location. With a horizontal resolution of 100 meters and a vertical resolution of just 1 meter, MOLA has been able to construct the first three-dimensional topological map of Mars. Also of major importance was the onboard Magnetometer/Electrometer, capable of measuring the magnetic properties of the planet and the movement of electrons in any detected planetary magnetic fields, telling us about Mars' past and present magnetic properties, and from which the internal structure and dynamics of the planet can be inferred.

Morphology

From its arrival at Mars in late 1997 to its eventual failure in late 2006, MGS's various instruments produced enormous datasets that will take decades to fully analyze. The MOC image set alone comprises in excess of 240,000 images—far greater than the Viking orbital image set in both coverage and detail. From Tharsis in the west to Elysium in the far-east, from the ancient southern highlands to the younger northern lowland plains, all have been photographed in unprecedented detail. Of major significance, for example, has been the discovery of vast tracts of deep sedimentary layering at hundreds of locations across the planet, and from the equator to the poles (Figures 42–44). While both Mariner 9 and the Viking orbiters had photographed sedimentary layering at several locations, MGS has revealed that it is a planetwide phenomenon, pointing to massive transport and deposition of material in Mars' distant past.

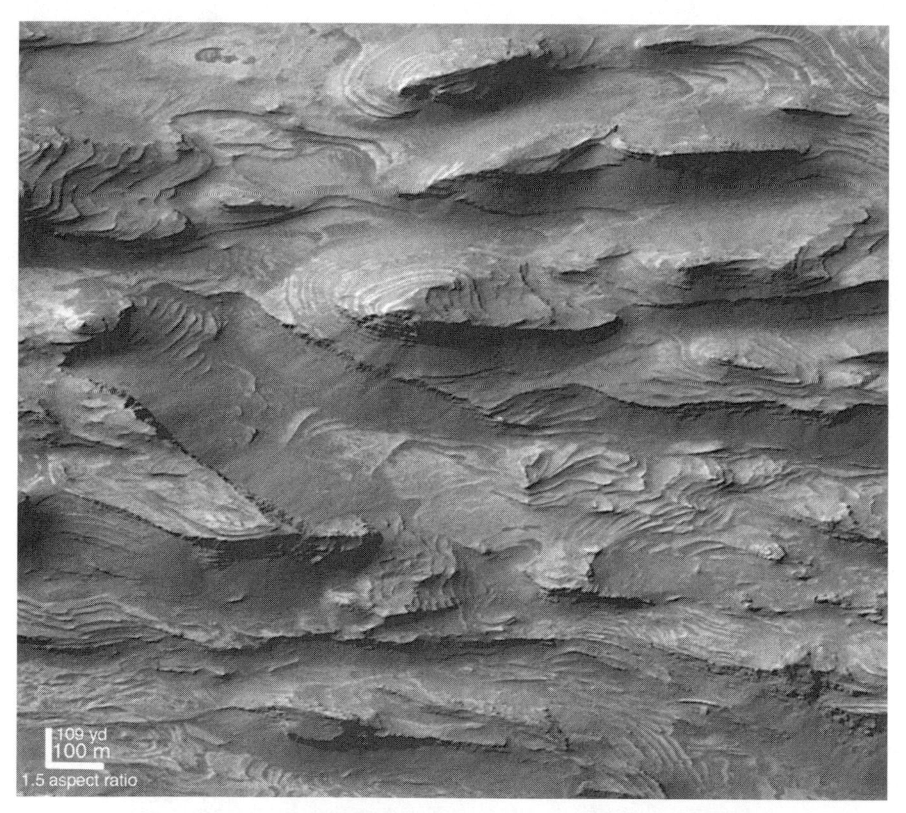

Figure 42: Mars Global Surveyor's Mars Orbiter Camera (MOC) confirms the presence of layered outcrops within Valles Marineris. This high-resolution image reveals extensive layering on the floor of western Candor Chasma. See Plate 16 in the color section. [Credit: NASA/JPL/MSSS]

Figure 43: MGS/MOC view of a scarp at the head of Chasma Boreale, a large trough cut by erosion into the Martian north polar cap and revealing layered material beneath the ice cap. The picture was taken using a resolution-enhancing technique called ''compensated pitch and roll targeted observation (cPROTO).'' [Credit: NASA/JPL/MSSS]

Three distinctly different types of sedimentation have been identified—light-colored layers that are in the order of tens to hundreds of meters thick, massive layers approximately 1 kilometer in thickness, and thin messa or darker thin layers found on top of the other types of layering. MGS has also identified four types of terrain within which sedimentation can be found: within craters, for example at Arabia Terra; on inter-crater terrain, as found at northern Terra Meridiani; upon chaotic terrains, such as at Margaritifer Terra; and within chasm interiors, most notably within the walls and on the floor of Valles Marineris.

Such widespread sedimentary layering constitutes substantial evidence of a once active planet. Often spread over hundreds of kilometers and several kilometers in depth, the processes involved must have occurred on a

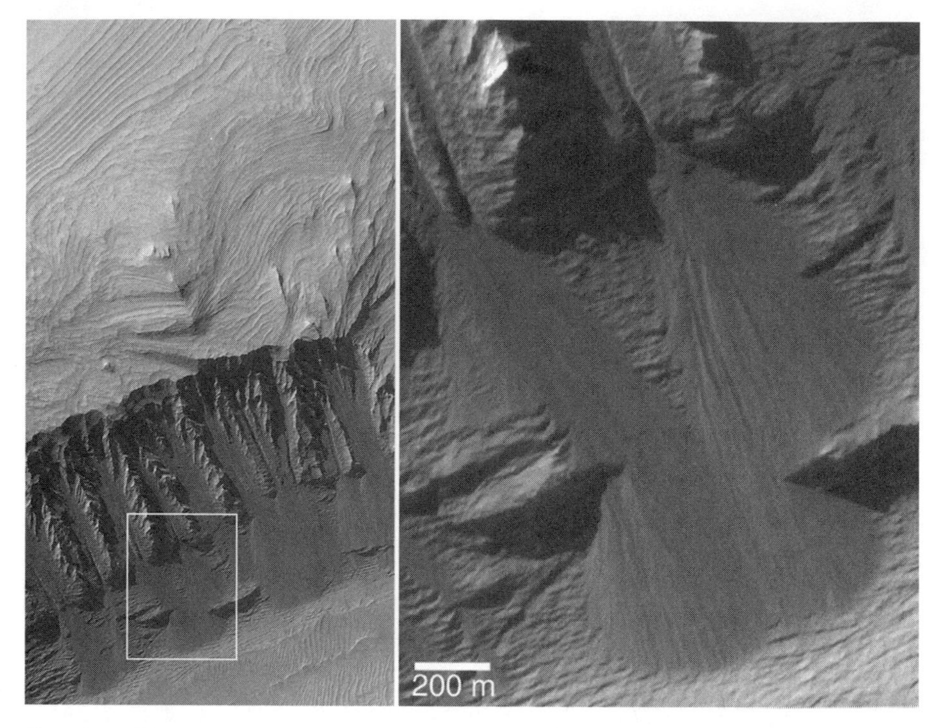

Figure 44: MGS/MOC/cPROTO image showing sedimentary layers in Candor Chasma, Valles Marineris. On the slope in the lower half of the image, fine-grained material has slid down the slope to create fan-shaped talus accumulations. [Credit: NASA/JPL/MSSS]

planetary scale that lasted for many millions of years. Also, in the particular case of the Valles Marineris rift, we know that the plain within which the rift resides was formed during the Hesperian period, therefore the several-kilometer deep sedimentary layering seen within its walls must have been laid down during the Noachian period, indicating that sedimentary layering occurred most prominently in Mars' earliest history. While there can be little doubt that much of the sedimentation was due to volcanic and tectonic activity, there are other possibilities. Eolian (wind) based erosion, transport and deposition, for example, may have been responsible for much of the observed layering, especially if Mars had a dense atmosphere at one time. There are also several types of aqueous activity—such as alluvial (flowing water), lacustrian (standing water as in lakes) and deltaic (running water flowing into a standing body of water)—that can give rise to sedimentation. At Hellas, Meridiani Planum, and the floor of Valles Marineris, for example, MGS/MOC has revealed extensive sedimentation that could well have been laid down by the action of water. Indeed MGS/MOC images provide

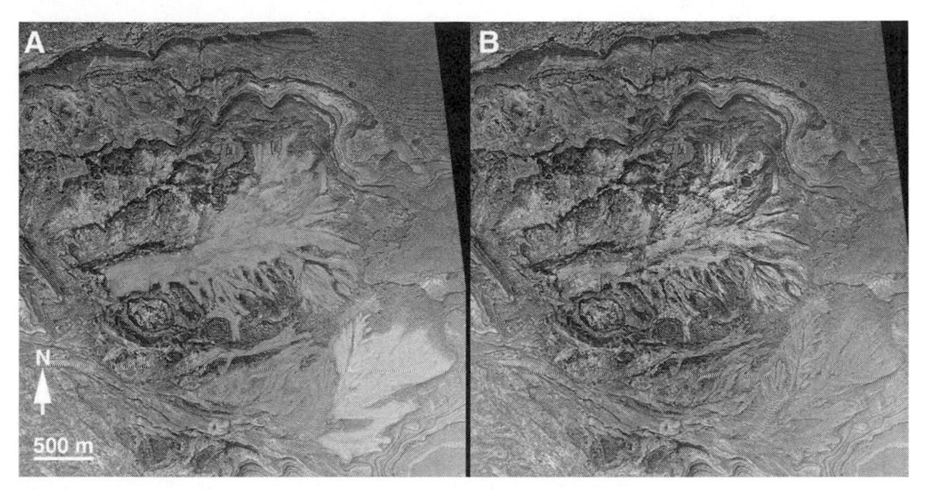

Figure 45: MGS/MOC image of alluvial sedimentation in Melas Chasma, Valles Marineris. The water-based "fan-shaped" sedimentation is highlighted in panel A (left). See Plate 18 in the color section. [Credit: NASA/JPL/MSSS]

intriguing evidence of alluvial sedimentation at Melas Chasma, one of the large troughs of the Valles Marineris system (Figure 45), revealing two sets of alluvial activity where the water sedimentation process created fans of debris with finger-like protrusions and with the channels through which the water flowed actually preserved to the present day. Among the most compelling evidence of persistent water flows and aqueous sedimentation was spotted by MGS at a site known as Holden Crater, in the Erythraeum region at 260°S, 340°W (Figure 46). The entire region covers an area of 4,000 square kilometers, with several valleys feeding into a fan-like layered landform covering about 100 square kilometers—unequivocal evidence of long-term, persistent surface water flows from river channels into a large standing body of water, within which a fan-shaped delta with layered sedimentation formed. The rhythmical layering also indicates episodic changes in environmental conditions.

Despite the extent of sedimentation identified by MGS, there are many questions that cannot be answered from MGS data alone. For example, many supposed dried river networks in the south of the planet show no evidence of sedimentary layering. And at many layered sites there is no obvious transport mechanism into or out of the region. Further, as sharp as the MOC images are, they are not quite good enough to definitively identify the processes that gave rise to the layering seen at many locations, and for this we need the sub-meter resolving power of the Mars Reconnaissance Orbiter with possibly lander missions being sent to those regions in the future.

Figure 46: MGS/MOC image of Eberswalde Delta, located northeast of Holden Crater—unequivocal evidence of surface liquid water activity on Mars in the past. [Credit: NASA/JPL/MSSS]

Whatever the processes involved however, the identification of planetwide sedimentary layering firmly indicates that Mars was active in its past, probably involving a range of deposition processes including tectonic, volcanic, eolian, and even aqueous activity over long periods of time.

Global Topology

Another of Mars Global Surveyor's crowning achievements has been the production of a three-dimensional (3D) map of the surface of Mars with its

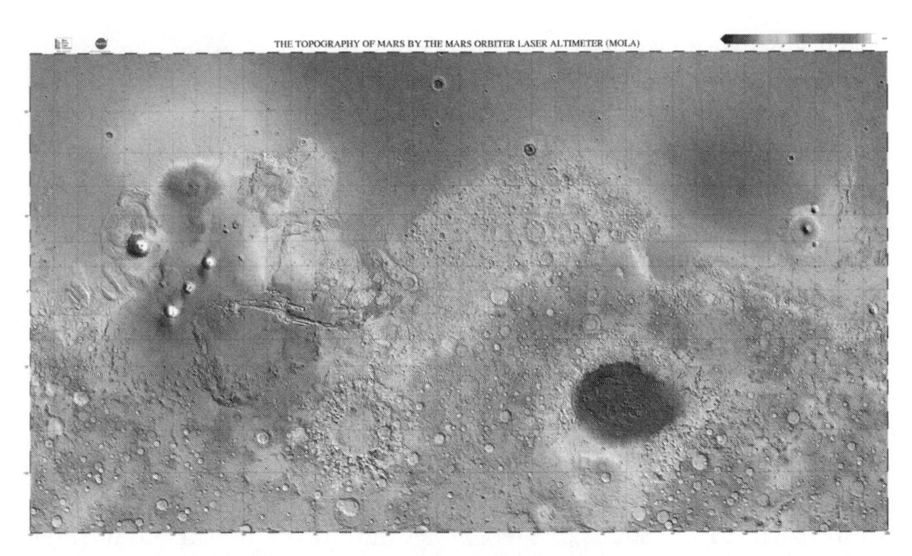

Figure 47: MGS MOLA 3D Topological Map of Mars. Of note is the north–south planetary dichotomy, with the northern hemisphere both lower and smoother than the cratered plains of the south. See Plate 17 in the color section. [Credit: NASA/JPL/USGS]

laser altimeter—MOLA. Between 1999 and 2001, MOLA fired over 670 million pulses at virtually every location on the planet, building a 3D topological map of the surface with a vertical resolution of just 1 meter (Figure 47). This has allowed the true relief of surface features to be seen for the first time, and greatly improves our understanding of the geological processes that give rise to them. The MOLA map is the best topological map of any world—including Earth, where the construction of such a detailed map of the entire surface is inhibited by a thicker atmosphere and the presence of trees, artificial structures, and water.

Among the most important results from MOLA is verification of a significant difference in the mean elevation of the northern and southern hemispheres; with the southern hemisphere about 3 kilometers higher. Such a difference points to significant yet currently unknown events in Mars' distant past that must be explained if we are to understand how such a dichotomy arose. MOLA has also indicated a general difference in geology between the north and south, where the south is dominated by older cratered terrain while the north is composed of a less cratered and younger surface. MOLA has also revealed surface roughnesses, which are important to determining geological processes such as erosion and deposition. Here it has shown that while the southern hemisphere is generally composed of rough terrain, the equatorial region is even rougher. By contrast, however, the northern lowlands have been shown to contain the smoothest surface in the entire Solar System.

MOLA has also provided a vital new approach to investigating possible water activity on Mars long ago. Because most of Mars' features were created during the Noachian and Hesperian eras, the topology seen today relates back to the time when we think water activity may also have occurred. Hence the MOLA map can reveal the ancient drainage patterns across the planet—identifying where water may have originated, as well as its final destination. Furthermore, because of the superlative resolution of the MOLA data, we can even determine water and flood channel capacities, the quantities of water involved and the length of time such activity may have persisted. Indeed MOLA suggests that, because of the drop in elevation from the southern to the northern hemisphere, the dominant drainage pattern would have been from the southern highlands to the northern lowlands. This seems to correspond with photographic evidence where, for example, possible water channels lead all the way from the giant Argyre Basin in the south to the Chryse outflow region in the northern hemisphere. Further photographic evidence shows that virtually all the gigantic flood channels seen on the planet also point northward. If substantial bodies of water resided on Mars in its early history, MOLA suggests that upwards of 90% of that water would have found its way to the northern hemisphere over time, prompting some scientists to consider that Mars may have possessed a vast northern hemisphere ocean in its distant past (although this idea is not unanimous, nor does the available evidence strongly suggest such an ocean). Intriguingly, MOLA also poses a Martian puzzle. From all combined estimates, we suspect that Mars currently retains enough water (in non-liquid form) that, if it were liquid on the surface, would cover the planet to a depth of about 30 meters; yet MOLA (among other surveys) suggests that there were sufficient quantities of water in the past to cover the planet to a depth of perhaps 500 meters. If this is correct, where has Mars' water gone?

MOLA has also greatly enhanced our understanding of the geological processes that shaped Mars' surface. It has been shown unequivocally that Tharsis and Elysium are gigantic continent-sized bulges on the surface. Intriguingly, it has also revealed that Olympus Mons is not part of the Tharsis bulge as was originally thought, but instead sits on lower terrain further west. And with Tharsis and Elysium so clearly seen as significant seats of past tectonic and volcanic activity, there can be little doubt that both would have contributed to the outgassing of volatile and biogenic materials. MOLA has revealed the full magnificence of many other surface features, also verifying, for example, that Hellas is a gigantic impact basin over 9 kilometers deep—the deepest impact basin in the Solar System. Apart from contributing substantially to our understanding of Mars' distant past, MOLA has also told us about the current state of Mars. For example, it has revealed the volume of water-ice locked into the north and south poles, and can even measure changes in elevation of the polar surfaces when carbon dioxide freezes during the winter, pointing to the quantities of carbon dioxide involved.

Mineralogy

Another important instrument on board Mars Global Surveyor was its Thermal Emission Spectrometer (TES), which photographed thermal or infrared emissions from the surface. Because different minerals emit heat in different ways, TES could detect various surface materials by observing their unique thermal emission properties, and by comparing such values to known heat emissions from minerals on Earth, identify those same minerals on Mars. While TES had a relatively crude resolution, it is only with such a broad pixel size that a comprehensive mineralogical survey of the entire planet could be completed. Of all the instruments on board MGS, TES has arguably provided the most puzzling results, not only because they provided a deeper insight into the material make-up of the planet but also because they have exposed a hitherto unseen complexity to the history of the planet.

TES has shown, for example, that the surface of Mars is composed primarily of silicates, as on Earth. It also reveals that the global dichotomy on Mars, identified from its surface terrain, is also broadly reflected through its surface composition. While the southern hemisphere is made up of ancient basalts, the northern hemisphere is largely composed of andesite, which on Earth is associated with younger surfaces produced through plate tectonic activity. The detection of andesite is therefore significant, suggesting some ancient tectonic process on Mars that is not obvious from optical or topological mapping. Also of significance has been the detection of olivine-rich feldspar on much of the northern surface. On Earth, olivine turns quickly into clay on contact with water moisture, suggesting that since none of the olivine on Mars has lost its integrity, Mars has been a very dry planet for at least several billion years.

In the search for water-altered minerals, TES has provided some intriguing results. First, TES has only detected trace amounts of carbonates (which form when precipitated from water). Such a stark discovery might suggest that Mars was never dominated by water, but this does not tally with other evidence that has suggested substantial water activity at one time. However, the absence of carbonates (at least as seen by TES) suggests that either they were never created in the first place or have subsequently disappeared. It may be, for example, that Mars was characterized by a dense carbon dioxide atmosphere that was not lost through precipitation as carbonates but was instead lost to space or by chemical reactions with the crust. Or, it may be that some carbonates did at one time exist and have been subsequently covered or chemically eroded. Whatever the scenario for water and carbon dioxide on ancient Mars, the TES carbonate measurements already tell us that we are far from understanding their ancient context and that more sensitive measurements are required. Almost in contradiction, TES has detected significant quantities of hematite at locations on Mars within Valles Marineris and across Meridiani Planum, verifying the existence of water at or near the surface for long periods in its past. As we will see in the next chapter, the Mars Exploration Rover Opportunity subsequently visited Meridiani Planum in 2004 (the first lander mission based on mineralogical evidence) and found precipitated salts, verifying the existence of an ancient shallow sea at that location.

Magnetic Fields

Because MGS required a full year to settle into orbit, it passed much closer to Mars than originally intended, approaching to within 150 kilometers of the surface on some passes. Critically, it temporarily passed underneath Mars' ionosphere, permitting MGS's magnetometer and electrometer to directly measure whether Mars' crust is magnetized and to infer the polarity (direction) of any detected magnetic field. Intriguingly, magnetic anomalies were found in the crust of the ancient southern highlands. The observations revealed stripes of magnetized crust, with the direction of the magnetic field within alternating stripes pointing in opposite directions. This was a fabulous result in itself because it verified, to some extent, that, in its distant past, Mars possessed an internal dynamo that drove a magnetic field and probably surface tectonic activity. Furthermore, the striping also looked superficially like the magnetic striping that is also found on Earth and is associated with its internal dynamo and plate tectonic movement. Here, Earth's internal magnetic field changes direction every few million years causing new material emerging from diverging tectonic plates to become magnetized and retain a record of Earth's internal magnetism of that time. By observing the alternating direction of the magnetic properties along stripes of the Earth's crust, we are witnessing a record of both the internal alternating magnetic field and of plate tectonic movement of the crust.

The initial magnetic striping on Mars seemed more elusive, however. It was not visibly associated with any feature that looked like a plate and indeed no evidence of tectonic plates could be seen across the planet.

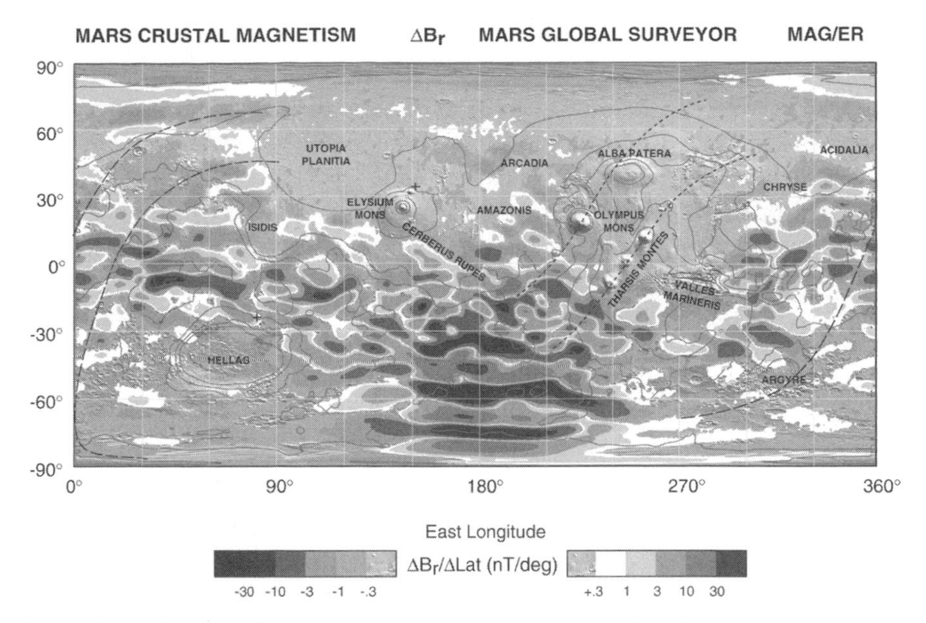

Figure 48: MGS map of the magnetic fields retained within Mars' crust. The widespread "striping" pattern suggests plate tectonic activity on Mars in its distant past. See Plate 19 in the color section. [Credit: MGS Magnetometer Team led by Mario Acuna at the Goddard Space Flight Center in Greenbelt, MD]

However, persistent and continuous magnetic mapping of the entire planet over the years by MGS has just recently yielded some of the most exciting results regarding Mars' distant past, and with far-reaching implications. As announced by J.E.P. Connerney, M.H. Acuna, and others at NASA's Goddard Space Flight Center in 2005, MGS has revealed significant magnetic striping across the entire globe of Mars, seen virtually everywhere except on younger terrain such as at Hellas in the south and Utopia Planitia in the north (Figure 48). The detection of such extensive striping suggests that Mars' surface was characterized by plate tectonic movement in it earliest history and that early Earth and Mars may have been quite similar in this respect. Indeed, this discovery may also clarify other features on Mars that, until now, have not been understood. For example, the three giant volcanoes on Tharsis all lie in a straight line. While this seemed curious, and unexplained, this new evidence points to the volcanoes having arisen on a moving tectonic plate sitting on an internal convective cell, as occurred with the Hawaiian Islands. Furthermore, while the gigantic scale of Valles Marineris has remained puzzling, the evidence of plate tectonics in Mars' distant past suggests that it may have been the result of two divergent tectonic plates.

Present Climate

Despite the importance of determining the nature of Mars in its early history, of emerging relevance is also the nature and dynamics of Mars today. For example, understanding the behavior of both water and carbon dioxide through their natural reservoirs—the polar ice caps, the planetary crust, and the atmosphere—will reveal much about the present climate as well as how it varies over periods of thousands or hundreds of thousands of years. And, of course, an up-to-date understanding of the planetary environment and climate are critical to the safety and success of all future orbital, aerial, lander, and human missions.

Once again, MGS has been pivotal in this respect. Through continuous monitoring on a daily basis and over several Martian years, MGS has provided valuable insight into Mars' climate. It has revealed, for example, that planetary wind patterns are dominated by a processes known as Hadley Circulation, where air from a cold winter hemisphere blows into the warmer summer hemisphere. Coupled to planetary rotation, trade winds similar to those on Earth, traveling at hundreds of kilometres per hour, blow eastward in the winter hemisphere and westward in the summer hemisphere. MGS has also provided some insight into one of the great mysteries about Mars: What is the source of its global dust storms? Dust storms on Mars can last for months or years and cover the entire planet, but their origin is currently unknown. MGS noticed, however, that local or regional dust storms most readily occur in the southern hemisphere during its summer and when Mars is closest to the Sun. It is possible, therefore, that a merger between local dust storms in the south and planetary trade winds may cause dust to circulate across the globe, while an amplification of solar heating within the accumulating dust storm may cause it to grow even larger.

MGS has also examined both polar regions of the planet in spectacular detail and over many seasonal changes. It has observed that both poles (but especially the North Pole) retain permanent caps of water-ice and ice-rich sediments that are several kilometers thick. It has also observed almost one-third of the planetary atmosphere freezing out as carbon-dioxide-ice on whichever pole is in winter, and being released again during spring thus radically affecting planetary winds and atmospheric pressure. MGS has also shown that each pole has a different climate, perhaps due to the fact that the South Pole resides at a higher elevation than the North Pole. Also seen, at the South Pole only, are permanent changes in the structure of carbon-dioxide-ice over an 8-year period, suggesting that Mars may be currently going through a period of climate change.

Yet another intriguing finding has been the discovery of near-surface gullies in tens of thousands of locations across Mars, especially on the flanks of crater walls in the southern hemisphere. Because liquid water cannot currently reside on or near the surface, some scientists are skeptical as to whether the phenomena involve running water and that flowing sand may be the cause. There is nevertheless growing consensus that liquid water containing salts that could lower the freezing point of water 60°C degree or more are the cause of these limited yet intriguing water flows. Furthermore, MGS has verified that they are possibly a current phenomenon by revealing new gullies in recent images where none was seen at the same location earlier in its mission (Figure 49). Irrespective of their underlying mechanism, the discovery of so many gullies is significant because they reveal surface activity that, hitherto, were not considered to occur on Mars today, and perhaps even point to micro-environments relevant to the search for evidence of life.

As already stated, Mars Global Surveyor failed in late 2006. It has been arguably the most successful mission to Mars to date and its legacy will certainly continue for decades to come. It has provided a far-reaching reconnaissance of the planet, the results from which have acted both as a vital reality check on our understanding of the planet and as a springboard to all current and future missions. Most of all, Mars Global Surveyor told us that Mars is far from understood and that there is a lot more to learn.

Figure 49: Gullies in the wall of a crater in Terra Sirenum, photographed in 2001 and 2005 by MGS/MOC, providing tantalizing evidence of recent water flows on or near the surface of Mars today. [Credit: NASA/JPL/MSSS]

Odyssey

As MGS was busily conducting its mission, NASA followed with a second reconnaissance orbiter in 1999—the Mars Climate Observer (MCO). It was not a coincidence that the Mars Climate Observer was timed to arrive at Mars two years after Global Surveyor. With the specific task of analyzing Mars' climate, atmosphere, polar, and volatile activity, MCO was timed to complement (and avail of) the significant survey carried out by its immediate forerunner. Similarly, in 2001, the aptly named Odyssey orbiter arrived after another two years, representing the third in a trio of orbiters set on comprehensively investigating the great mysteries of Mars. Odyssey was tasked with yet another unique series of analyses—to photograph the entire surface at a different and intermediate resolution between those of Viking and MGS, to conduct follow-up thermal mineralogical analyses of many specific sites at thousands of times the resolution of TES, to determine the abundances of over 20 elements upon the surface and, critically, to determine the global distribution of water in the top layer of Mars' crust.

With the loss of the Climate Observer and Polar Lander, the arrival of Odyssey in April 2001 was monitored very closely by all concerned. It is perhaps no exaggeration to say that the future of the Mars program rested on the success of Odyssey. Should a problem arise *en route* or while entering orbit, there would be serious repercussions. This time, however, everything worked to perfection. Odyssey used the same aerobraking procedure as MGS, but this time it worked perfectly and on October 24, 2001, Odyssey settled into its final science orbit about Mars, just a single kilometer off its targeted position. And to celebrate the mission, NASA even commissioned an original composition from the Greek composer Vangelis, whose music had been used to breath-taking effect in Carl Sagan's TV series "Cosmos." Titled *Mythodea*, Vangelis composed a major work for synthesizer, orchestra, choir, and two solo parts sung by world-renowned sopranos Kathleen Battle and Jessye Norman—exploring both the mythological legacy and the current space odyssey to the Red Planet. The work was performed live in Athens in June 2001 and was a resounding success, heralding the success to come for the space probe that had inspired the work.

Odyssey's Science Mission

Odyssey's science objectives are numerous and unprecedented. First, Odyssey has been tasked with photographing the planetary surface at a resolution of about 18 meters—which is of immense value when attempting to understand the planet's geology. While Viking had photographed some regions at a resolution of about 8 meter, general global mapping was at resolutions of between 150 to 300 meters, often insufficient to identify the underlying geological processes at work on the surface. Mars Global Surveyor's MOC addressed this problem in part with an ability to image interesting features to about 4 meters resolution; but even with its lengthy mission it has only been possible to image a small portion of the surface in such detail. And while in wide-angle mode, MOC has photographed virtually the entire planet at 250 meters resolution, but again these images are often of too low a resolution to provide defining answers on surface geology. So, with Odyssey photographing a far greater portion of the planet at an 18-meter resolution, it provides unprecedented views of much of the planet, finally providing us with the capability of interconnecting local-scale to regional and planetary-scale geology (Figure 50).

Using its onboard Gamma Ray Spectrometer (GRS), another of Odyssey's goals has been to map the global abundance and distribution of approximately 20 chemical elements, including hydrogen, silicon, iron, potassium, thorium, and chlorine, as well as the global distribution of water to a depth of about 1 meter. GRS is actually three instruments combined: two neutron detectors and a gamma-ray detector. Each instrument works by observing gamma rays or neutrons emitted from the Martian soil when

Figure 50: From high above Valles Marineris, this 2001 Mars Odyssey derived view looks down upon a sight resembling parts of the desert west of the United States. Here the canyon is 150 kilometers wide, with the floor composed of rocks, sediments, and landslide debris. Within the canyon walls lie hundreds of layers revealing Mars' geologic history. See Plate 20 in the color section. [Credit: NASA/JPL/Arizona State University]

bombarded by cosmic rays that reach the surface through Mars' tenuous atmosphere. With each element in the soil reacting uniquely to incoming cosmic rays and subsequently emitting its own particular gamma-ray or neutron signature, GRS can infer the composition of the soil by analyzing the patterns of the radiation emitted. In particular, the detection of gamma rays indicates the presence of hydrogen bound up in water within the soil, pointing to where water-ice or hydrated minerals may reside across the planet.

Another significant experiment being conducted by Odyssey is a planetary mineralogical survey using an onboard thermal imaging camera called THEMIS, also designed by Phil Christensen. Where MGS's TES camera had a low pixel resolution suited to global mapping, THEMIS' resolution is just 100 meters—1,800 times that of TES and capable of producing detailed local and regional level mineralogical images. Such resolution, coupled with excellent sensitivity and a large number of thermal frequencies has made THEMIS one of the defining instruments in our quest to understand the complex history of Mars and the search for life on that planet. With TES having already produced an exquisite global mineralogical map, THEMIS can now follow up on the multitude of interesting regions that TES identified. Further, THEMIS can identify a myriad of minerals—carbonates, sulfates, phosphates, silicates, oxides, and hydroxides, among others—all to near-trace levels. It is a powerful tool for determining the specific mineralogy and hence the geological evolution of each region examined.

Apart from determining surface mineralogy, THEMIS can also monitor surface reflectance and thermal inertia, from which surface compositions—sand, pebbles, boulders, solid rock, and so on (Figure 51)—can be determined. "Thermal inertia" means the resistance of a material to changes in temperature, and each surface type has its own thermal inertia. For example, dust and sand have a high thermal inertia as they heat up quickly during the day and cool down quickly at night. Conversely, large boulders and solid rock have a low thermal inertia as they are slow to warm up and equally slow to cool down. Hence, if THEMIS takes, for example, two or more thermal images of a given location during the day and at night, changes in temperature and hence the thermal inertia of the various surfaces in that location can be determined, from which their composition and petrology can be inferred.

Odyssey has also been tasked, with a radiation experiment called MARIE, to monitor solar and cosmic radiation both *en route* to Mars and in Mars orbit. Such measurements are critical to the wellbeing of all upcoming missions and most especially for any human mission. Finally, Odyssey has acted as an excellent communications satellite for both of the Mars

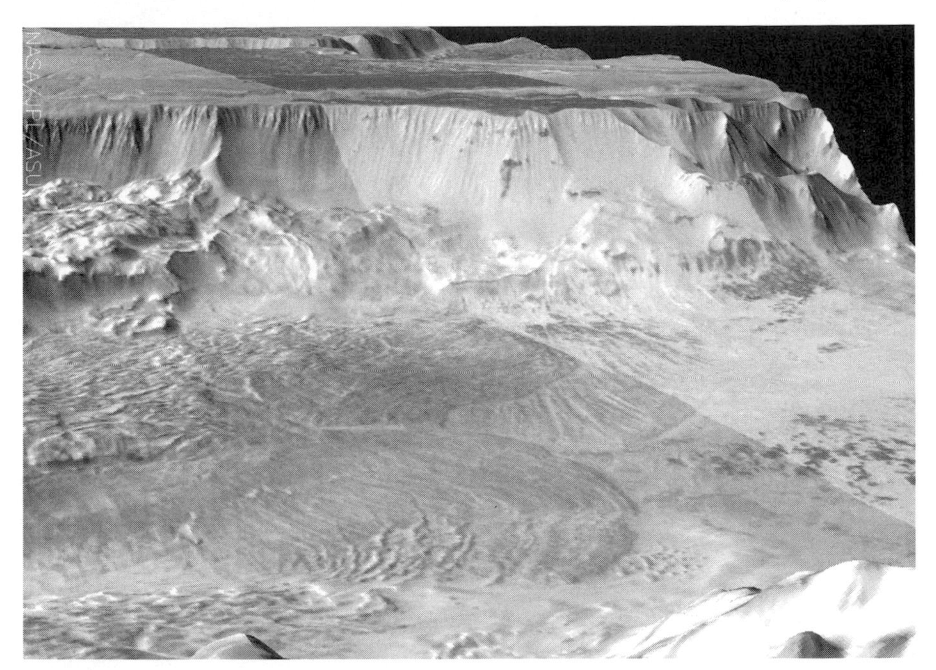

Figure 51: Odyssey/THEMIS infrared image of Melas Chasma at night, revealing the abundance and distribution of surface materials. Rocks retain their heat at night and stay warm, while dust and sand cool more rapidly. See Plate 21 in the color section. [Credit: NASA/JPL/Arizona State University]

Exploration rovers, relaying instructions from Earth to the rovers and scientific data from the rovers back to Earth.

Odyssey's Findings

Central to our investigations on Mars is the detection of water. While ancient river and flood channels indicate that water once flowed there, many fundamental questions regarding its ancient and current context remain unanswered. In particular, we have been uncertain of the amount of water, if any, that remains on Mars today, of its chemical and physical state, or of its location on the planet. While Viking verified the existence of water-ice at the poles, it was not until the arrival of Odyssey that the full extent of water on Mars could be robustly addressed. In true spectacular fashion, within just 10 days of operation, Odyssey's GRS experiment put beyond all doubt the existence of vast quantities of water-ice at both the North and South Poles. Furthermore, it found that at above 60 degrees latitude in each hemisphere, the topmost layer of the crust is composed of upwards of 50% water-ice, suggesting that vast quantities of Mars' water has been locked into the crust as a permanent permafrost lasting billions of years (Figure 52). Indeed, even

Figure 52: Mars Odyssey map showing the lower limit of the water content of the upper meter of Martian soil. The highest water–mass fractions, exceeding 60% in places, are found in the high latitudes and polar regions. See Plate 22 in the color section. [Credit: NASA/JPL/Los Alamos National Laboratory]

in the equatorial regions GRS found that the crust consists of several percent water—quantities too great to be in equilibrium with the current environment—perhaps suggesting that Mars' climate is gradually changing. GRS has also been busy determining the global distribution of various chemical elements, from which we will be able to better determine the composition and evolution of the crust as well as initial stocks of volatile and biogenic materials in Mars' distant past (Figure 53).

Odyssey's THEMIS instrument has also provided significant results to date. It has transmitted back to Earth spectacular optical and thermal maps of literally tens of thousands of sites on Mars. Many of the optical images are the best available of those regions, while the thermal images are proving to be of importance in studying Mars' ancient geology (Figure 54). First, THEMIS has verified TES findings, showing that the southern highlands are dominated by ancient basalt rocks while much of the northern lowlands are composed of andesite. Significantly, however, THEMIS has also identified isolated andesite in the southern hemisphere, indicating that even during Mars' earliest history at least some crustal recycling was occurring. THEMIS has also turned to the sites at which hematite was found and has constructed detailed maps of each region. These are critical if we are to determine the precise nature and extent of the

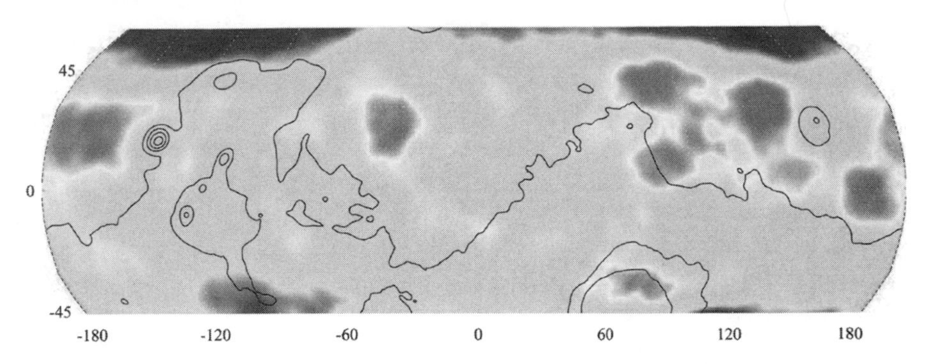

Figure 53: A Mars Odyssey gamma-ray spectrometer map of the mid-latitude region of Mars for the element iron. Regions of highest iron content are found at latitudes above 45°N and below 45°S. Regions of highest iron content are found nearer the Equator. Similar maps have been constructed for other elements. See Plate 23 in the color section. [Credit: NASA/JPL/ Arizona State University]

hematite and, hence, the specific water activity from which it originated. The true power of THEMIS is in its detailed examination of the surface petrology and mineralogy of a given region. Although it will require the analysis of thousands of sites to gain a global context into the ancient processes shaping Mars, there have already been some significant results. For example, not only has THEMIS verified the presence of hematite at Meridiani Planum, but it has also been able to identify widespread sedimentary layering across the region, strengthening the case for past surface water activity at that location. No less than eight types of material have been identified within the sedimentary layers across the region, from ancient volcanic materials to hematite. Through THEMIS we can also see their stratigrapic relationship—that is, the chronological order in which they were laid down. Also, the nature of the layering in the region shows episodic erosion and deposition, as if Mars' early environment changed in a cyclical fashion (at least in that region). In particular, rhythmic variations in the layering observed suggest that one layer in particular was laid down by burial or cementation by percolating fluids, perhaps compacted by about a 1-kilometer-thick overburden that was subsequently eroded, pointing to significant activity in the region over a very long time.

Another important yet paradoxical discovery by THEMIS has been the identification of unweathered olivine layers at a depth of 5 kilometers within the walls of Valles Marineris. Since olivine is reduced to clay on contact with water, such a discovery would seem to suggest that Mars was incredibly dry even during the Noachian period. Such apparently contradictory evidence is an indication of both the incompleteness of our surveys

Figure 54: Ganges Chasma, part of the Valles Marineris, showing layering in the canyon walls to a depth of 5 kilometers. Outcrops of olivine, seen along the base of the canyon, suggest that the region has been dry for billons of years. See Plate 24 in the color section. [Credit: NASA/JPL/Arizona State University]

Figure 55: 2001 Mars Odyssey THEMIS image of near-surface water-ice in a region on Mars approximately 70°N. The instrument is able to reveal ice lying at depths from 4 centimeters to more than 18 centimeters. See Plate 25 in the color section. [Credit: NASA/JPL/Arizona State University]

of Mars and of our understanding of its complex history. None the less, we are now confident that, as with the spectacular analysis the Meridiani Planum, THEMIS will allow us to strip away the layers of time at site after site and reconstruct the detailed history of the planet.

THEMIS is also providing unrivaled insight into the present nature Mars. It can, for example, closely monitor changes in frost and ice covering both poles and as the seasons pass, allowing for accurate monitoring of changes in carbon dioxide and water at sub-kilometer resolutions. In particular, THEMIS images, coupled to advanced computer modeling of the thermal inertia of rock, dust and ice on the surface of Mars, have allowed Phil Christensen and his colleague Joshua Banfield at Arizona State University Tempe, to determine the depth of near-surface ice in the polar regions over scales of just hundreds of meters, revealing that water-ice is likely to reside closer to the surface when covered by dust than by rock (Figure 55). Such measurements will prove critical to future missions, in particular for the Phoenix Lander, whose 2008 mission involves landing in the far north in search of near-surface ice. Furthermore, THEMIS thermal inertia measurements of near-surface ice reveal the precise ground patterns of water-ice retained under the surface, providing insight both into quasi-periodic climate change that is now thought to occur on Mars over timescales of tens of thousands of years, and into the cycling of water between the atmosphere and the crust over various timescales. Finally, with its extraordinary temperature-measuring capability, THEMIS can already tell us, at least to the limits of its resolution and sensitivity, that there are no high-temperature anomalies such as volcanoes or hydrothermal vents on Mars today. Fumaroles and hot springs may well exist, but we can now be confident that, if they do, they are on a small scale. Overall, THEMIS has turned out to be a monumental instrument for studying the mineralogy, past geology, and current environment on Mars (Figures 56 and 57).

Mars Express

With a legacy of unsuccessful Soviet Mars missions and a more limited European Space Agency (ESA) space science program, Mars exploration remained exclusive to the United States of America through the second half of the twentieth century. That changed in 2003, however, with the launch to Mars of ESA's reconnaissance orbiter Mars Express.

Mars Express, so named because it was designed, built, and launched in record time, was just one of a number of missions commissioned by ESA as part of its new Cosmic Vision program of space-science exploration.

Figure 56: 2001 Mars Odyssey THEMIS image of a landslide in Juventae Chasma, part of the vast Noctis Labyrinthus (the Labyrinth of Night) in west Valles Marineris, perhaps created when tectonic faults opened and allowed subsurface water to escape, causing the ground to collapse. See Plate 26 in the color section. [Credit: NASA/JPL/Arizona State University]

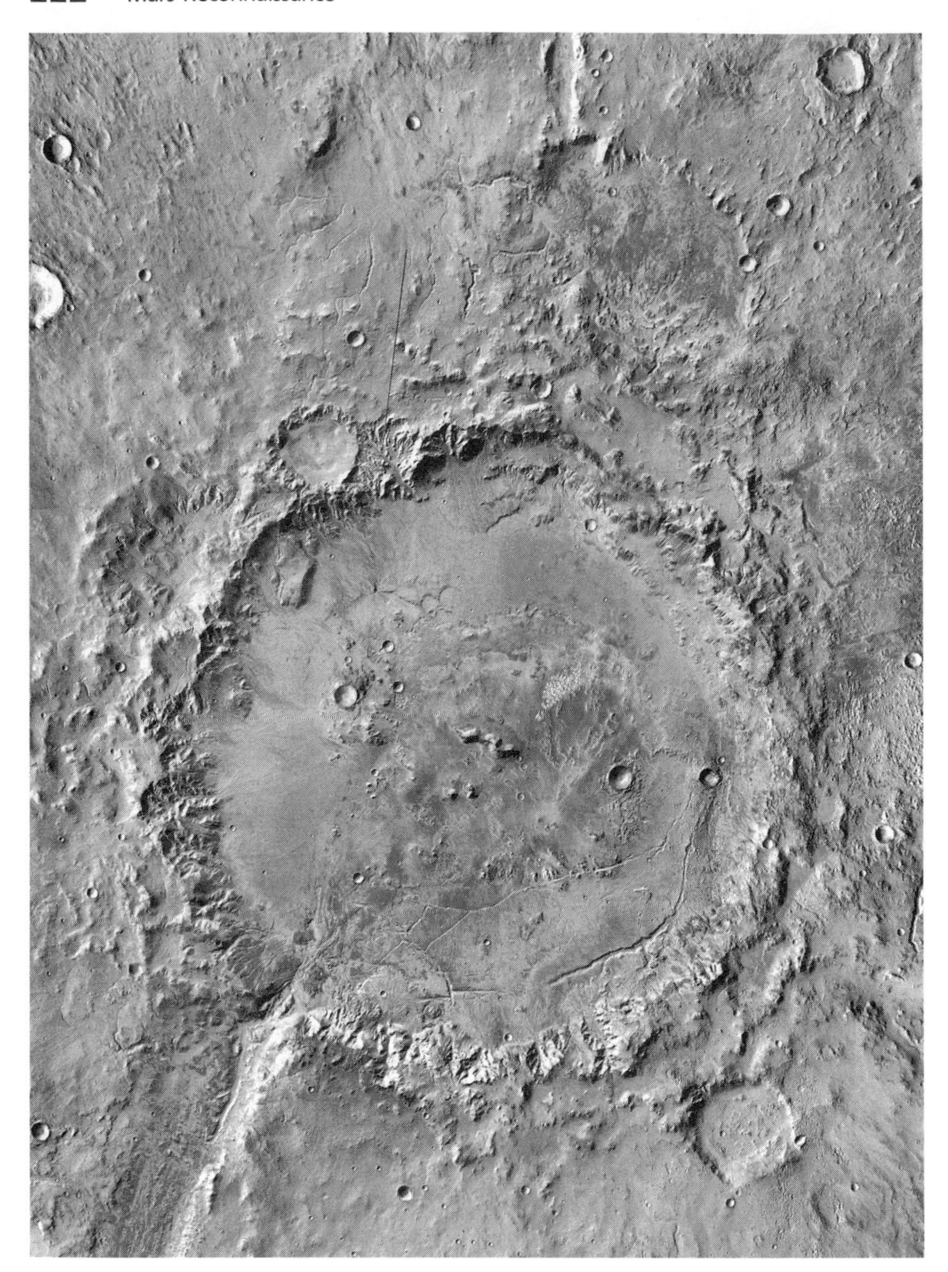

Figure 57: 2001 Mars Odyssey THEMIS image of Holden Crater, which formed on what may be the longest watercourse on Mars—the Uzboi–Ladon–Margaritifer valleys—stretching from the Argyre Basin in the south to Chryse Planitia in the north. See Plate 27 in the color section. [Credit: NASA/JPL/Arizona State University]

Mindful of (and learning from) the US Mars program, Mars Express was built both to complement US efforts by carrying out follow-on surveys based on MGS and Odyssey findings, as well as to conduct surveys specific to ESA's own emerging Mars program. The result was yet another exquisite robotic space probe equipped with an array of instruments designed to pursue new investigations at Mars. Mars Express—accompanied by its ill-fated passenger, the Beagle 2 Mars lander—was launched in June 2003 to avail of the extraordinarily close opposition of Mars that year; and with NASA's two Mars Exploration Rovers also *en route*, no less than four space probes were heading toward Mars at the same time. In December 2003, Beagle 2 was unfortunately lost on arrival, although Mars Express itself successfully settled into orbit and immediately began its reconnaissance mission. With the successful landing of the two Exploration Rovers in January 2004, no less than five robotic explorers were at work on and around the planet.

As with MGS and Odyssey, Mars Express is a multifaceted orbiter. First, its spectacular optical camera—the High Resolution Stereo Camera (HRSC)—is capable of mapping the planet at about 10 meters resolution (surpassing Odyssey), while a special super-high resolution mode can even photograph selected sites to 2 meters resolution. Furthermore, the HRSC can also take true stereo images, providing unprecedented three-dimensional views of Mars that are radically improving our understanding of the planet's topology, morphology, and geology.

Mars Express is also tasked with conducting an extensive mineralogical survey with OMEGA, its infrared spectrometer. While the spatial resolution of OMEGA is lower than that of THEMIS, its sensitivity is higher, meaning that it can identify even lower levels of minerals that are crucial to a complete picture of Mars' mineralogy and the processes by which they originated. For example, OMEGA can identify the various classes of silicates—whether feldspar, olivine, pyroxene, etc. OMEGA can also identify the iron content and oxidation state of iron-based minerals, locate clays and hydrated minerals, and detect carbonates and nitrates to abundances of just a few percent. While the mantra for NASA has become follow the water, ESA have aptly decided to prioritize tracing the history, evolution and current planetary cycling of carbon—and in this search, OMEGA should provide many answers.

Unique to Mars Express are a number of instruments whose aim is to determine the make-up and circulation of the planet's atmosphere, as well as to investigate how the atmosphere interacts with the crust and solar wind. Such information is critical to understanding Mars' current climate, the nature of its ancient atmosphere, and the mechanisms that may have caused it to dissipate over time. The energetic neutral atom analyzer, ASPERA, examines how the uppermost region of the atmosphere interacts with the solar wind, revealing the escape (or leakage) of the atmosphere to space over time. A Planetary Fourier Spectrometer (PFS) measures the global atmospheric distribution of water vapor (among other atmospheric constituents); an optical and ultra-violet camera called SPICAM can look through Mars' atmosphere to determine its composition; while a radio transceiver called MARS uses radio waves to measure both the temperature and pressure within the atmosphere and on the surface.

Mars Express carries another enigmatic instrument called MARSIS (Figure 58). This radio-sounding instrument consists of two radio booms, each 20 meters in length, which transmit radio waves at the planet and subsequently detect their reflections. The characteristics of each reflection are analyzed to reveal the underground structure to a depth of about

Figure 58: Image depicting Mars Express' MARSIS Radio Sounding instrument scanning for underground reservoirs of water and other buried geological features. See Plate 28a in the color section. [Credit: ESA]

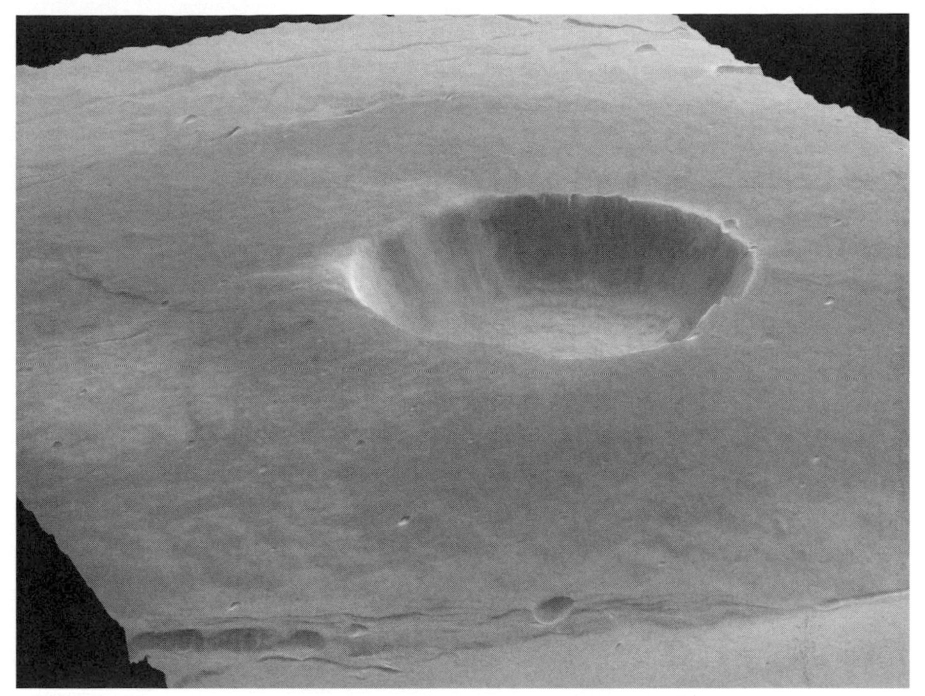

Figure 59: Mars Express/HRSC 3D view of the summit of Albor Tholus, a volcano on Elysium. The caldera is 30 kilometers across and 3 kilometers deep. On the rim, a bright "dust fall" triggered my Martian winds flows from the surrounding plateau into the caldera. [Credit: ESA/DLR/FU Berlin (G. Neukum)]

5 kilometers. In a similar manner to how a submarine uses sonar to detect otherwise invisible structures, MARSIS uses radio waves to reveal both ancient geological structures now covered over, as well as any remaining underground reservoirs of water or water-ice.

Data through Mars Express

The findings of Mars Express have been dramatic. The HRSC, for example, is providing new ways of looking at the planet through high-resolution color perspective images as well as true stereo 3D images (Figures 59–61; see also Figures 62 and 63). Mars Express scientists can now look at the relief of Mars in all three dimensions, gaining a unique insight into the underlying geology. MGS's MOLA also provides spectacular 3D perspectives of course, though its resolution is lower and was designed to produce a global topological map. With the HRSC's stereo images being of higher lateral resolution and revealing incredibly fine relief of individual surface features, both instruments complement each other in providing exquisite regional and local 3D views.

Figure 60: Mars Express/HRSC 3D perspective view of the Martian north polar ice cap, showing layers of water-ice and dust. The cliffs are 2 kilometers high. See Plate 29 in the color section. [Credit: ESA/DLR/FU Berlin (G. Neukum)]

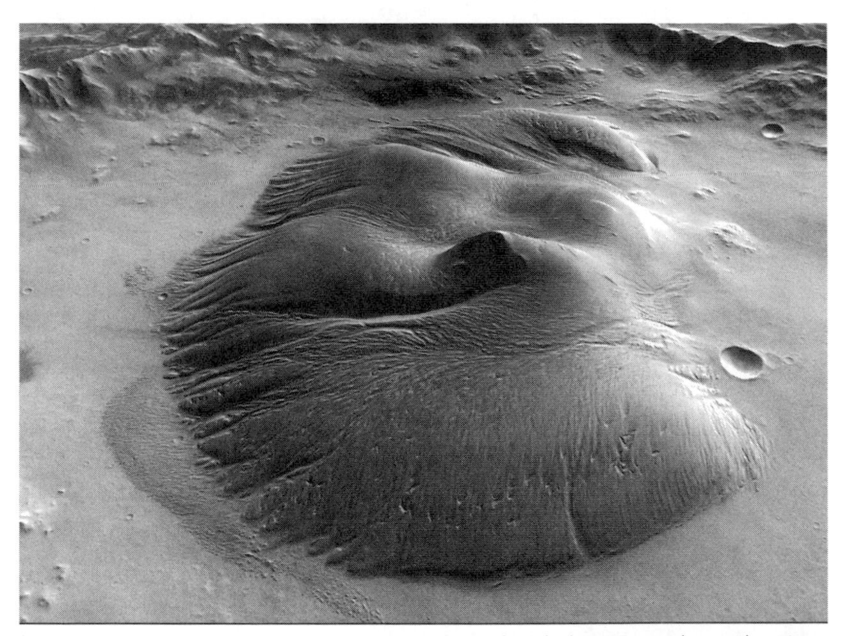

Figure 61: Mars Express/HRSC 3D perspective view of Nicholson Crater located at Amazonis Planitia. This view shows the central part of the crater, measuring 100 kilometers across. See Plate 30 in the color section. [Credit: ESA/DLR/FU Berlin (G. Neukum)]

Figure 62: Mars Express/HRSC 3D perspective view of Coprates Chasma and Catena in Valles Marineris, with a ground resolution of approximately 48 meters. The main trough, appearing in the north (top half) of this image, ranges from 60 kilometers to 100 kilometers wide and extends 9 kilometers below the surrounding plains. See Plate 31 in the color section. [Credit: ESA/DLR/FU Berlin (G. Neukum)]

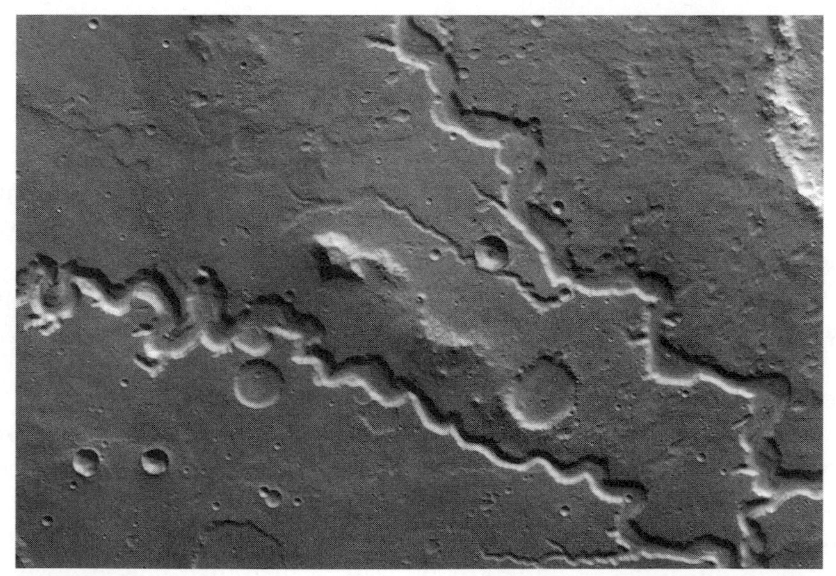

Figure 63: Mars Express/HRSC image of Nanedi Valles valley, extending 800 kilometers diagonally across Xanthe Terra, southwest of Chryse Planitia. The valley exhibits meanders and a merging of two branches in the north. Erosion was perhaps caused by ground-water outflow, flow of liquid beneath an ice cover or collapse of the surface in association with liquid flow. See Plate 32 in the color section. [Credit: ESA/DLR/FU Berlin (G. Neukum)]

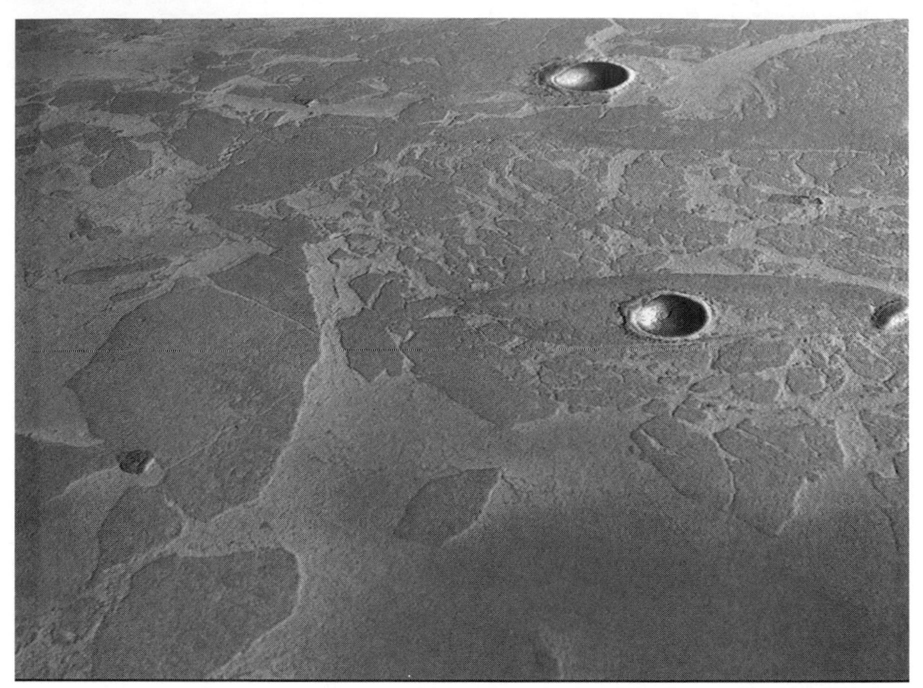

Figure 64: Pack-ice on Mars? This image, taken by the HRSC, shows what appears to be a dust-covered frozen sea at Elysium Planitia near the Martian equator. This scene is tens of kilometers across, centered on latitude 5°N and longitude 150°E. See Plate 33 in the color section. [Credit: ESA/DLR/FU Berlin (G. Neukum)]

One of the most dramatic results has been the discovery in 2004 by John B. Murray, Jan-Peter Muller, Gerhard Neukum, and others of what looks like a frozen sea on Mars today (Figure 64). A region south of Elysium, known as Cerberus Fossae, had previously been identified as having undergone tectonic-based lava and water flows within the last five million years or so. It was assumed that all water in the region would have long since disappeared. However, on photographing the area, HRSC images revealed what appears to be a frozen sea actually residing there today, measuring approximately 800 by 900 kilometers (about the size of the North Sea); and about 45 meters deep. Also seen within the region is what looks like broken glaciers that drifted apart before the sea froze and locked them into their current positions. The sea may have arisen due to tectonic and/or volcanic activity within Cerberus Fossae fissures, releasing enormous quantities of underground water which then flowed down for hundreds of kilometers, filling a vast area that then froze. The water may originally have come from underground ice that melted during the event, or it might even have originated as liquid water residing underground where the geothermal gradient prevented it from freezing and it burst onto the surface during the cracking of the fissures. Subsequent to the creation of the sea, volcanic ash falling on the freezing sea may have formed a protective layer, inhibiting all of the ice from sublimating into the atmosphere. Even so, it is clear that the water level in the region has dropped about 20 metres since it first formed due to sublimation. This discovery is significant. It reveals for the first time dramatic events occurring on Mars in the current geological era, indicating that the planet is certainly not dead. Also, in the search for past and present life, this discovery tells us that environments conducive to basic life as we know it may have persisted on Mars for the past four billion years and until several million years ago. If underground water reservoirs still exist, life could survive well there, protected from ultraviolet radiation and oxidation and availing of geothermal energy and geochemical nutrients.

HRSC has also provided unprecedented insight into the movement of water and ice across the surface of Mars both in its distant past and perhaps even today. For example, images of the Kasei Valles outflow channel which connects the southern Echus Chasma to Chryse Planitia in the east over a distance 2,500 kilometers reveal evidence of a vast flood on Mars billions of years ago, followed by the rapid freezing of much of the flood waters into a colossal glacier that subsequently carved out much of the valley and which may have persisted for as long as a thousand million years (Figures 65 and 66).

James Head of Brown University and Ernst Hauber of the German Aerospace Center have also used HRSC to identify recent glaciation on the planet, even in its equatorial regions. For example, glacial deposits can be seen at the base of Olympus Mons, pointing to glaciers in that region about four million years ago. Above 7,000 meters, glaciers may currently exist on the flanks of Olympus Mons and be disguised by a layer of dust. Further evidence of glaciation has been found at five other volcanoes. For example, HRSC images also reveal that Hecates Tholus (which erupted explosively about 350 million years ago) contained glacial deposits inside depressions within its caldera about four million years ago. And evidence of snow-ice accumulation, flow, and glaciation can also be found on the eastern side of Hellas Basin. Here, numerous concentrically ridged lobate and pitted surface features could only have been created by glaciation in the last few million years.

The detection of recent glaciation on Mars is important as it points to significant climate change occurring on the planet over timescales of millions of years or less, where we suspect that periods of greater obliquity affect the planet's hydrological cycle and mobilize polar ice to micro-environments at lower latitudes. In response to the HRSC findings, a team

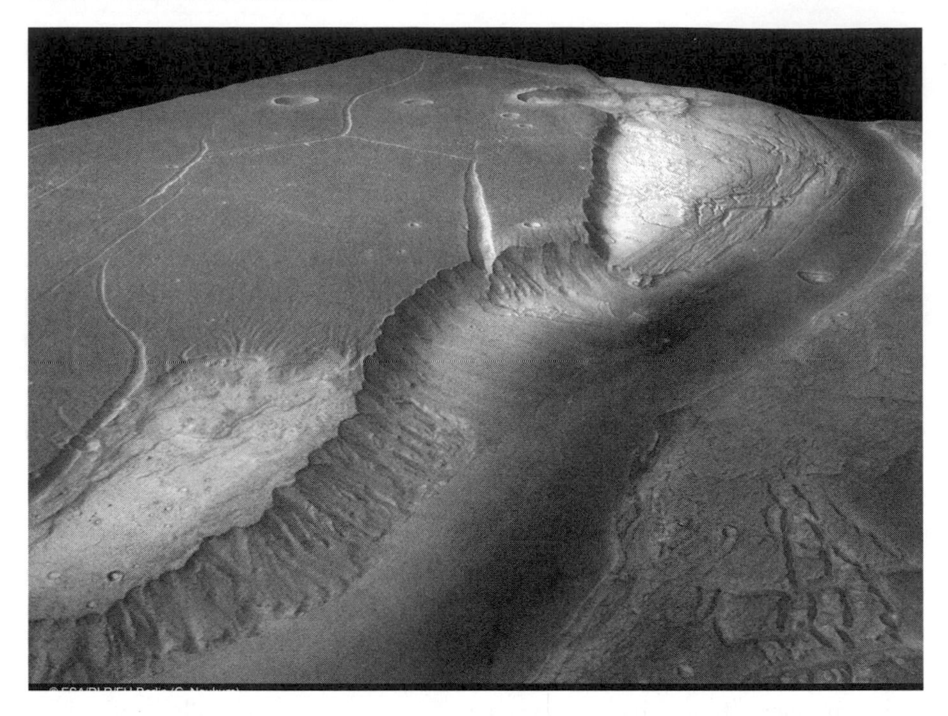

Figure 65: Mars Express/HRSC 3D perspective of the Kasei Valles outflow channel – a gigantic channel 2,900 kilometers long and 500 kilometers wide that may have persisted on Mars for upwards of a billion years. See Plate 34 in the color section. [Credit: ESA/DLR/FU Berlin (G. Neukum)]

led by Mars Express scientist François Forget at the University of Paris have even applied Mars meteorological computer simulations to the problem, yielding compelling results. By turning the clock back several million years and assuming an axial tilt of 45 degrees, Forget's simulations produce climatic conditions leading to glaciation at precisely the locations identified by the HRSC. With a greater tilt, increased solar illumination in the north polar summer increased the sublimation of the polar ice, leading to a more intense water cycle. The simulations show water-ice accumulating at a rate of 30 to 70 millimeters per year in localized areas on the flanks of the Elysium Mons, Olympus Mons, and the three Tharsis Montes volcanoes. After a few thousand years, the accumulated ice forms glaciers up to several hundreds of meters thick. Comparisons between the simulated and actual glacier deposits give excellent agreement. For example, simulations predict maximum deposition on the western flanks of the Arsia and Pavonis Montes of the Tharsis region and this is where the largest deposits are actually observed. Forget's simulations even reveal the manner in which ice accumulates on the flanks of mountains in the Tharsis region during such

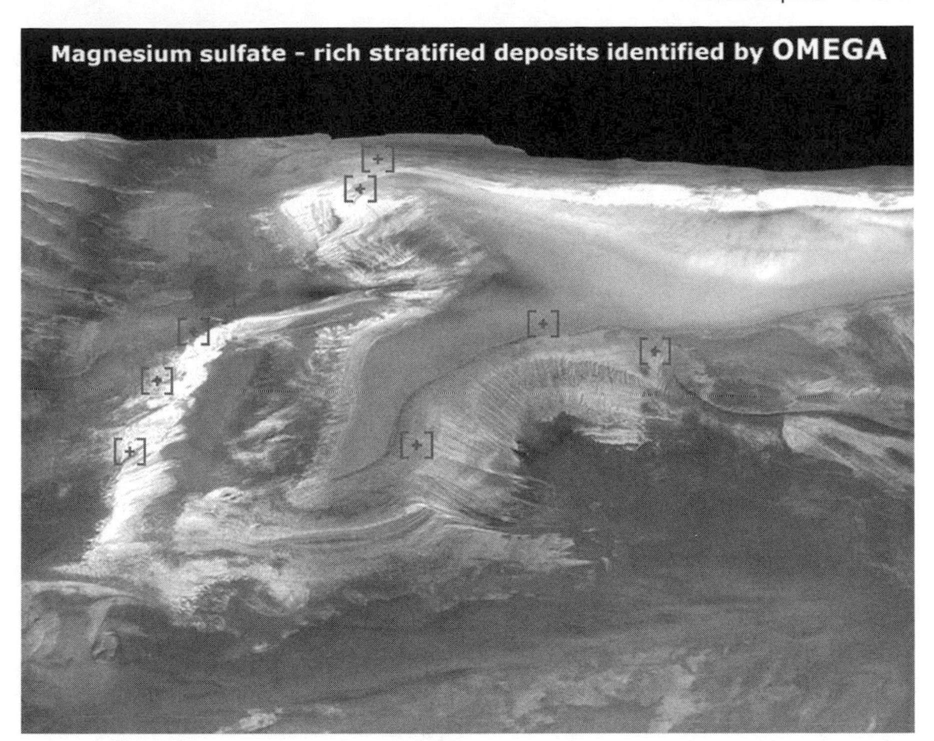

Figure 66: Mars Express/HRSC 3D perspective view of Candor Chasma with a superimposed infrared image from OMEGA. This image shows deposits of the mineral kieserite (hydrated magnesium sulfate). See Plate 35 in the color section. [Credit: ESA/OMEGA/HRSC]

periods of high obliquity. With year-round winds similar to Earth's monsoons, movement of moist air occurs particularly up the slopes of Arsia and Pavonis Montes; it then cools, leaving water condensation on the slopes as ice particles. Olympus Mons, on the other hand, experiences less deposition (as observed) because, even during high obliquity, monsoon-type winds and water condensation only occur during the northern summer and not all the year round. Simulations also reveal that water from the south polar cap accumulates at east Hellas Basin, which is so deep as to create a northward wind flow on its eastern side carrying water vapor sublimating from the south polar cap during summer. When the water-rich air meets the colder air mass over the eastern Hellas, the water condenses, precipitates, and forms glaciers over time.

Mineralogy

The OMEGA mineralogical detector has complemented the findings of both TES and THEMIS while also delivering unprecedented results of its own regarding Mars' distant past. First, OMEGA has further verified the broad presence of mafic materials (ancient and pristine basalts) in the south of the planet and felsic (andesite) to the north, corroborating the idea that the south is composed of the most ancient surface while the north is much younger. OMEGA has also found significant evidence of water-altered minerals. For example, it has found an area in the north polar region measuring 60 by 200 kilometers that is dominated by gypsum (calcium sulfate), revealing that water played a major role in altering the surface of the region. Also, hydrated minerals such as kieserite, gypsum, and polyhydrated sulfates have been found within the sedimentary layers of Valles Marineris, at Margaritifer Sinus, and Terra Meridiani, constituting significant evidence of ancient aqueous activity on Mars (Figures 67–69).

OMEGA has also re-examined the eastern Meridiani, where TES, THEMIS, and Opportunity have already verified the presence of hematite and precipitate salts. OMEGA has found widespread presence of water molecules within etched terrain in the region, overlying heavily cratered terrain but underlying the hematite-bearing plains explored by Opportunity.

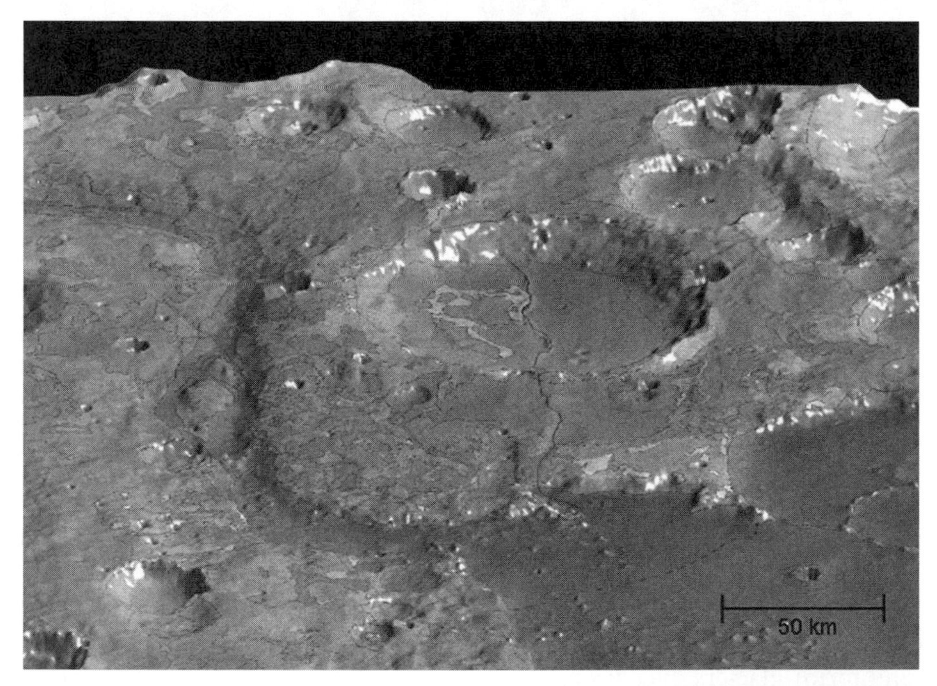

Figure 67: A Mars Express/HRSC 3D perspective view of Marwth Vallis, with a superimposed OMEGA map showing water-rich minerals. While no hydrated minerals are detected in the channel, the outflow was so violent as to expose ancient hydrated minerals, revealing an earlier era when water was present. See Plate 36 in the color section. [Credit: ESA/OMEGA/HRSC]

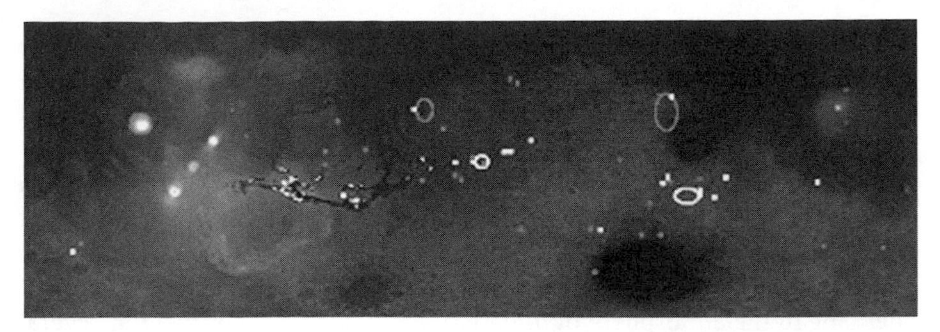

Figure 68: The global distribution of hydrated (water-rich) minerals as discovered by Mars Express/OMEGA. The map is superimposed on an MGS/MOLA map. Such maps indicate the distribution of phyllosilicates, sulfates, and other hydrated minerals. See Plate 37 in the color section. [Credit: IAS/OMEGA/ESA]

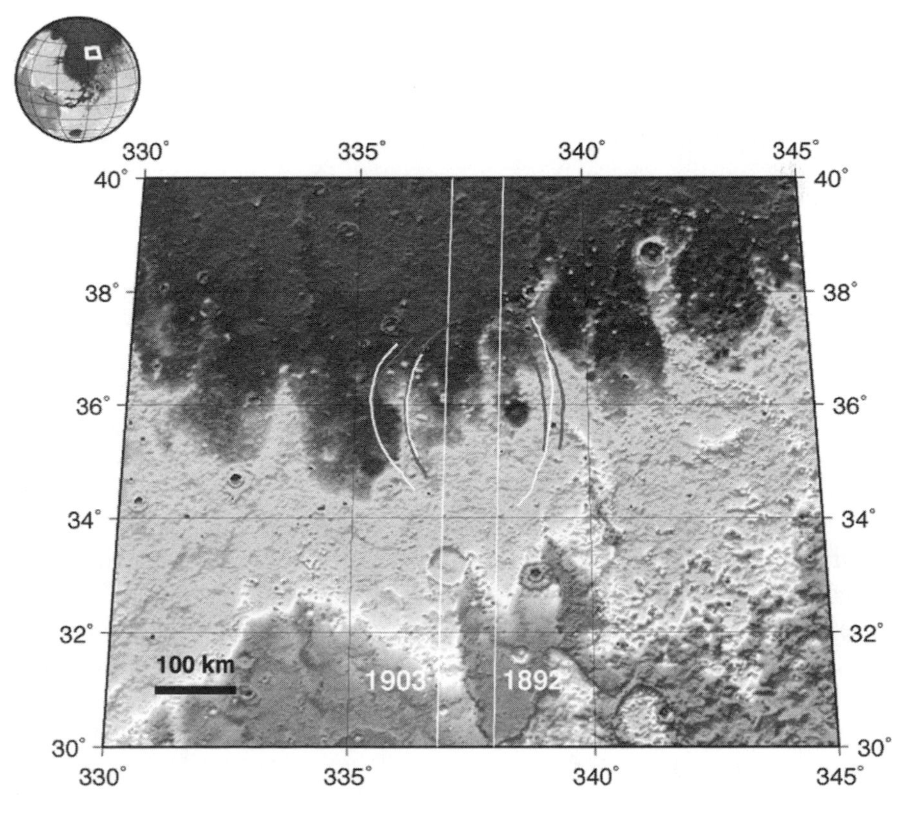

Figure 69: Mars Express/MARSIS identification of an ancient crater 250 kilometers in diameter buried under Chryse Planitia, superimposed on a MGS/MOLA topology map. See Plate 38 in the color section. [Credit: NASA/JPL/ASI/ESA/Uni. of Rome/MOLA Science Team]

This indicates that ancient aqueous environments inferred by Opportunity indeed extend over a large area, and corroborates the idea that an ancient sea once existed there. Of direct interest to the search for ancient prebiotic and life-related activity on Mars, OMEGA has also identified layered and water-altered terrain on Syrtis Major, suggesting the interaction between volcanism and volatile rich deposits and pointing to the possibility of ancient hydrothermal activity in the region.

Among some of the most intriguing discoveries has been the detection of methane in the atmosphere by the Mars Express PFS experiment at several specific locations on the planet, most notably from the Hellas Basin, from within Valles Marineris, and from Elysium. Since methane cannot reside within Mars' current atmosphere for long periods of time, it must have entered the atmosphere only recently. Measured at levels of about 10 parts per billion, an indigenous production rate of about 150 tonnes per year is suggested. Furthermore, the detection of formaldehyde—perhaps produced when methane is oxidized—points to a limit of methane production upwards of 20 times greater.

Currently, three natural mechanisms of methane production are envisaged—from active volcanism, from delivery by comet impacts, and as a by-product of life activity. Given how quiescent Mars is today, it is difficult to see how volcanism could give rise to such quantities; and while comet impact could deliver the detected methane, its localized nature suggests an indigenous source. Hence, while the detected methane may arise from a geochemical process, a biological source cannot be ruled out.

Finally, the Mars Express MARSIS radio sounding experiment has also probed below the surface to depths of up to 5 kilometers, identifying large impact basins, including one up to 250 kilometers across, buried beneath the smooth northern lowlands near Chryse Planitia. Such discovery suggests that even though the northern lowlands are younger than the southern highlands they are none the less extremely old and point to planet-wide water activity occurring only in Mars' earliest history. MARSIS radar maps have also detected what appear to be underground reservoirs of water-ice, while radar maps of the south polar region reveal the structure of the water-ice cap in exquisite detail, unequivocally verifying that enough water resides in the southern ice-cap alone to cover the entire planet to a depth of 11 meters (Figures 70 and 71).

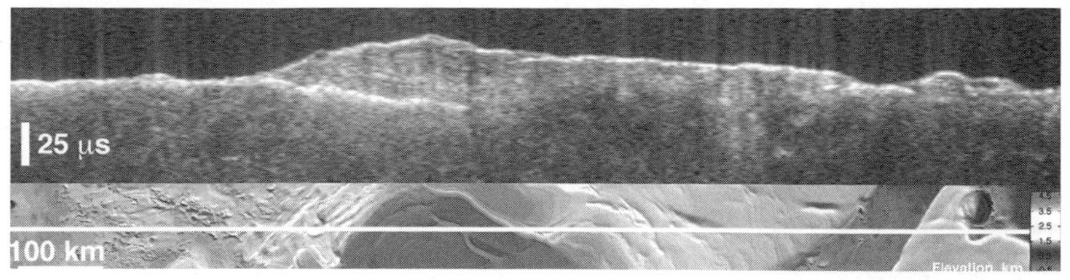

Figure 70: Mars Express/MARSIS subsurface map of the water-ice of the south polar region of Mars. The amount of water-ice in the region is equivalent to a layer 11 meters deep covering the entire planet. See Plate 28b in the color section. [Credit: NASA/JPL/ASI/ESA/Uni. of Rome/MOLA Science Team]

Figure 71: Mars Reconnaissance Orbiter (MRO), among the most sophisticated planetary probes ever sent to another planet. It will pave the way for continued Mars exploration well into the twenty-first century. [Credit: NASA/JPL]

The Future: Mars Reconnaissance Orbiter

In March 2006 yet another orbiter, NASA's Mars Reconnaissance Orbiter (MRO) arrived at Mars. This orbiter is set on gathering more data than all previous missions to Mars combined. MRO's onboard stereo camera—called HiRISE—is no less than half the diameter of the Hubble Space Telescope and has an extraordinary pixel resolution of just 30 centimeters, allowing it to resolve features on the surface less than 1 meter across. MRO will also use revolutionary navigation techniques to guide future landers to the surface of the planet with 10 times current precision, and will provide images of the surface of sufficient quality to allow safe human landings to be planned.

The Mars Reconnaissance Orbiter represents a significant step forward in orbital reconnaissance, housing the most sophisticated computers and instruments ever sent to another planet. With its HiRISE camera capable of sub-metre resolutions and in stereo, MRO will finally bridge the gap between orbital imaging at the planetary level and lander-based imaging at centimeter and smaller scales. Such a seamless connection is already providing defining views of the planet that allow for underlying processes behind geological features to be determined. As an example, sub-meter resolution HiRISE images of layering within an exposed scarp at the head of Chasma Boreale in the north polar region reveal unprecedented detail of geologically recent climate variations recorded in the layers at 40 centimeters resolution (Figure 72). And with the prospect of obtaining tens of thousands of images of the surface at such resolutions, we will finally be able to unravel much of the vast and intricate geological history of the planet. Views of the surface from HiRISE will also reveal thousands of candidate sites for follow-up exobiological missions such as NASA's upcoming Phoenix and Mars Science Laboratory rover, and ESA's ExoMars rover.

MRO is also engaged in a mineralogical survey of the planet at no less than 10 times the resolution of THEMIS or OMEGA. Here, MRO's thermal imaging camera, called CRISM, will specifically look for water-altered minerals such as clays, carbonates, and precipitated salts beyond the detection threshold of THEMIS and OMEGA. In tandem with HiRISE, CRISM will reveal the history of water activity of the planet in unprecedented clarity, as well as unveil hitherto invisible small-scale recent or current activity. Both instruments will become the eyes through which an entirely new planet, currently invisible to us, will be revealed over the coming years. Indeed, results already received are proving to be extremely revealing. For example, sedimentary deposits in Candor Chasma (Figure 73), examined both by HiRISE and CRISM, reveal for the first time ancient fluid flows through cracks in the rocks of that region on the sub-meter level, suggesting possible ancient hydrothermal sites as habitats for microbial life.

MRO will also survey the planetary subsurface using the same radio sounding technique employed by MARSIS on board Mars Express. While

Figure 72: High-resolution MRO/HiRISE images of layering within an exposed scarp at the head of Chasma Boreale in the north polar region, revealing unprecedented detail of climate variations recorded in the layers. Pixel resolution is 38 centimeters. [Credit: NASA/JPL-Caltech/Univ. of Arizona]

Figure 73: Sedimentary deposits in Candor Chasma examined by MRO's HiRISE and CRISM reveal ancient fluid flows through cracks in the rocks on the sub-meter level, suggesting a possible ancient hydrothermal site as habitats for microbial life. See Plate 39 in the color section. [Credit: NASA/JPL-Caltech/Univ. of Arizona]

MARSIS can probe to a depth of 5 kilometers and has a vertical resolution of 100 meters, MRO's SHARAD radio sounder will probe to a depth of just 1 kilometer but with a vertical resolution of 10 meters. With both instruments working in tandem, the subsurface structure of the planet can be probed in excellent detail, revealing whether underground reservoirs of water reside there today. Here again results already relayed back reveal a layered structure to Mars' southern ice cap that suggests past climate change and clearly defines the boundary between the underlying Martian surface and the ice cap above (Figure 74).

MRO will also conduct a number of other studies of the planet-atmospheric radio sounding to determine the three-dimensional structure of the atmosphere, seasonal and yearly climate monitoring, and a color full-planet imager providing daily planetary weather reports. Subsequent to its two-year primary science mission, MRO will enter a unique second two-year *relay mode* phase of operations, specifically designed to test a range of new technologies and help with the precision landing requirements of Phoenix, the Mars Science Laboratory, and ExoMars. Through its unique UFH communications facility called Electra, MRO will communicate with and guide all future spacecraft when approaching the planet, allowing them to accurately determine their celestial coordinates and trajectory. Similarly, landers and rovers on the surface will use Electra to better determine their position. A technology demonstration optical-navigation-camera will

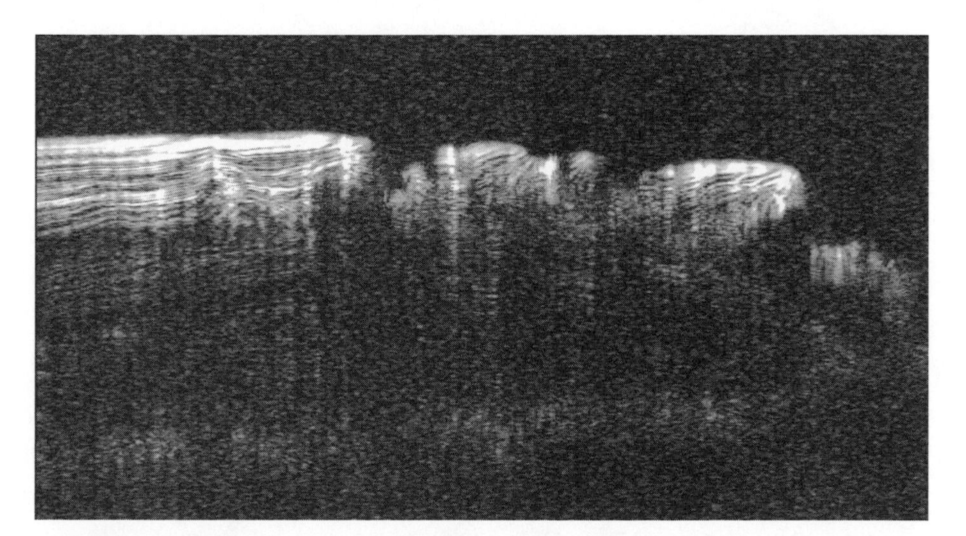

Figure 74: MRO/SHARAD radio 'sounding 'image' subsurface section of Mars' northern polar cap. The ice cap is approximately 2,000 meters deep, with horizontal layers in the upper 600 meters (brighter) of the ice cap suggesting cycles of climate change. [Credit: NASA/JPL-Caltech/ASI/University of Rome/Washington University in St. Louis]

accurately track the position of Mars' two satellites Phobos and Deimos, comparing their predicted positions with actual positions and hence provide improved navigation data. Finally, MRO will test a new mode of communications with Earth called Ka-Band Communications which demands less power than traditional X-band communications.

Combined, all four reconnaissance orbiters have radically altered our understanding of Mars, revealing it to be a complex world where significant volatile and water activity occurred both in its earliest history and episodically over billions of years. Our understanding is far from complete, but the picture emerging is of a planet with a vibrant ancient past in ways perhaps relevant to understanding the early history of our own planet and the emergence of life.

15 Spirit and Opportunity

Within just a few years of reconnaissance from orbit, hundreds of sites were identified on the surface as relevant to advancing our understanding of Mars' ancient history and its potential for life. Underpinned by the successful Pathfinder mission, NASA saw fit in 2003 to commence Phase II lander exploration by sending robotic rovers to two locations of past water and geochemical interest. Two identical rovers, called the Mars Exploration Rovers (MERs), would be sent to sites on opposite sides of the planet. Whereas Sojourner had been essentially a test rover that remained close to its landing site, the Mars Exploration Rovers would be far more capable—physically larger, autonomous, capable of traveling up to 100 meters per day over a 90-day period, carrying a sophisticated science payload.

True to the phased strategy, the Mars Exploration Rovers would not look for life; they would instead conduct precursor investigations for evidence of past water activity, as well as examine the morphology, geology, mineralogy, and surface chemistry at each site. Only subsequent to such analyses could we develop appropriate biological exploration missions with any chance of success. To maximize this opportunity, NASA invited the entire Mars-science community to propose possible sites for the two MERs. By January 2002, no less than 185 sites had been proposed, from which 12 were then selected. Many of the rejected sites were of equal or greater interest to those chosen, but posed insurmountable technical challenges or too high a risk. With further analysis of the 12 sites then carried out by MGS and Odyssey, only four were chosen by October 2002. Valles Marineris and Meridiani Planum were selected because of the detection of hematite at each location, suggesting past water activity; Athabasca Vallis (south of Elysium) was chosen because it is among the youngest of tectonically induced flooding regions on the planet; and Gusev Crater made the list because it is a shallow crater with a valley called Ma'adim Vallis intersecting it, suggesting that it may once have contained a lake. With several more months of geological and logistical analysis, the two final locations selected—Meridiani Planum and Gusev Crater—were announced in January 2003. While Vallis Marineris and

K. Nolan, *Mars, A Cosmic Stepping Stone*,
DOI: 10.1007/978-0-387-49981-9_15, © Praxis Publishing, Ltd. 2008

Athabasca Vallis were equally compelling, Meridiani Planum and Gusev Crater represented the best balance of safety, adequate solar power (being within a required 25 degrees of the equator) and scientific interest.

Rover Characteristics and Science Mission

The Jet Propulsion Laboratory (JPL) has spent decades developing the types of technology required to carry out remote science on Mars. Among the systems to be developed for the Mars Exploration Rovers included power generators, propulsion and navigation systems, robotic arms, robust and flexible suspension and wheel systems, computer hardware and software, and a multifaceted scientific payload—all to be developed under the constraints of restricted mass and required robustness, maneuverability, and so on. Furthermore, such technology would have to be able to survive being launched into space, traversing hundreds of millions of kilometers across freezing and irradiated interplanetary space, entry and landing onto another planet and then to operate in an uncharted and alien environment. Although vital experience and know-how had been gleaned from the Pathfinder mission, the requirements of the Mars Exploration Rover program brought a different order of challenge. Each MER would weigh in at more than 10 times that of Sojourner, be required to travel 10 times further, last longer, and carry an onboard scientific payload, much of which had never been tried in the field.

Despite all of this, JPL responded in spectacular fashion. As with the early Mariner program, the task of developing and building two MERs was completed in record time and with a budget of just several hundred million dollars—about one-tenth the budget of the Viking program. Both rovers are identical. Each has a mass of 185 kilograms, is composed of a central body (called the Warm Electronic Body or WEB) and six wheels attached to flexible legs that incorporate a sophisticated *rocker-bogie* suspension system that allows the entire rover to swivel and the WEB to rock back and forth in response to the tricky Martian terrain. Power generation is via solar panels that can provide up to 900 watts of power. Unlike Sojourner, which communicated to Earth via the landing station, each MER is equipped with two antennas that communicate autonomously back to Earth via one of the orbiting reconnaissance satellites. Each rover is completely self-contained, capable of operating, communicating, and navigating independently of the landing craft that delivered it to the surface.

The scientific objectives of the MER program have been to build upon the reconnaissance effort still in operation today. Specifically, the program contributes to the four overarching goals of NASA's current Mars Exploration Program:

- *Determine whether life arose on Mars.* Since life as we know it depends on water, knowing the history of water on Mars is critical to finding out if it was ever conducive to life. In this respect, one of the primary goals for each MER is to search for evidence of water activity. While liquid water cannot reside on the surface of Mars today, both rovers were sent to sites of possible past water activity. Each could verify the presence of any water activity by looking for geological features such as sedimentary layering and mineralogical evidence such as water-altered rocks, the presence of hematite as well as precipitated minerals and salts.
- *Characterize the geology of Mars.* Each MER is equipped with several specific instruments to analyze the geology of their respective sites. The onboard stereo camera, called Pancam, sits upon the Pancam Mast Array (PMA) and can image the landscape exquisitely over 360 degrees and in stereo—directly revealing the geology of the region from kilometer to just centimeter scales. Pancam can reveal, as never before, the distribution of rocks, soils, sand, and bare rock over an entire region and from different perspectives, allowing us to identify geological processes such as water and wind erosion, sedimentation, volcanism, cratering, and past hydrothermal activity. Also, with an onboard rock abrasion tool (called the RAT) and an optical microscope, individual rocks can be cleaned, cut, and imaged to micrometer scales, revealing the types of rocks in the region as well as their small-scale morphology, texture, strength, and alteration over geological timescales. And with the ability to rove about the surface and analyze many rocks, each MER can provide a detailed analysis of even the most complex aspects of the site geology. A suite of mineralogical analysis tools also allows for chemical and mineralogical analyses. A mini-TES similar to that on board the Mars Global Surveyor can identify both the mineralogy and petrology of the site. An APXS similar to that on board Sojourner can determine elemental abundances, while a Mössbauer Spectrometer can determine the composition of iron-bearing minerals to high accuracy. Combined, these instruments represent a sophisticated arsenal from which to determine the chemistry, mineralogy and geological processes, and evolution of the region.
- *Characterize the climate of Mars.* The climate of Mars today is affected by many factors including its distance from the Sun, changes in its celestial motion and orientation about the Sun, the

nature of its atmosphere and seasonal changes that affect polar cap and atmospheric carbon dioxide levels, water vapor in the atmosphere and on the surface and dust movement across the planet. While the MER program does not specifically incorporate weather and climate monitoring instruments, the quality of the MER Pancam is such that it can reveal a great deal about the atmosphere by simply looking at the sky! So useful is this technique that regular photographs of the sky are taken and transmitted back to Earth for analysis. Further, chemical, mineralogical, and geological analyses can reveal a great deal about both past and present climatic conditions; while imaging of sand accumulations and local dust-devils (small atmospheric vortices visible because of the dust they raise from the surface) can indicate past and current prevailing wind patterns. Comparisons between the chemistry of sand and rocks can tell if the sand is locally derived or has traveled in from elsewhere. Imaging of sedimentary layering can reveal patterns of past water behavior such as the duration and episodic nature of the activity, the depth of water involved and speed of flow—all of which can help to determine the climatic conditions in which they occurred.

- *Preparation of Human Exploration of Mars.* If we are to send people to Mars, a myriad of preparations must be made. In this regard, the MER program is extremely useful in characterizing surface and atmospheric conditions, identifying chemical hazards such as the super-oxidizing nature of the soil, the characteristics of Mars' ubiquitous dust, radiation dosages from space, daily and seasonal changes, and wind behavior. MER can also provide valuable insight into entry, descent, and landing (EDL) systems, communications and navigation and efficient modes of long-term remote-controlled exploration.

Journey to Mars

The Mars Exploration Rovers were launched in June 2003. As with ESA's Mars Express, this launch window took advantage of Mars' August 27, 2003, closest approach to Earth for almost 60,000 years, providing a *short hop* across to Mars in only six months. The first MER, named *Spirit*, was launched on June 10, while the second, *Opportunity*, was launched on July 7. Both probes traversed the distance between the planets to near-perfection, with just several minor course corrections needed *en route*. Upon arrival, an uncontrolled entry, descent, and landing procedure—very similar to that of

Pathfinder—was carried out, involving the rover (enclosed in a protective capsule) decelerating through the atmosphere, followed by deployment of a parachute and small retrorockets to slow the lander and, finally, airbags inflating around the lander to cushion the impact. Once safely on the surface, the airbags deflated and retracted, the lander petals opened, the rover opened its solar panels, checked its systems and descended to the surface (Figure 75).

For each MER, part of their descent and landing phase could be monitored via the transmission by the rover of a set of 36, 10-second tones during descent and landing, revealing its progress. On arriving at Mars, each MER entered the atmosphere at 12,000 kilometers per hour (kph) with the aeroshell reaching a maximum temperature of 1,447°C. Within just four minutes the capsule had slowed to 1,000 kph and reached an altitude of 10,000 meters, whereby the parachute opened. Just 20 seconds later the probe had descended to 7,000 meters where the aeroshell was jettisoned. Ten

Figure 75: A Mars Exploration Rover (MER) enters the Martian atmosphere. With the parachute deployed the retrorockets fire, suspending the lander at 12 meters from the surface. Protected by airbags, the lander is dropped and bounces along the surface. The lander's petals then unfold and the rover egresses. [Credit: NASA/JPL]

seconds later the landing pod descended from the protective capsule along a Zylon rope attached to the parachute, and five minutes into the mission, the onboard altimeter switched itself on to determine the precise height above the ground. A descent imager also took photographs of the surface to provide orientation information and during this time the rover continuously transmitted its status-tones back to Earth via the Mars Global Surveyor. At 100 meters from the surface the airbags inflated and the retrorockets fired, bringing the probe to *zero vertical velocity* at just 12 meters from the surface. The airbag assembly was then released, where it dropped and hit the ground, bouncing a number of times before coming to rest about 1 kilometer downrange. After 14 minutes the lander tones resumed to let Mission Control on Earth know that all was well. After 80 minutes the airbags were retracted and the lander petals opened. Finally, the lander opened its solar panel arrays and carried out a systems check before entering a *safe mode* and going to sleep for the night! It had taken six months to reach the planet but just six minutes to traverse the hazardous descent to the surface. With critical events occurring just seconds apart while slowing from thousands of kilometers per hour in violently changing conditions, everything had to operate to perfection or the entire mission would be lost. However, both Spirit and Opportunity worked to perfection, arriving safely at their respective destinations with no problems—an unprecedented achievement in planetary exploration. Despite the fact that Pathfinder had landed in similar fashion, the larger physical size and sophistication of the Mars Exploration Rovers had brought new challenges that were all accurately managed and overcome. Also unprecedented was the live broadcast of the EDL phase on many of the world's TV networks, where, for the first time, the public at large could share in the tension during entry and descent and the elation of the successful landings.

Before driving safely on to the Martian surface, a choreographed sequence of steps were first required to prepare each rover, to be conducted over a several day period. First, the Pancam Mast Array and Hi-Gain Antenna were raised into position and activated. The Pancam then took some panoramic images to characterize the immediate vicinity for the first time. Subsequently, the rover raised itself up onto its six wheels and a calibration of all science instruments was conducted. Finally, a suitable egress path from the landing deck was chosen, whereupon the rover drove onto the Martian surface.

With each MER expected to travel far from its landing site, surface operations have presented a unique and unprecedented opportunity (and challenge) to Mission Control at JPL, heralding a new mode of space exploration never before engaged. The challenge is unique, because of the time lag between Earth and Mars, combined with the interactive nature of the expedition, making instantaneous interaction with the rover or with its environment impossible. Hence, whenever the rover is driving across the surface or conducting field investigations, it is doing so completely on its own. As a result, scientific investigations of the surface need to be intimately connected to navigation of the rover and managed within a well-defined program of operations maintained on a daily basis from Earth. Only in this way could JPL hope to ensure steady scientific progress while also maintaining the safety of the rover. A maximum traveling distance of about 100 meters per day was set to occur over a four-hour period around noon. With sophisticated onboard hazard avoidance and navigation capability, each rover can literally drive itself across the Martian surface to a predetermined destination and avoid obstacles *en route.*

Each rover begins its day after a night's sleep by being awoken by an alarm call. Instructions uploaded the previous day from mission control configure the rover adequately to carry out its operations for the day without any further communication from Earth. Instructions are sent the previous day simply because it cannot be done overnight when the rover must enter *safe mode* to maintain its systems during the freezing Martian nights. The rover proceeds on its own to the instructed destination and carries out the required science investigation. In the afternoon it transmits its data back to Earth, where both mission scientists and operation engineers perform a quick analysis of the data and determine the next sol's work for the rover. The instructions, which include navigation, destination, and science operations, are transmitted to the rover in late afternoon (on Mars), after which the rover goes into safe mode for the night, but is ready to carry out the received instructions the following morning and early afternoon.

This highly orchestrated sequence of events has been in operation for several years to date and has literally provided NASA and JPL with a new methodology of planetary exploration barely envisaged a few years ago. With the unprecedented daily cooperation of mission scientists and operation engineers, award winning rover control software (called Maestro, which can be downloaded by the public) and the autonomy and intelligence of the rovers themselves, a new era of far-reaching scientific exploration on other planets using robots has by now begun.

Spirit's Journey

Spirit landed on January 4, 2004, in a gigantic crater 150 kilometers wide called Gusev, situated just south of the equator at 15°S, 185°W. With

Ma'adim Vallis cutting into the crater from the southeast, there is a good chance that Gusev contained a lake at one time in its distant past. The hope was that Spirit would find evidence of that lake by identifying sedimentary layering within the rocks, or perhaps mineralogical evidence of past water activity. It was also clear to mission scientists, however, that lava flows in the region could well have subsequently covered over any evidence, so there were no guarantees.

Subsequent to landing and conducting system checks and instrument calibrations, Spirit's egress eventually took place on Sol 12. The rover drove off its landing deck to properly begin its 90-sol primary mission to look for evidence of past water and to characterize the geology and mineralogy of the region. Pancam's first panoramic images revealed an extremely flat terrain that was much smoother than the Viking or Pathfinder sites. Spotted some 300 meters to the northwest was a crater called Bonneville; while a set of hills rising over 100 meters above the plain could be seen to the southeast, over 2.5 kilometers away. After only three sols of investigating the landing site, it became clear that the immediate vicinity was not, as was hoped, dominated by past water activity. Rather, all of the rocks examined were seen to be volcanic basalt in nature, indicating past volcanic and impact activity only. But this was early in the mission and too soon to draw conclusions. For Spirit's first trek, it was decided to send it to Bonneville Crater to the northwest to see if it might reveal more about past activity in the region. Mission scientists decided on a combined strategy of conducting both rapid stop-and-go as well as some longer detailed analyses of any interesting features encountered *en route*, according to the situation. Two rocks, named Humphrey and Mazatzal, were examined in detail and at last revealed some signs of what we were looking for—alteration of the rocks due to past water activity. Humphrey for example, was grinded by Spirit's Rock Abrasion Tool (RAT) and examined by its microscope to reveal a fine-grained structure indicative of water alteration. However, while some evidence of past water activity had been detected, the extent of that activity did not point to a region once dominated by water. There had been water there, but to what extent and under what conditions remained unclear.

After a journey lasting 68 sols, Spirit arrived at Bonneville, and spent a further nine sols analyzing rocks and soils, but the journey had been in vain. As no further evidence of past water activity could be seen, it was decided on Sol 77 to turn Spirit round and attempt a crossing of the plain to the hills nearly 3 kilometers to the southeast. If past water activity had occurred and was subsequently covered by volcanic and impact deposits, then perhaps the hills rising above the plain had avoided the worst of the deposition and retained some evidence of water activity. By Sol 90 and *en route* across the

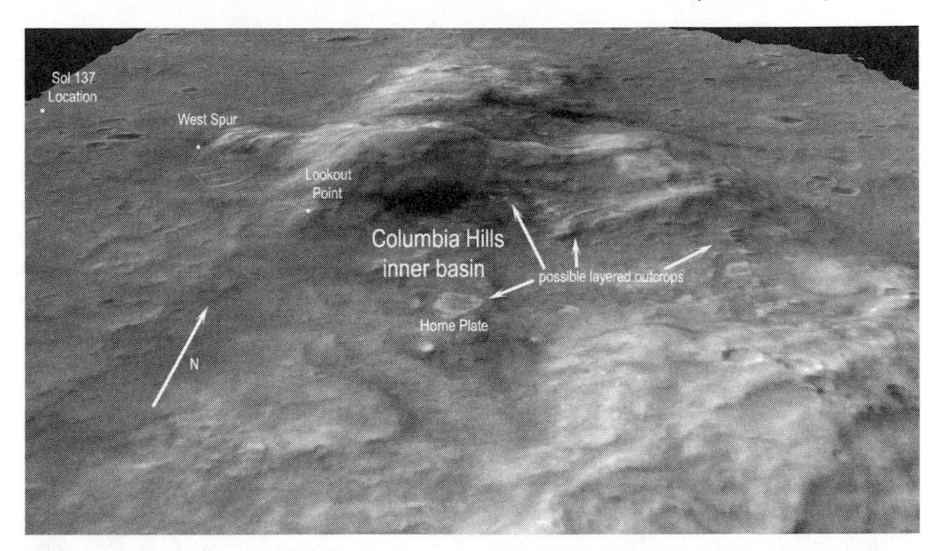

Figure 76: View of Columbia Hills in Gusev Crater from MGS/MOC, showing Spirit's journey toward West Spur by Sol 146 and to Larry's Lookout by Sol 165. [Credit: NASA/JPL/MSSS/USGS]

Figure 77: Mosaic image acquired by MER Spirit having left Bonneville Crater, approximately 600 meters from the base of Columbia Hills. See Plate 43a in the color section. [Credit: NASA/JPL/Cornell]

plain, Spirit's primary mission formally ended, but the rover was in exquisite health and showed no significant signs of wear and tear. It had traveled a total of 630 meters, greatly exceeding expectations. With plenty of life left, NASA initiated Spirit's first extended mission, instructing the rover to continue across the plain toward the hills, now named Columbia Hills (Figure 76) in honor of Space Shuttle Columbia that was tragically lost on re-entry to Earth's atmosphere on February 1, 2003, with the loss of the entire crew.

As Spirit headed toward Columbia Hills, it set another first in space exploration by completely traversing the ejecta-field of Bonneville Crater (Figure 77). Finally, on June 9, 2004 (Sol 152), it arrived at the foot of the hills where it quickly became apparent that the hunch of the MER scientists had been correct. The terrain of Columbia Hills looked significantly different to that of the surrounding plain, showing signs of water alteration (Figure 78). One of the first rocks to be investigated, called Pot of Gold, revealed the

Figure 78: MER Spirit image of a rock titled "Tetl" that provides evidence of past aqueous sedimentary activity in the region. [Credit: NASA/JPL/Cornell]

presence of hematite—which was a strong indication of the existence of water at one time (Figures 79 and 80). Other rocks seemed to have rotted, while others contained thin mineral *rinds*, again all constituting signs of past water activity. From June until late August 2004, Spirit proceeded up Columbia Hills toward a point called West Spur; and with every investigation indicating significant water activity billions of years ago affecting all of Columbia Hills, it became clear that Spirit had indeed landed in a crater that was once significantly affected by water and subsequently filled in by violent volcanic deposition. Around this time, Spirit experienced its first technical problem when its front right wheel stopped working. To continue, the rover had to be turned around and driven backwards up the hills while dragging its front wheel. The rover was otherwise in excellent health and when, in

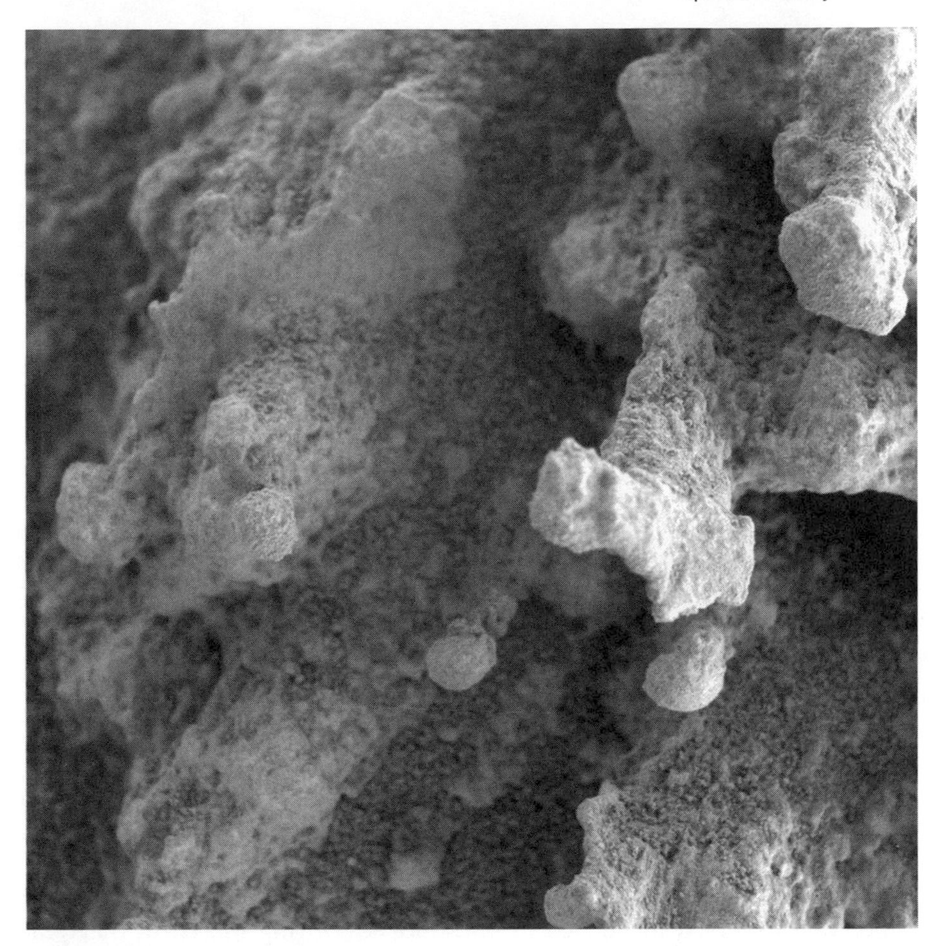

Figure 79: Close-up image taken by Spirit of a rock called "Pot of Gold." Nuggets appear to stand on the end of stalk-like features containing the mineral hematite, which can form in the presence of water. [Credit: NASA/JPL/Cornell/USGS]

October 2004, it completed its first extended mission, NASA promptly announced a second. The journey through the hills continued—backwards—but with significant discoveries being made at almost every stop and investigation. One of the most significant finds occurred in November 2004, where the mineral goethite was detected in a rock called Clovis, providing compelling evidence of past water activity.

On January 4, 2005, Spirit celebrated one year on Mars. Incredibly, it had lasted over four times longer than originally planned, had traveled 3.79 kilometers and was showing little signs of wear (apart from its front wheel). During February 2005, Spirit continued westward and upward toward a point called Larry's Lookout, where it was able to look north into a neighboring

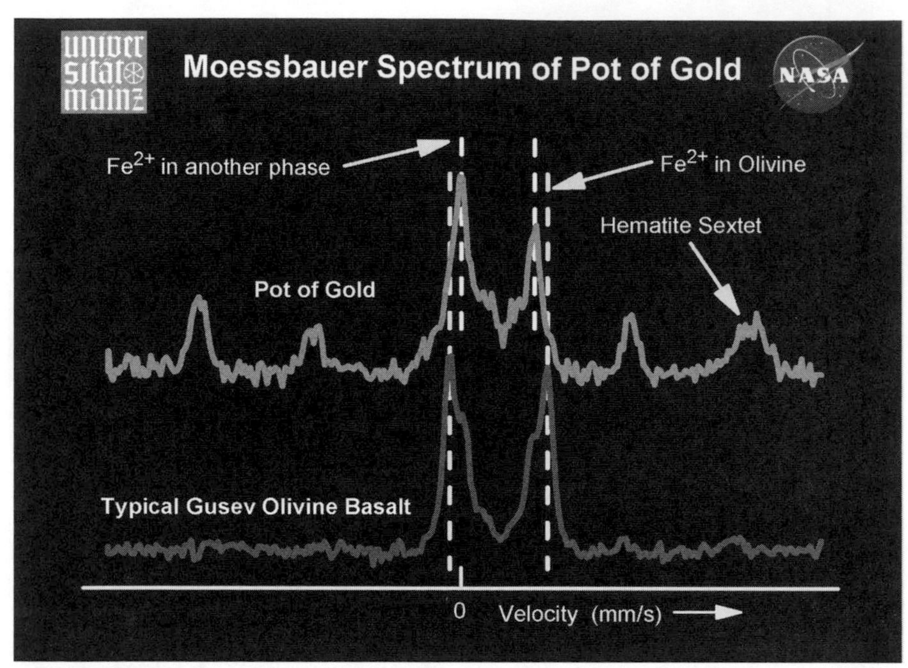

Figure 80: Graphical evidence for the mineral hematite in "Pot of Gold." Image. [Credit: NASA/JPL/Cornell/University of Mainz]

valley to see if there were any features of interest. Yet further discoveries around this time provided more evidence of past water activity. A rock named Peace showed precipitated magnesium sulfate salt within. Dust in the area revealed iron sulfate salts with water molecules bound to them, while other rocks seemed to be rich in phosphorus. During March of that year Spirit had incredibly good fortune. First, dust that had previously gathered on the solar panels and reduced the power output to about 400 watts was blown off by high winds, giving the rover a welcome boost to about 700 watts. Second, the faulty front right wheel started to function once again, giving Spirit back its full mobility! Having finished its second extended mission and received a clean bill of health, NASA announced in April 2005 yet another extended mission, but this time for no less than another 18 months, predicting that the rover could last until late 2006. Spirit continued on and upward toward the summit of one of the hills, now named Husband Hill (Figure 81). All seven peaks of the Columbia Hills were named in honor of the seven Columbia crew, with Husband Hill named in honor of the Columbia commander Rick D. Husband. Spirit reached the summit of Husband Hill on August 21, 2005, having traveled a total distance of 4.287 kilometers. While there, the rover received yet another boost when another gush of wind blew the remaining dust form the solar panels, restoring full power at 965 watts—

Figure 81: MER Spirit panoramic view of the impact feature called East Basin to the northeast of Husband Hill, taken on Sol 653 (November 3, 2005).The rim of "Thira" Crater is visible on the horizon, 15 kilometers away. See Plate 43b in the color section. [Credit: NASA/JPL-Caltech/Cornell]

just like new! On looking back down the slope it had ascended, MER scientists could eventually decipher the terrain—Columbia Hills are indeed made of layered rocks, but parallel to the incline of the slope, which had made it near-impossible to identify on the way up.

On November 20, 2005, Spirit celebrated one Martian year (or 669 sols) in spectacular style by photographing a natural fireworks display in the form of a meteor shower caused by dust from the tail of Halley's Comet burning up in the atmosphere. A decision was taken from the summit of Husband Hill to send Spirit down toward a point called Home Plate. But Spirit's front wheel became problematic once again, finally failing completely on Sol 779 in March 2006. With reduced mobility and the prospect of the oncoming Martian winter, it became a priority to move and park Spirit on the north-facing slope of McCool Hill (named in honor of Columbia pilot William C. McCool) to conduct a "winter campaign" of investigations at just one site (Figures 82 and 83). Because Spirit is on Mars' southern hemisphere, the north-facing slope would point Spirit toward the Sun to the north to maximize the solar power needed to survive the harsh Martian winter. And although Spirit would remain stationary for much of the remainder of 2006, MER scientists could avail of a long-awaited opportunity to conduct long-term and detailed surveys at just one location, yielding different type of scientific data to that achieved when constantly on the move. The MER team devised innovative ways of carrying out research that was not originally envisaged. Although the metal teeth on Spirit's RAT were completely worn and could no longer grind into rocks, its wire cleaning brush was still functioning well. Hence, the rotating wire brush could be programmed to remove soil in a layered fashion, revealing the structure of the near subsurface, layer upon layer.

Having endured the harsh Martian winter well, Spirit celebrated its one thousandth sol (1,026 Earth days) on Mars on October 25, 2006, and with the onset of spring was ready once again to resume mobile exploration activities.

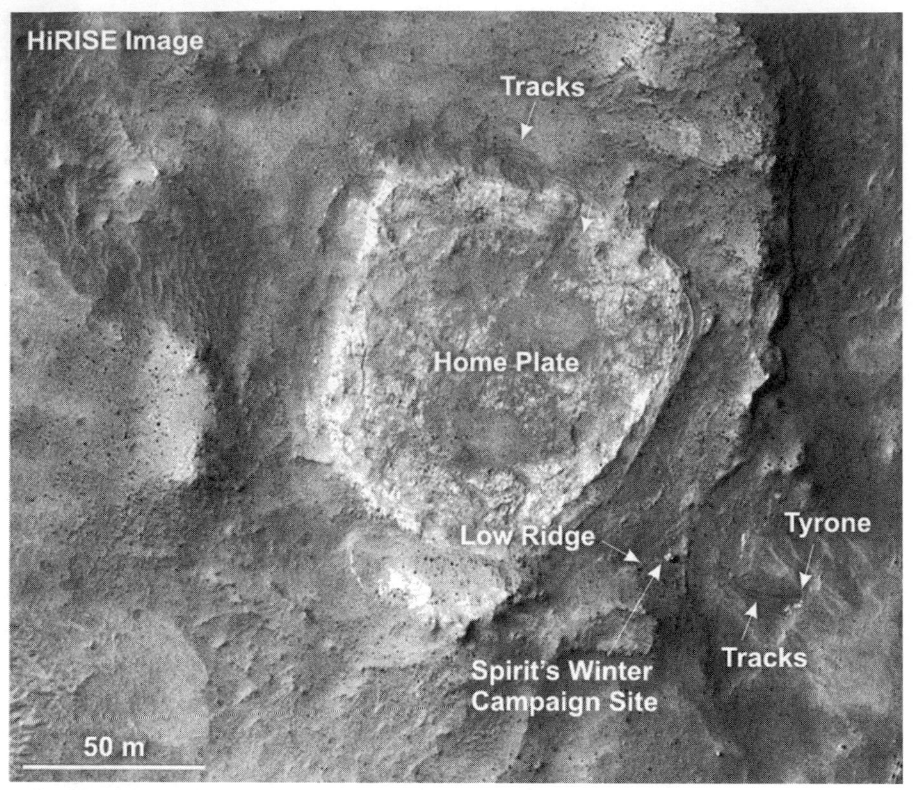

Figure 82: MRO/HiRISE image of Home Plate in Columbia Hills. The MER Spirit rover is visible at its winter campaign site. Image. [Credit:N ASA/JPL-Caltech]

Figure 83: On February 19, 2006, Sol 758, Spirit took this panoramic view of the interior of Home Plate, a circular topographic feature within Columbia Hills with exposed layered rocks. [Credit: NASA/JPL-Caltech/USGS/Cornell]

Furthermore, NASA had by then developed and uploaded sophisticated new software to both rovers, improving their capabilities to what they called "thinking spacecraft." From here on, each rover would be more autonomous than ever, capable of recognizing many Martian surface features (including potentially hazardous "dust devils"), track targets while moving, avoid significant hazards including backing out of cul-de-sacs, and implement "go and touch" maneuvers where the rover could think several steps ahead, approach a target, and gather data with the appropriate instruments, all without instruction from Earth. This new capability, coupled to radically improved orbital support from the newly arrived Mars Reconnaissance Orbiter, has brought about yet again another significant advance in rover-based exploration.

Resuming its course toward Home Plate, Spirit, still dragging its faulty front wheel, uncovered perhaps the most compelling evidence to date of past water activity in Gusev Crater, by inadvertently scooping out a shallow trench from the surface soil and uncovering power-like silicate salts of more than 90% purity. Such concentrations most likely arose from past water activity such as from outgassing of volcanic steam or from a hot spring (Figure 84).

Spirit, in exploring the surface Mars for many years, has far exceeded all expectations and has been a resounding success (Figure 85). Yet, despite such an enigmatic journey, it has taken virtually all of that time to barely begin to understand the past history of Gusev Crater as that of an ancient impact site apparently influenced by water activity.

Opportunity's Journey

Opportunity, the second Mars Exploration Rover, arrived on Mars 20 days after Spirit, on January 24, 2004. It landed on Meridiani Planum, a vast plain just south of the equator on the far side of the planet to Gusev Crater. Opportunity was sent there because of mineralogical evidence from both MGS and Odyssey showing the presence of hematite across the plain, suggesting that there had been a large standing body of water at that location long ago. This was the first time ever that a space probe was sent to investigate direct chemical and mineralogical evidence of past water activity on another planet—a milestone in planetary exploration.

Having being extremely careful and taking 12 sols to allow Spirit to drive onto the surface, MER scientists felt confident enough to conduct Opportunity's egress to the surface after only 7 sols. It became immediately clear that something extraordinary had occurred—Opportunity had landed inside a 20-meter-wide crater (named Eagle Crater). More startling was that the subsurface bedrock of the surrounding plain was clearly exposed on the side walls of the crater. The rover had truly been appropriately named, as here was one of the most fantastic opportunities ever presented to science exploration—a chance to study the subsurface of Mars with no extra effort

Figure 84: Spirit inadvertently exposes bright, silica-rich clay while driving toward Home Plate, showing what may have been at one time an ancient hydrothermal setting. [Credit: NASA/JPL/Cornell]

by the rover whatsoever. A mission specifically designed to drill underground is a future objective, yet here was an unprecedented opportunity to study the subsurface of Mars!

This was nothing, however, compared to the findings that Opportunity transmitted back to Earth upon analyzing the exposed sides and floor of the crater. First, the walls were seen to be composed of sedimentary layered rock, laid down by a standing body of water; and with multiple layers, it looked as if the region had undergone multiple episodes of water inundation over long periods of time in the distant past. Also clearly seen, spread out across the floor and the sides of the crater, were small blue spheres of hematite, while a chemical analysis of the rocks revealed numerous precipitated salts that

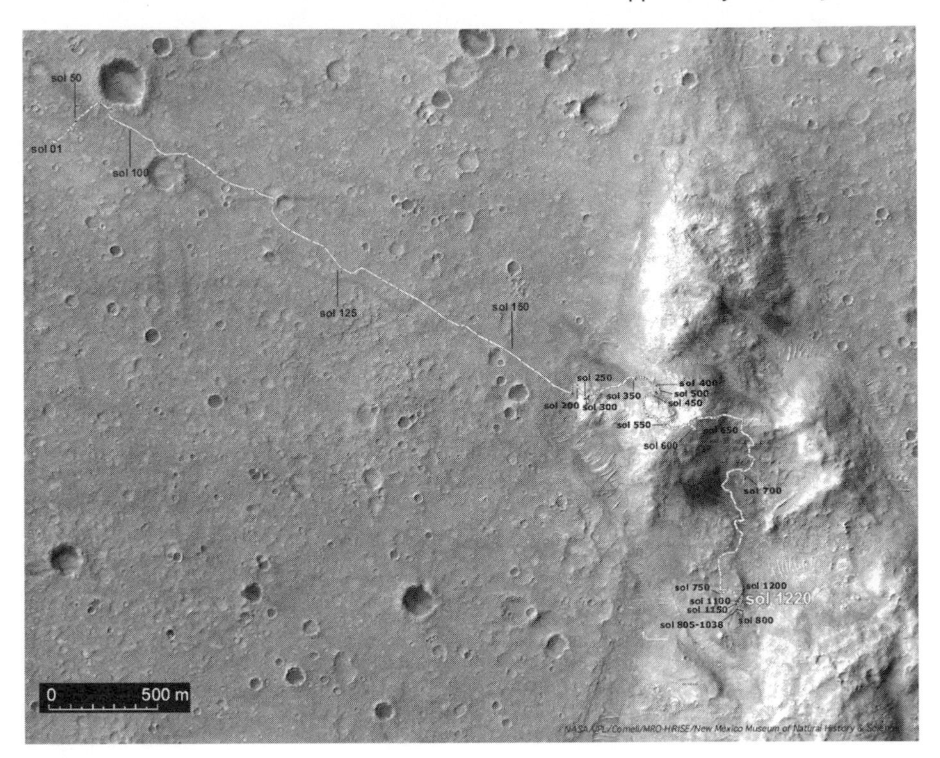

Figure 85: MGS/MOC Map of Spirit's long journey as of Sol 1220 (May 2007) having traveled 4,276 meters. [Credit: NASA/JPL/MSSS/NMMNH]

could only have been produced and then laid down by water: gypsum, bassanite, jarosite, epsomite, kieserite, glauberite, magnetite, goethite as well as magnesium sulfate, calcium sulfate, and numerous oxides and hydroxides (Figures 86–89).

It had taken Spirit over 150 sols to conclusively verify the influence of water at Gusev; but Opportunity had hit the bull's eye on day one. Not only did it unequivocally verify the presence of hematite in the region, but it landed within a crater whose walls bore exposed bedrock with sedimentary layers and precipitated salts, constituting dramatic evidence that the region had been repeatedly saturated by water in its distant past. Without moving from its landing site, Opportunity had fulfilled its purpose spectacularly. And despite the fact that the evidence pointed to shallow and slow-moving water, and to water in the region with an "activity" (a measure of the amount of water in a substance and a good indicator of its suitability to life-related activity) too low to have ever supported life; the discovery has none the less radically affected our perception of the planet as a whole. Perhaps life may not have arisen at that specific location, but conclusive evidence of liquid

Figure 86: ''Shoemaker's Patio,'' near Opportunity's landing site, showing finely layered sediments. Spherical wet sediment grains or concretions (nicknamed ''blueberries'') align with individual layers on the outcrop. See Plate 40a in the color section. [Credit: NASA/JPL/Cornell]

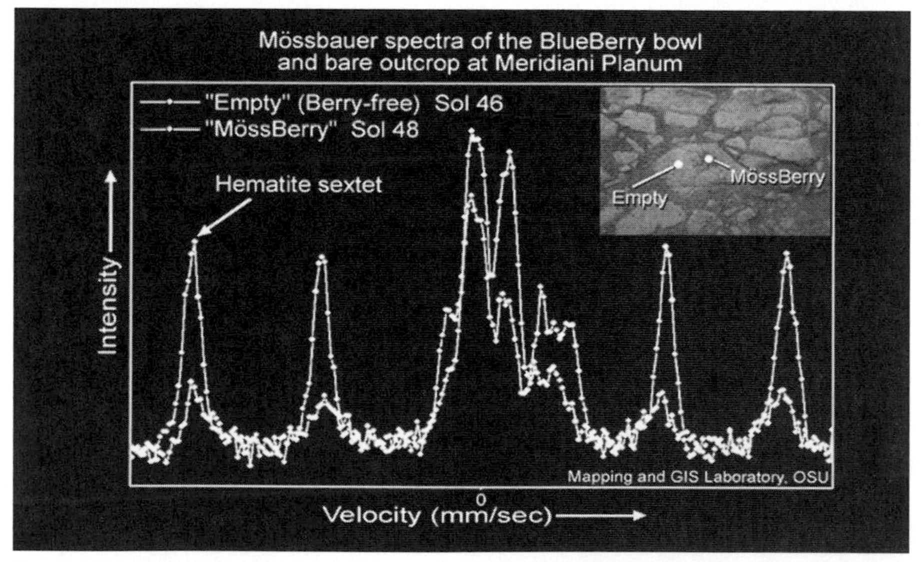

Figure 87: Spectra of an outcrop at Shoemaker's Patio. Two lines are shown: one reveals the presence of hematite in an area called ''Mössberry,'' while the other represents an area called ''Empty'' that is devoid of hematite. See Plate 40b in the color section. [Credit: NASA/JPL/Cornell/University of Mainz]

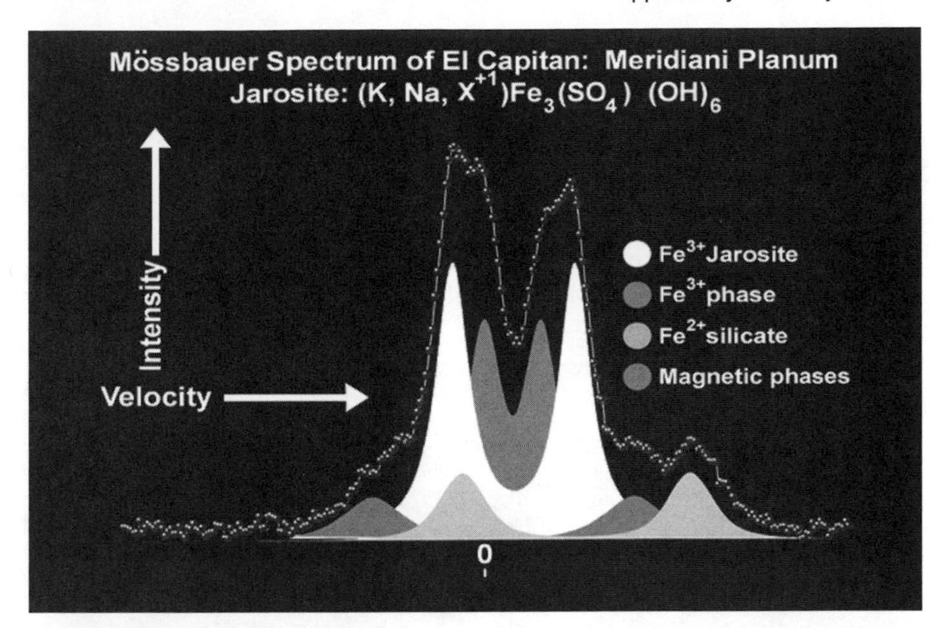

Figure 88: Mössbauer spectrum reveals the iron-bearing mineral jarosite at "El Capitan" in Eagle Crater, suggesting water-driven processes in the region at one time. See Plate 41 in the color section. [Credit: NASA/JPL/Cornell/University of Mainz]

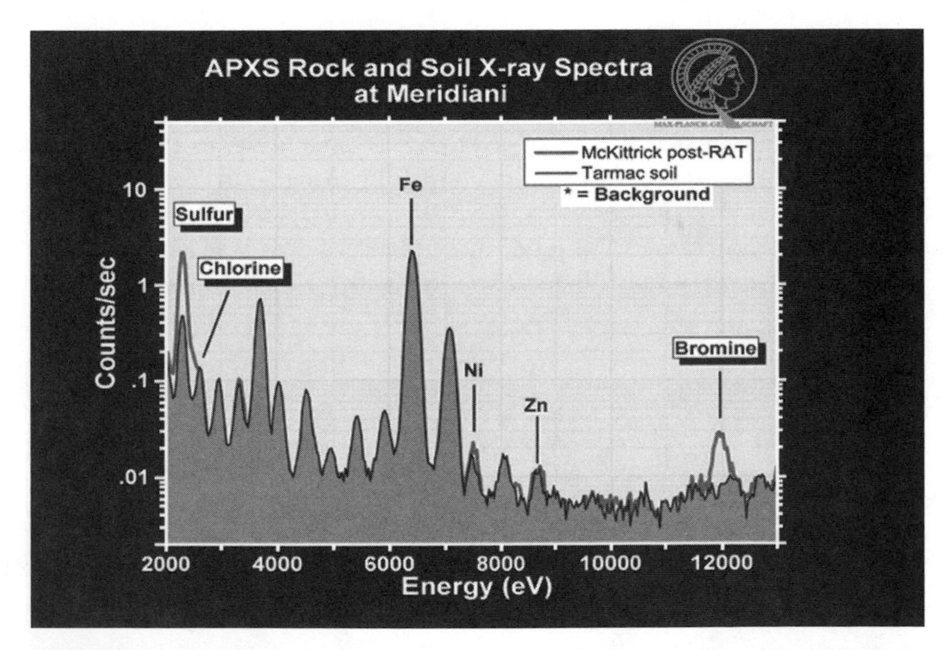

Figure 89: Spectra of high concentrations of sulfur and bromine, arising when watery brine evaporates, in a rock dubbed "McKittrick" near Opportunity's landing site. See Plate 42 in the color section. [Credit: NASA/JPL/Cornell/Max Planck Institute]

water on the surface in its distant past opens up many other possibilities with regard to both surface and underground water and life-related activity.

In total, Opportunity spent 57 sols exploring just Eagle Crater. Such was the magnitude of the find that it warranted that scale of effort. Several outcrops along the crater walls were given names: El Capitan, Guadalupe, and Last Chance, all comprising sedimentary layers and revealing the presence of oxides, sulfates, and other precipitated salts (Figure 90). Eventually, Opportunity drove up one of the shallow side slopes of the crater

Figure 90: The Mars Exploration Rover Opportunity's view of its landing-site—Eagle Crater—on Sol 56. [Credit: NASA/JPL]

and onto the incredibly flat Meridiani Planum, finding it to be covered from horizon to horizon in tiny blue spheres of hematite (labeled "blueberries" by the Opportunity team). It was decided to send Opportunity to a much larger crater called Endurance, which would hopefully provide further evidence of past water activity (Figure 91). On Sol 90 (April 26, 2004) Opportunity's primary mission came to an end, but because it was well on its way to Endurance Crater and still operating perfectly, NASA initiated its first

Figure 91: The heat shield impact site of MER Opportunity, imaged on Sol 330 (December 28, 2004). On the left the main heat shield piece is inverted and reveals its metallic insulation layer, glinting in the sunlight. [Credit: NASA/JPL/Cornell]

Figure 92: Endurance Crater and the surrounding plains of Meridiani Planum as seen by Opportunity's panoramic camera. See Plate 43c in the color section. [Credit: NASA/JPL/Cornell]

extended missions for a further 90 sols. On Sol 117 (June 4) Opportunity arrived at Endurance, stopping at the rim to reveal a spectacular view of a deep, stadium-sized (130 meter) crater (Figure 92). Once again, a bounty of evidence immediately pointed to significant past water activity. And because Endurance is much larger and deeper than Eagle Crater, its side walls revealed no less than three distinct sets of sedimentary rocks. With careful analysis of each set. Opportunity could conduct a stratigrapic analysis of the region, providing unprecedented insight into the long-term history of the region with far-reaching implications for the planet in general.

Opportunity spent six months (June to December 2004) conducting a thorough investigation of Endurance Crater. Not only did it find similar widespread evidence of long-term saturation, but it found that the deeper into the crater it went, the more water alteration had occurred. While layers of rock at the same level in both Eagle and Endurance were very similar—suggesting that they were formed at the same time during a region-wide wet period—deeper rocks and layers were different, with greater concentrations of chlorine and less magnesium and sulfur, for example (Figure 93). All clues pointed to a second period of saturation subsequent to the impact that created Endurance. By early 2005—with similar chemistry at similar depths in Eagle and Endurance, and different chemistry at greater depths in Endurance and the widespread presence of hematite on the surface of the entire region—Opportunity had accumulated enough evidence to suggest

Figure 93: Exposed layered sedimentary rocks at a location within Endurance Crater named "Burns Cliff," revealing repeated aqueous activity in the region billions of years ago. [Credit: NASA/JPL/Cornell]

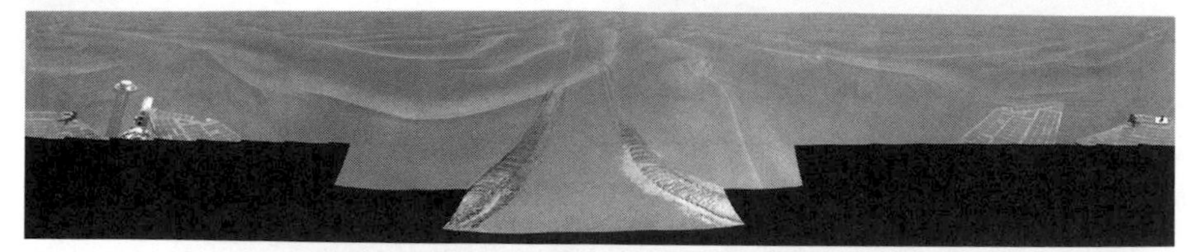

Figure 94: Panoramic image taken by Opportunity of a region dubbed "Rub al Khali," 2 kilometers south of Endurance Crater. The rover became stuck in one dune titled "Purgatory Dune" for more than a month. [Credit: NASA/JPL/Cornell]

that a shallow body of standing water had covered tens of thousands of square kilometers across Meridiani Planum for at least hundreds of thousands of years—a finding subsequently verified by Mars Express. Meridiani Planum was once a vast shallow sea, and Mars once supported surface water, at least regionally, over significant periods of time.

As with Spirit, Opportunity continued to function so well that in 2005 it, too, was given an 18-month extension until the end of 2006. The ambitious decision was taken to send Opportunity to an even larger crater called Victoria, over 4 kilometers south of Endurance. The route was certainly not without incident or excitement. On March 20, 2005, Opportunity set a new distance record by traveling 270 meters during a single sol. As it traversed some incredibly smooth terrain utterly devoid of loose rocks and covered by very shallow sand dunes (about 20–30 centimeters high), Opportunity suddenly became stuck. For over five weeks, from April 26 to June 6, the rover struggled to maneuver itself free, sometimes managing less than a few centimeters per sol. On Opportunity's escape, MER scientists appropriately named that particular sand trap *Purgatory Dune* (Figure 94)! They used the rover's mini-TES to build a thermal image of the region and noticed that the sand within which Opportunity had become lodged was slightly cooler than the surrounding terrain, providing a vital clue on how to avoid similar hazards in the future. Opportunity also encountered some intriguing terrain, rock outcrops and bedrock *en route*. In November 2005 it encountered some rocks that suggest cyclical variations of the environment in the region. Also between December 2005 and April 2006 it had identified some of the most important items of evidence of past water activity in the region (Figure 95). In December 2005, for example, Opportunity had to make a service stop at a rocky outcrop called Olympia, where MER engineers needed to attempt a remote repair of the rover's failing robotic arm. Having outlived its expected age by over six-fold, the joints on the rover's robotic arm were beginning to stick, with the arm not stowing away in the required position for travel. Having stopped, however, MER scientists noticed a feature called *festoon*

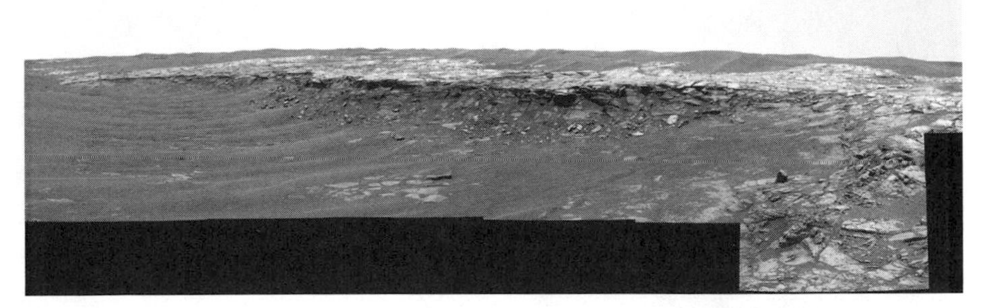

Figure 95: MER Opportunity panorama of the Payson outcrop on the edge of Erebus Crater on Sol 744 (February 26, 2006). Layered rocks are observed in the crater wall, while to the left a thin layer of spherule-rich soils overlies outcrop materials. See Plate 43d in the color section. [Credit: NASA/JPL-Caltech/USGS/Cornell]

cross-bedding within the rocks of Olympia. Such small-scale cross-bedding reveals itself as small curved (smiling) ripples within the rock, and can only occur in the presence of surface running water. Here, finally, was unequivocal evidence of *surface* liquid having once flowed on Mars. Such a discovery would have far-reaching implications regarding the planetary environment during that period of water activity. Such was the importance of the find that Opportunity took more photographs of Olympia than of any other location on its journey. Thankfully, MER engineers managed a partial solution to the problem of the robotic arm, and Opportunity could safely continue on its journey. Further along the route, a small crater called Erebus provided even better examples of festoon cross-bedding as well as pavement type rocks formed from water percolating through cracks from regolith above (Figure 96).

Opportunity's Recent Progress

After four months of investigations at Erebus, Opportunity was directed in March 2006 to complete the final 2-kilometer journey to Victoria Crater. Although winter for the southern hemisphere was due, such was Opportunity's proximity to the equator that it did not need to stop and bank on a north-facing slope, and could, instead, continue with its journey. Within about a kilometer and in sight of Victoria Crater however, Opportunity encountered a vast field of sand dunes not unlike Purgatory Dune. Thankfully, however, with the experience gained in escaping Purgatory Dune, MER scientists were able to plot a safe "zigzag" path toward Victoria, with Opportunity finally arriving at the rim of the 750-meter crater in late September 2006 at a point named "Duck Bay." At precisely the same moment, the enigmatic Mars Reconnaissance Orbiter was

Figure 96: Exquisite example of centimeter-sized layering called ''cross-lamination''—nested sets of concave layers in sedimentary rocks that are remnants of tiny underwater sand dunes formed waves in shallow water on the surface of ancient Mars. Opportunity acquired this image of a rock called ''Overgaard'' at the edge of Erebus Crater on Sol 690 (January 2, 2006). [Credit: NASA/JPL-Caltech/Cornell]

settling into orbit and was by then able to use its HiRISE camera to take exquisitely detailed images of Opportunity arriving at the rim of the crater (Figures 97, 98, and 99). Orbital and surface reconnaissance were finally brought together, for evermore providing the possibility of seamless analysis of the planet from the planetary through to microscopic scales. MRO could for evermore help Opportunity (and Spirit) to navigate more carefully and successfully across the Martian surface, while the rovers would provide vital "ground truth" information to MRO, allowing it to analyze the entire Martian surface more precisely.

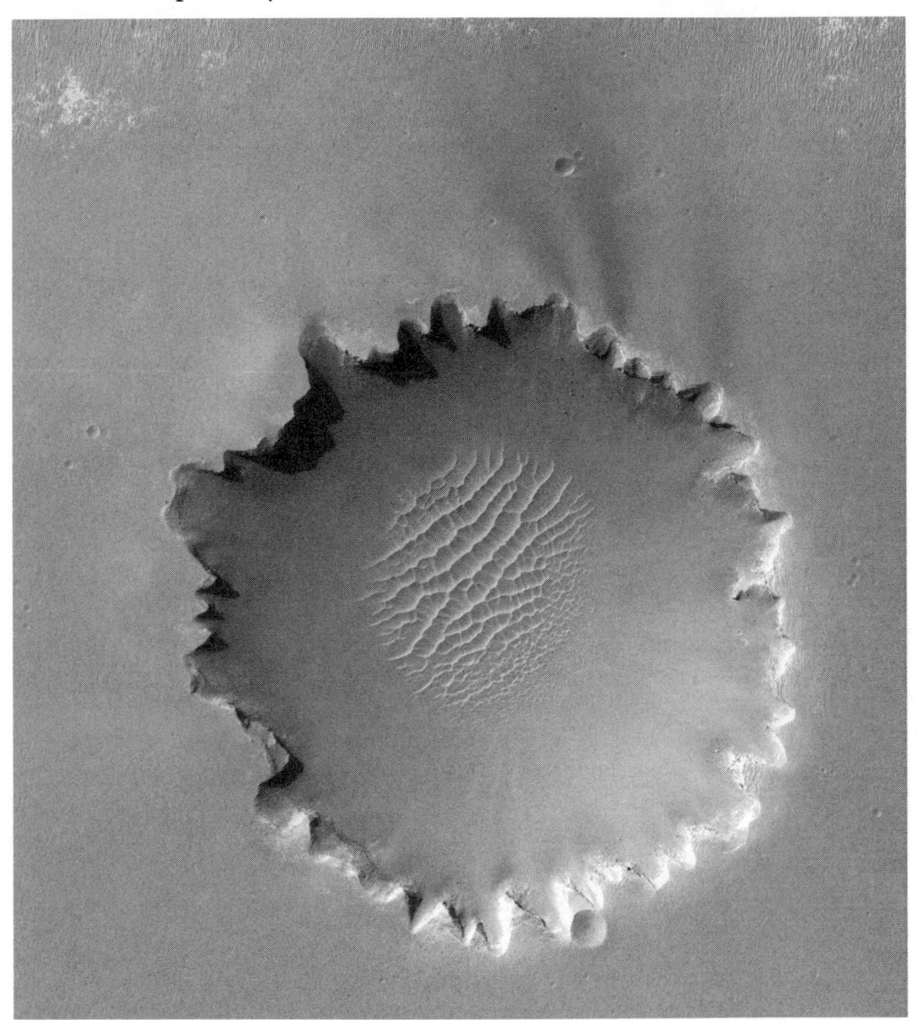

Figure 97: MRO/HiRISE image of Victoria Crater, taken as Opportunity arrived. Such images bridge the gap between orbital and lander reconnaissance. See Plate 44 in the color section. [Credit: NASA/JPL-Caltech/Univ. of Arizona]

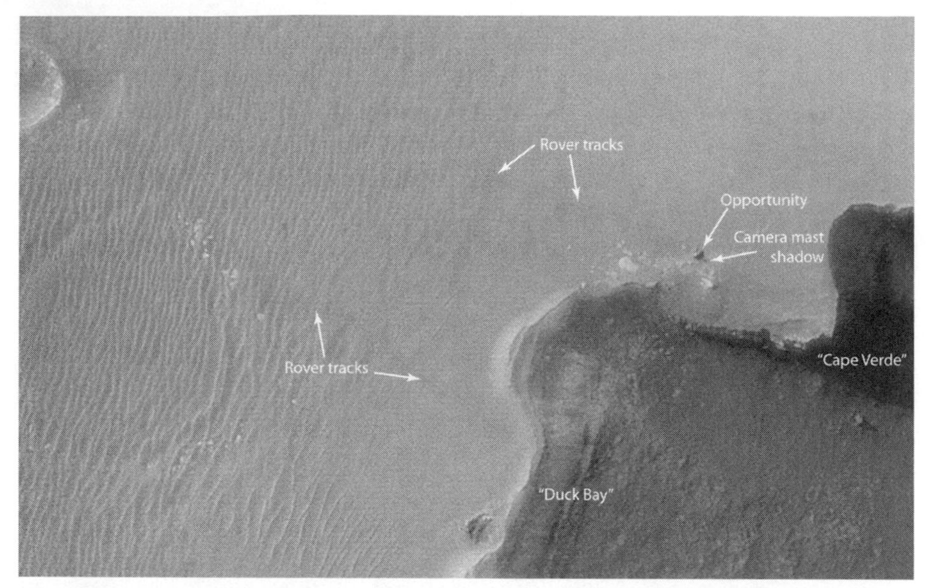

Figure 98: MRO/HiRISE image of Victoria Crater, showing the arrival of MER-Opportunity. See Plate 45a in the color section. [Credit: NASA/JPL-Caltech/Univ. of Arizona]

Figure 99: Opportunity image taken at the rim of Victoria Crater, with a simulated image of the rover superimposed. The walls of Victoria Crater reveal approximately 15 meters of exposed sedimentary layers for Opportunity to investigate. See Plate 45b in the color section. [Credit: NASA/JPL-Caltech/Univ. of Arizona]

Over the following eight months Opportunity navigated the rim of Victoria Crater in a clockwise direction, and in February 2007 had traversed 10 kilometers across the Martian surface (Figure 100). In May 2007, it turned back to Duck Bay having revealed Victoria Crater to be of significant scientific interest. First, the crater bares approximately 15 meters of layered rocks—far deeper than in either Eagle or Endurance—and could therefore reveal details of Mars over a greater period of its past and further back in history. Indeed the initial findings reveal that deeper layers are less acidic than those found within Endurance, suggesting that the area was less hostile to life-related activity during wet periods earlier in its history.

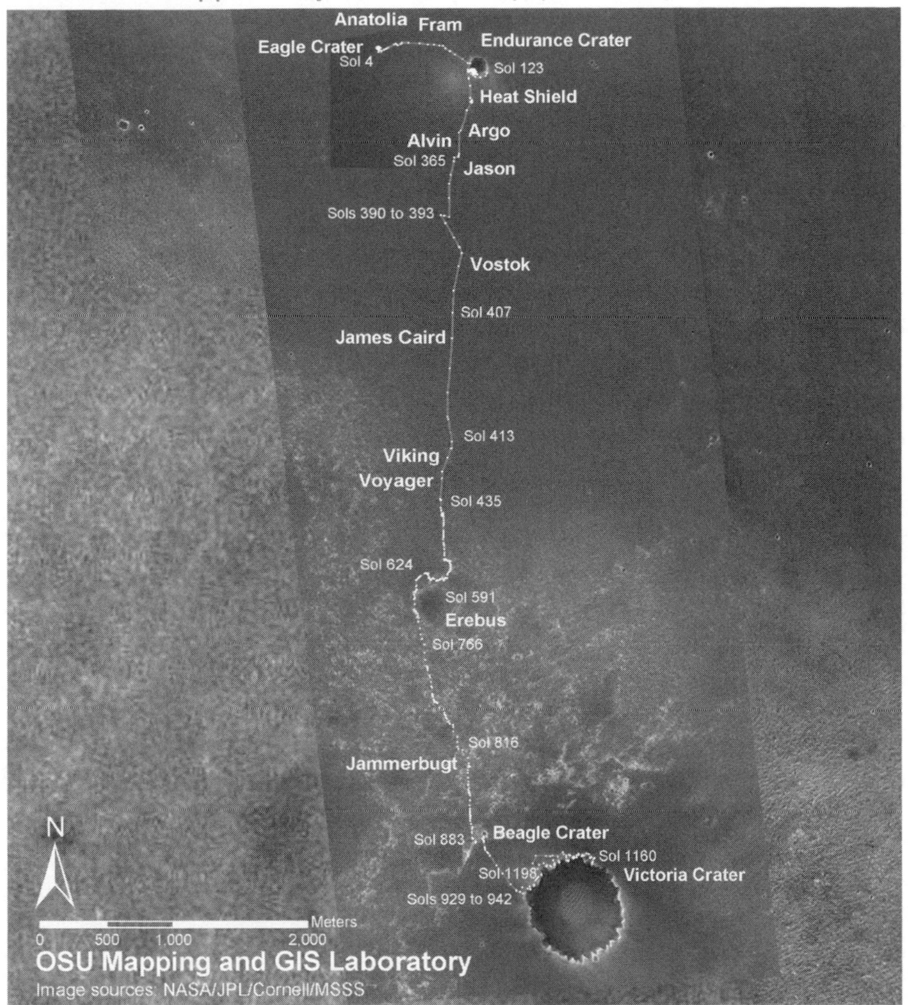

Figure 100: This image from MGS shows the route that Opportunity has taken from its landing site inside Eagle Crater to its position on Sol 1198 at the rim of Victoria Crater, having traveled more than 10 kilometers across the surface. [Credit: NASA/JPL-Caltech/MSSS/OSU]

On rare occasions, a space exploration mission will greatly surpass our hopes and expectations. Sputnik 1, Apollos 11 and 13, and Voyagers 1 and 2 are examples of historic space missions whose impacts still resonate today; and now Spirit and Opportunity join that illustrious short list of human endeavors that define the essence of space exploration. The Mars Exploration Program is a pinnacle of human achievement.

Spirit and Opportunity embody everything that their names mean to us.

Both rovers are now aging and at some stage they will cease to operate. Irrespective of how they finish their active duty on Mars, their mission will not have come to an end. Both rovers performed so far beyond expectations that we will spend years analyzing the data they have presented. Astoundingly, they have provided unequivocal evidence of past surface water activity on Mars. This confirmation has been the aspiration of every Mars and planetary scientist for 200 years, and it has been delivered by Spirit and Opportunity.

The Mars Exploration Rovers have provided so much data that it will take us many years to analyze, absorb, and understand. Furthermore, what we have learned about doing planetary exploration is significant. For several years, hundreds of scientists and engineers have spent their lives focused on Mars. They have had to live their daily routines in part set by the calendar of Mars. They have had to live with daily logistical, technical, and scientific issues associated with long-term and long-range robotic exploration on another planet. Problems we thought might not be encountered until well into the next decade have now been met and overcome. Such exploration is now routine, and such experience is priceless. JPL have learned so much about how to design, build, and operate better rovers, and are so prepared to accomplish similar or even greater feats in the future that Mars exploration will never be the same. Generation after generation has wished for a time when humanity could pursue uncompromising space exploration, and with regard to robotic exploration of Mars that time is now. Through Spirit and Opportunity the planet Mars—and its exploration—will in future years be vastly more engaging.

A Paradigm Shift 16

At last, the nature and history of Mars is beginning to reveal itself. The current era of exploration is providing such depth of insight into the planet that it has literally become a new place to us. It is no longer a distant and inaccessible place. Rather, it has become a real and tangible place, one in which we now have a long-term and vested interest. A paradigm shift has happened regarding our perception and understanding of the Red Planet.

That shift may best be characterized by a new realization of the complexity of the planet's natural history. The enormity of the surface, the immensity of its history and the intricacy of planetary processes occurring across time has conspired to weave a planetary landscape that still largely confounds us. We now realize that no singular or simple answer will reveal the history or nature of Mars. Its story is one of many events at many locations and over billions of years; and only when we retrace all those events will we have a reasonable understanding of the planet. And while that continues to be hugely challenging, we are fortunate to have such a complex place to explore, because what currently perplexes us is neither arbitrary nor abstract—it relates to a world where activity necessary for the origin of life may have actually occurred. It would have been an enormous disappointment to have found a planet that is too easily understood, and where nothing complex ever happened. The hidden nature of Mars pertains to our quest, and the more complex Mars turns out to be, the more we are likely to learn. Finding a planet of such depth and diversity tells us that we are on the right track.

Our efforts of late have not only brought about a new perception of the planet; they have also gone some way toward helping to decipher its natural history. And what we have learned, while far from definitive, confirms that activity occurring in Mars' early history involved water and other volatile materials in environments that were possibly conducive to prebiotic and life-related chemistry. Everything we are currently learning validates the history implied by Viking orbital data and our sense that Mars has much to say about a planetary context for life. Although we have a long way to go, we

K. Nolan, *Mars, A Cosmic Stepping Stone*,
DOI: 10.1007/978-0-387-49981-9_16, © Praxis Publishing, Ltd. 2008

know that the results of our efforts will be important to us. So let us now attempt to bring together what we have learned about Mars, and attempt to determine the type of place it was in its early history, what it is like today and what it might eventually tell us about the planetary context for life.

The Evolution of Mars: An Emerging Picture

The Age of Mars' Surface: A Key to its Evolution

Determining the chronological sequence of indigenous surface activity on Mars is a key to understanding its internal dynamics, the resulting surface environment, and how they evolved. Only when we know the nature, extent, timing, and duration of volcanic and tectonic activity, for example, can we determine the nature of water, volatile behavior, and surface conditions prevailing in the planet's early history. So complex is the geological history of Mars, however, that many years will pass before a comprehensive picture finally emerges. None the less, the quantity and quality of data accumulated from the Mariners to the Mars Reconnaissance Orbiter provide a sufficient resource with which to commence a detailed investigation.

We have determined that five major types of terrain (called *geological units*) dominate the surface: the southern highlands, the northern lowlands; tectonic and volcanic regions such as Tharsis and Elysium; polar and high-latitude layered terrain; and, finally, the multitude of dry water channels found across the globe. Unlike on Earth where the surface has been regenerated many times, Mars has not been active enough to erase all previous surface features, even from its earliest history, and it is now clear that the five geological units relate to distinct periods in Mars' past. Indeed, we can even infer the approximate sequence of events from the nature of the geological units themselves, although a far more intricate chronology is needed to infer the specific environmental conditions prevailing during each era. Currently, the best way to attempt this is by a *crater impact stratigraphic analysis*. We do this by mapping the density and distribution of craters of various sizes across the planet, where regions with high densities of both large and small craters must be the oldest, and those with relatively few craters are younger, having wiped clean previous craters with more recent lava flows or weathering, erosion and deposition processes.

While crater-based stratigraphy is excellent at revealing relative ages, it is less effective in providing absolute ages of surface features because the current technique relies upon lunar crater fluxes, which work excellently for the Moon but not so well for Mars. In the case of the Moon, the age of its surface can be accurately determined through radiometric dating techniques applied to rocks returned to Earth by the Apollo astronauts. Relating those known ages to observed crater patterns provides an accurate *lunar cratering chronology relationship* which can then be extended to all parts of the Moon. When using this technique to date the surface of Mars, however, we must modify that same lunar cratering chronology relationship, taking into consideration differences in the nature and density of impactors, Mars' larger size, its stronger gravity and so on. Unfortunately, those modifications lead to an uncertainty in dating Martian features by as much as a factor of 2.

Interestingly, this uncertainty has little impact on our interpretation of both Mars' earliest and most recent history. We are already confident that the most ancient surfaces on Mars are those saturated by craters, and the uncertainty in determining the age of very recent geological features by a factor of 2 is inconsequential to our understanding of the long-term evolution of the planet. The uncertainty is most critical, however, when trying to date events in the intervening periods of Mars' long history where, for example, an event thought to be one billon years old could actually have occurred two billion years ago. The use of crater-based aging techniques with lunar fluxes as a reference is therefore valuable only to a degree. While ultimately we will resolve the uncertainty through radiometric means, acquiring Martian samples to achieve this is still some way off. As neither Spirit nor Opportunity was equipped to make such measurements, we must await the observations of new landers in the coming decade.

Despite this limitation, cratering impact stratigraphy has been sufficient to reveal that there have been three great geological epochs in Mars' history: the Noachian period, which began with the birth of the planet and ended some time between 3.8 and 3.5 billion years ago; the Hesperian period, which ended some time before 2.9 billion years ago; and the Amazonian period, which has lasted from the end of the Hesperian period to the present day. The Noachian period is broadly identified through the most ancient terrain on the planet such as that found at Noachis Terra in the southern hemisphere. The Hesperian period occurred later and is characterized by vast tracts of lava that overlay older Noachian surface, as found at Hesperia near the equator. Finally, the Amazonian period is best exemplified by the Amazonia Planitia, Tharsis, Elysium, and the polar regions, spanning nearly three billion years from the end of the Hesperian period to the current day. Though only crudely determined, the identification of these three geological ages has provided a valuable foundation upon which we can now investigate the planet's geological evolution more closely, because we now know where on the planet to examine each epoch in finer detail.

The Tectonic and Volcanic Evolution of Mars

The four reconnaissance orbiters recently analyzing Mars have provided significant evidence of tectonic and volcanic activity during the Noachian and Hesperian periods, and even of continued lava flows up to the current geological era.

The oldest remaining surfaces have been shown by TES, THEMIS, and OMEGA to be composed of pristine basalt rock produced by widespread volcanism, and crust formation during its earliest history, that have mostly retained the same composition since that time. Evidence suggests that Mars' first crust would have been largely in place by four billion years ago. Conversely, many of the lava-filled regions from the Hesperian and Amazonian periods are composed of a more advanced form of igneous rock called andesite. Although on Earth andesite is typically associated with plate tectonic subduction and melting, the presence of plate tectonics on Mars has until recently remained elusive. As previously mentioned, as far back as the Vikings we have been aware that tectonic activity occurred on Mars—but not *plate* tectonics. The radial faults emanating from Tharsis, the similar fossae, pits, and grabens south of Elysium, and the gigantic Valles Marineris are just some examples of significant tectonic activity in Mars' past. However, because no trace of tectonic plates is visible on the surface today, we have traditionally regarded Mars' surface, like that of Venus, to have remained stationary throughout history, acting as a stagnant lid that has frustrated the cooling of the planet and inhibited surface activity. However, as already pointed out in Chapter 14, with nine years worth of data available from MGS about Mars' surface magnetic properties, space scientist John E.P. Connerney and his colleagues at NASA's Goddard Center recently produced intriguing new evidence of plate tectonic activity occurring on Mars during the Noachian and early Hesperian periods. Tentatively identified early in MGS's mission, but now seen clearly across the planet, are a multitude of magnetic lineations in the crust which suggest moving tectonic plates and a reversing dynamo in the interior of the planet, billions of years ago. In particular, the lineations are most evident on Noachian and Hesperian terrain and noticeably missing from the younger terrains at Tharsis, Elysium, and the gigantic impact basins such as Hellas in the south and Utopia Planitia in the north. We can interpret the lack of lineations at these regions as having been subsequently covered over by lava flows or literally wiped from the surface during the force of asteroid and comet impacts.

These findings suggest that Mars' surface, as on Earth, was once composed of numerous moving plates, but only during its earliest history. At divergent boundaries where new material emerged from beneath, the planet's internal magnetic dynamo left a permanent magnetic imprint in the newly forming crust. Whenever the dynamo reversed, subsequent magnetic imprinting of the then forming crust would be correspondingly oppositely magnetized. Over millions of years, the alternating dynamo resulted in permanent magnetic stripes within the crust. Subsequent to the dynamo collapsing and plate tectonics halting, activity such as lava flows during the late Hesperian and Amazonian periods, as well as comet and asteroid impacts, all reworked much of the surface to such an extent that no obvious trace of the plates are visible today. But with magnetic evidence prevailing, close examination by Connerney and his colleagues may reveal where some of the original plates actually resided. Cerberus Rupes, for example—part of a fracture system southeast of Elysium—is found to align perfectly with a magnetic contour for about 2,000 kilometers along its length. Contours also align with Valles Marineris on its eastern end; and where they are not detected on the western end, visual and mineralogical evidence reveals younger terrain possibly covering the more ancient magnetized crust. Furthermore, although not visible from the images of the surface, regular *shifts* in magnetic contours through the Meridiani region suggests two ancient parallel faults lines with properties unique to plate tectonics—evidence of transform faulting (plate slippage), symmetry about a ridge axis, and coherence over a long distance. If this is the case, then these faults may even explain the formation of Valles Marineris, where the direction of crustal tensile forces needed for its creation are consistent with those implied by the position of the two Meridiani fault lines.

The alignment of the three great volcanoes at Tharsis (and possible alignment of Olympus Mons and Alba Patera) also suggests the movement of a plate—called the Tharsis plate—sitting over convective cells within the planet that gave rise to its giant volcanoes. Throughout the Noachian period, the Tharsis volcanoes released no less than 300 million cubic kilometers of magma, creating the continent-sized Tharsis bulge. Many river networks in the Tharsis region that are known to have formed in the Noachian period are oriented in accordance with the bulge, indicating that it, too, must have formed during the Noachian period. Indeed, so heavy was the Tharsis bulge even then that, as revealed in MOLA 3D topological maps, a vast ringed deformation of the surface, of radius 5,000 kilometers, encircled it entirely. The formation of Tharsis was therefore a singular and significant event in Mars' early history, affecting the topology of much of the planet's surface and setting the direction of flow of many of the planet's river valleys and the majority of the great flood outflow channels.

If, as current evidence suggests, plate tectonics occurred, it would have lasted throughout the Noachian period and possibly into the early

Hesperian. Eventually, Mars' internal dynamo collapsed, bringing plate tectonic movement to a halt. Although we are unclear on how precisely the dynamo came to a halt, one possible reason could have been that the thicker crust of the southern hemisphere acted as a partial stagnant lid, curtailing Mars' internal activity and halting the dynamo. During the Hesperian period, it appears that the entire crust came to a halt, forming a permanent and sealed lid over the surface. Tectonic and volcanic activity persisted, though plate tectonics was now gone forever. The period was characterized by the formation of many of the currently visible features—Valles Marineris, volcanism across Syrtis Major, and continued significant but static volcanism at Tharsis and Elysium. Over hundreds of regions across the planet, volcanic deposition produced layered terrain many kilometers thick, resurfacing much of the northern lowlands and the north polar region. Such activity continued across much of the equatorial and northern regions well into the Amazonian period, extending the bulges at Tharsis and Elysium, filling the northern lowlands with lava and creating the gigantic Olympus Mons. The past three billion years have been characterized by the virtual cessation of all tectonic activity, but with occasional volcanism persisting in the vicinity of both Tharsis and Elysium. Although today Mars is essentially tectonically quiescent, some tectonic activity seems to have occurred south of Elysium perhaps as recently as five million years ago. Geologically, recent volcanic activity has also been detected on several volcanoes indicating that Mars, while extremely dormant by comparison to its early history, is not totally so.

Water on Mars

The question of water on, and within, Mars is of paramount importance to determining the planet's internal structure, resulting surface environment, and potential for life. The amount of water in the accretion disk from which Mars formed would have governed the amount of water contained within the planet and hence whether it was internally wet enough to allow its mantle to flow upon heating up. If the planet's water content was too low, the mantle would remain viscous even when hot; but with sufficient water it could undergo gigantic heat-induced convective cycling, releasing water and other volatiles to the surface and producing a surface environment perhaps conducive to life.

Among the important discoveries by Mariner 9 and the Viking orbiters was a multitude of geomorphologic features suggesting water activity on the planet in its distant past. Image after image revealed what looked superficially like dried river valleys, giant flood channels, and ancient lakes within craters. The reconnaissance orbiters currently orbiting Mars are also providing significant new evidence. They have imaged many of the supposed water features detected by Viking at higher resolutions, providing greater clarity on their nature and evolution, while a myriad of unprecedented data such as three-dimensional (3D) topological imaging, mineralogical mapping, and identification of sedimentary layering all contribute to a picture of water activity across Mars spanning millions of years, and possibly even occurring (on rare occasions) to the present day. All reconnaissance craft ever to have visited the planet—without exception—have provided increased evidence of past water activity.

While the valley networks are seen mostly on the oldest terrain, it is likely that they arose across the entire planet in its early history. Indeed, the Mars Global Surveyor has revealed about 10 times more river networks than was seen by Viking, suggesting significant fluvial activity in Mars' early history. Two broad types of valley have been identified: valleys hundreds of kilometers long with few tributaries, such as Nanedi Valles in the Xanthe Terra region, and smaller valley networks tens of kilometers long with complex branches of tributaries as seen in the Thaumasia region. Although they look superficially like dried river valleys on Earth produced from precipitation, many of the valleys on Mars reveal different morphologies when examined close up. Some may have arisen from precipitation, but others suggest water sapping from beneath the ground as the water source. Determining their origin is important. If the valleys were created from precipitation, that suggests a dense atmosphere and warm climate at that time. While their nature is not yet resolved, the increased number of valleys and drainage densities identified by MGS point to at least some of them arising from precipitation, suggesting atmospheric and climatic conditions capable of supporting liquid water on the surface for some time during the Noachian period. None the less, the evidence is equivocal; and even where valleys were created through precipitation, we cannot yet determine whether they emerged during a long and stable climate, or during brief wet periods on an otherwise cold and bleak planet. Some of the most captivating evidence has shown several ancient river delta features: two in the Nili Fossae region and one northeast of Holden Crater. In all three cases, evidence of ancient rivers can clearly be seen flowing into a standing body of water where deltas were created. So extensive are these features that they must have been created over periods of tens of thousands of years or more, suggesting relatively long periods of surface liquid water activity.

Whatever environmental conditions gave rise to the river networks, they provide important insight into Mars' internal structure and early surface activity. First, since river valleys in the Tharsis region all follow the orientation of the bulge, they reveal that Tharsis itself must have formed before the river valleys and, therefore, during the Noachian period. Such significant lava flows from Tharsis occurring during Mars' early history, together with the quantities of water indicated by the large number of subsequent rivers, suggest that Mars had a wet mantle. Coupled with evidence of plate tectonic movement, a picture is beginning to emerge of a wet and active mantle producing convective flows that led to widespread tectonics, volcanism, and the outgassing of water, carbon dioxide, and other volatiles.

Intriguingly, while the majority of the valley networks occur on the oldest Noachian terrain, some are seen on younger Hesperian and even Amazonian terrain near Alba Patera, raising questions about the stability of near surface water and climate change on Mars throughout its history. Equally significant is the existence, across the central latitudes of Mars, of gigantic flood channels that all flow from south to north. Many of the channels are tens of kilometers wide and thousands of kilometers long. They all originate from chaotic terrain found within canyons and grabens, such as at Hydaspis Chaos near the equator; and virtually all are found on terrain from the Hesperian era, indicating that they occurred much later—perhaps a billion years or more—and under different environmental conditions to the river networks. Although their origin is uncertain, the release of gigantic quantities of underground water must have been involved, suggesting that large areas of Mars retain underground reservoirs of water as liquid ice or chemically bonded to the crust. The orientation of many of the outflow channels also seems to be governed by the Tharsis deformation, suggesting a connection between the formation of Tharsis and water activity. For example, the giant Argyre impact basin in the south seems to have filled with water during the Noachian period, overflowing northward into the Tharsis trough. It appears that the direction of flow of great floods to the eastern end of Valles Marineris were also governed by the trough, flowing north into Chryse Basin and then into the northern lowlands.

One hypothesis for their origin considers the melting of underground ice by geothermal activity, where the container of the reservoir (whether composed of ice or of crustal material) breaches, releasing the water in an enormous flood, causing the overlying surface to collapse. A second scenario considers vast tracts of hydrated minerals similarly released through heightened geothermal activity. Whatever their origin, the Hesperian era floods were of colossal proportions, with some releasing tens of thousands of times the discharge of the Mississippi river, for example. Understanding their origin, nature, and timing is critical to understanding the evolution of the planet, especially during the Hesperian period. Evidence already points to vast quantities of water retained within the crust, and where episodes of tectonic and/or climate change instigated the floods of such proportion that the released water traveled great distances across the Martian surface without freezing over.

As we have gained knowledge of Mars, the question of lakes, seas, and even oceans residing on the planet have also arisen. Certainly many of the river channels flow into craters, suggesting crater ponding, and this indeed was the basis for sending Spirit to Gusev Crater, where the great valley Ma'adim Vallis intersects and breached the crater from the southeast. And as we have seen, Spirit has indeed verified the substantive influence of liquid water within the crater at one time, supporting the idea that a lake once resided there. Other craters are seen from orbit to exhibit similar characteristics, suggesting that lakes may have been common during the Noachian period. Working in combination, Odyssey, Opportunity, and Mars Express have provided unequivocal evidence of an ancient shallow sea, about the size of the Baltic Sea, at Meridiani Planum. With the detection of hematite, precipitated salts, and aqueous sedimentary layers tens or hundreds of meters deep across the region, it appears that a free-standing body of water existed at that location for hundreds of thousands of years.

One of the most controversial hypotheses regarding water on Mars has been the proposal of a great ocean in the northern hemisphere in its early history. This idea gained credence during the 1980s when Timothy Parker of JPL used Viking orbital data to identify what could be interpreted as an ancient shoreline circling the planet at the division between the southern highlands and the northern lowlands. And as the northern region is also largely devoid of craters, it must be relatively young and has been shaped by forces hitherto unexplained. Hence the idea of a great northern ocean that may have arisen in the late Noachian or Hesperian era seemed at least plausible, but closer examination by MGS and Odyssey did not support the hypothesis. First, MOC images of the supposed shoreline reveal no evidence of coastal morphology. Second, MOLA has revealed that the proposed shoreline features do not all reside at the same elevation, varying by as much as 8 kilometers from region to region. Finally, neither TES nor THEMIS has identified carbonate sediments associated with large standing bodies of water.

Despite this, other evidence seems to indicate that the region was influenced by water to some degree. A second inner shoreline-type feature circling the planet sits at a single elevation around the planet. It is also clear that river networks of the south come to an abrupt halt at the division between the south and the north, while most of the great flood channels of the Hesperian period flow northward into the region. Also, sedimentary layering along the flood channels and across the northern lowlands could have been created through aqueous sedimentation. We must also remember that the lack of identifiable coastal features could be a result of weathering and erosion over the past two billion years or more, while the observed variations in the elevation of the proposed coastline may have occurred subsequently through tectonic activity. The identification at Meridiani Planum of an ancient shallow sea with an abundance of iron and sulfates has also provided new insight into the possibility of a great northern sea or ocean. Victor Baker of the University of Arizona has indicated that if an ocean had formed, containing similar iron and sulfate salts to those found at Meridiani Planum, it would have become acidic and inhibited the precipitation of carbonates. Hence, the detection of mere trace levels of carbonates on Mars (expected in significant quantities wherever liquid water exists in a carbon dioxide atmosphere) might equally be a consequence of a sulfate-dominated chemistry rather than an indication that liquid water never existed in the region.

A significant development in the quest to determine the extent of water on early Mars has arisen from the first complete mineralogical survey of the planet from the Mars Express OMEGA thermal imaging spectrometer. OMEGA has revealed two distinctly different types of hydrated minerals on the oldest terrains of all, suggesting two different eras of water-based activity and an early geological history hitherto unconsidered. As detected by Jean-Pierre Bibring and his colleagues of the Mars Express OMEGA group, particular minerals called philosilicate clays have been identified at numerous locations, including Arabia Terra, Terra Meridiani, Syrtis Major, Nili Fossae, and Marwth Vallis. The nature and pattern of the minerals across the oldest surfaces suggest the presence of a large body of standing water and a warm, moist climate lasting for millions of years prior to four billion years ago in what is now called the Phyllosian era. Subsequently, from about 4.2 to 3.8 billion years ago in what is called the Theiikian era, the planetary climate may have changed radically, resulting in the loss of those standing bodies of water. But a later era of widespread volcanism seems to have released acid rain produced from water and volcanic sulfur and iron, leading to a second period of surface water activity that deposited the hydrated sulfate minerals seen by OMEGA. Finally, during what is called the Siderikian era between 3.8 and 3.5 billion years ago, Mars lost all capacity to retain surface liquid water, and became a cold, dry planet.

While this hypothesis is far from verified, the philosilicate clays in particular provide significant new evidence of an early period where surface liquid water must have played a significant role, possibly providing opportunity for complex and even life-related chemistry. Hence the identification of these clays represents a significant find, not only in providing further evidence into Mars' distant past but also suggesting that, if microbial life did arise, it is within such clays that we would expect to find their fossilized evidence.

Mars' Early Atmosphere and Climate

Mars formed about 4.6 billion years ago, yet its surface geology does not go back much beyond four billion years. Hence the nature of the planet just after its formation cannot be directly inferred from its geology alone. It is quite likely, however, that an early atmosphere arose, produced from outgassing as differentiation occurred within the planet. Indeed the evidence produced by the OMEGA team of surface water prior to four billion years ago suggests such a scenario. But the early atmosphere would not have survived for long. From crater densities seen on the oldest parts of Mars, we can infer that between 50% and 90% of the original atmosphere would have been blasted into space during the period of mass bombardment, turning the surface into a cold, desolate environment incapable of supporting volatile activity.

Despite this set back, the rise of tectonic and volcanic activity throughout the Noachian period—especially at Tharsis—seems to have given rise to a second atmosphere later in the Noachian period. Given that the Tharsis bulge was mostly formed within this period, we can infer that Tharsis alone would have released enough carbon dioxide to produce a new atmosphere with a pressure of one and a half bars (one and a half times Earth's current atmospheric pressure) and enough water to create a surface hydrological cycle including the formation of the river valleys. Uncertainty persists, however, about the prevailing climate within which the rivers and standing bodies of water arose. While the planet possessed ample volatiles to produce a warm climate, its distance from the Sun, and the fact that the Sun was then dimmer, suggests that Mars' climate would have remained cold without the influence of a mild greenhouse effect, and currently we do not know if one occurred. Irrespective of how Mars' climate played out, we are now confident that the planet was active enough for water and other volatiles to be cycled through the atmosphere, crust, and liquid reservoirs. The hydrated salts detected by OMEGA and Opportunity, the vast tracts of possible aqueous sedimentary layering, and the ancient river valleys, deltas, and shallow sea at Meridiani Planum all point to the cycling of volatiles and to the possibility of a complex chemistry system. Furthermore, SNC meteorites reveal the presence of atmospheric gases having being absorbed into the crust, most likely via interactions with liquid water. The presence of carbonates in the meteorites also supports this, and some of the SNCs even suggest that interactions occurred at hydrothermal vents. Overall, despite the considerable uncertainty about the prevailing climate during the Noachian period, it seems that Mars was then an active planet with a significant volatile and complex chemistry.

Eventually, Mars lost its second atmosphere. From an examination of the properties of the SNC meteorite ALH84001, for example, we can infer that the meteorite was created about 3.9 billion years ago—during the height of the Noachian period. Isotopic fractionation is minimal within this meteorite in particular, indicating that the atmosphere at the time was not affected by external forces such as stripping by solar and cosmic radiation. Also, the magnetic properties of the meteorite, as well as the magnetic striping seen across the oldest crust by MGS, indicates that the planet's internal dynamo and resulting magnetic field were then active and probably protecting the atmosphere. But about 3.7 billion years ago the planet underwent a massive change, with the disappearance of rivers and other free-standing bodies of water. As seen by MGS, many V-shaped river valleys became U-shaped, indicating that the flow of liquid water was replaced by glacial movement as surface water turned to ice. The timing seems more than coincidental with the simultaneous cessation of Mars' internal dynamo and of tectonic and volcanic activity on the surface. From that point on, Tharsis would no longer replenish the atmosphere and the loss of the planetary magnetic field would leave it exposed to stripping from solar radiation. Indeed isotopic analyses of the atmosphere strongly suggest that up to 99% of that atmosphere and over 60% of the planet's surface water have been lost to space through solar radiation stripping since that time. The examination of hydrogen, carbon, nitrogen, and argon isotopic fractions in Mars' current atmosphere also indicates that the atmosphere and the water supply were both bombarded by solar radiation and energized preferentially to allow the lighter isotopes to escape, leaving only a trace atmosphere with the excess of heavy isotopes seen today. The end of the Noachian period spelt disaster for Mars. The planet's interior essentially came to a halt, Tharsis became relatively quiescent, the planet's surface water froze and the atmosphere was eroded over the ages to the tenuous state we see today.

A New Mars

Although the end of the Noachian period brought about a dramatic downturn in the planet's internal, atmospheric, and volatile processes, it did not mean a complete collapse of all activity. The planet's internal dynamo and plate tectonics had ceased, but volcanic and tectonic activity persisted through the Hesperian period, especially at Tharsis and Elysium. And although the planet's surface and atmospheric volatile inventories were increasingly lost to space, vast amounts of water became trapped in the crust, and at the poles, as underground water, permafrost, and hydrated minerals. The gargantuan floods that took place throughout the Hesperian period indicate the existence of such crustal reservoirs and of occasional periods of greater internal and climatic activity long after the Noachian period.

But even the activity of the Hesperian period came to an end, and Mars became a fundamentally different place. No longer would it be dominated by indigenous processes and from that point onward the surface would be affected as much by the planet's celestial context as by any remaining internal activity. And while we might expect (and have often imagined) that Mars has remained essentially dormant in the intervening billions of years to the Amazonian period, evidence from Mariner 9 and the current orbiters reveals a somewhat different picture. We have now come to realize that even today Mars is not totally quiet. The spectacular discovery of a recent sea at Cerberus Fossae south of Elysium, created by tectonic-based lava and water flows within the last five million years—together with the detection of volcanic activity at numerous volcanoes within the past half billion years, and the explosive volcanic eruptions at Hecates Tholus and the young surfaces at both Tharsis and Elysium—all point to a planet on which tectonic and volcanic activity have persisted, albeit sporadically, until the present geological era.

There is also evidence that Mars' celestial context has had, and continues to have, a significant effect on the planetary climate and ensuing volatile activity. Unlike Earth whose motion about the Sun is somewhat stabilized by our planet-sized Moon, Mars endures the full force of gravitational perturbations from the other planets of the Solar System, inducing quasi-periodic but somewhat chaotic changes to the obliquity and precession of its axis and the eccentricity of its orbit. For example, while Mars' obliquity is currently about 25 degrees, it has reached 45 degrees within the last 10 million years, and may even lean over by as much as 60 degrees on occasion. Such a radical variation in the planet's orientation to the Sun produces equally radical variations in its climate and volatile behavior. So significant is the effect that Mars undergoes greater climate change than any other planet, including Earth, over periods of tens or hundreds of thousands of years.

Evidence of such climate change has been identified across the planet. As first discovered by Bruce Murray from Mariner 9 data, the polar regions are surrounded by hundreds of kilometers of layered terrain that suggest climate change. The MOC camera on board Mars Global Surveyor has imaged the layers in exquisite detail, showing intricate layering all the way to its limiting resolution of just 1.5 meters. And the GRS instrument on board Odyssey has simultaneously revealed that the upper meter is composed of more than 50% water, suggesting a vast permafrost across the polar regions. Although it is not conclusive, we now suspect that the layers are laid down by changes in carbon dioxide, water, and dust in response to climate change. It has been suggested that, in the future, we may be able to directly track the climatic history of Mars over billions of years by examining the individual layers—similar to the way in which we can decipher the climate during the life of a tree by examining its rings. As already mentioned, significant and dramatic evidence produced by MGS, Odyssey, and Mars Express also point to climate change. Geologically, recent glaciation in the equatorial regions, glacial deposits at the base of Olympus Mons from four million years ago, possible glaciers on the flanks of Olympus Mons above 7,000 meters and evidence of glaciation at five other volcanoes and on the eastern end of Hellas can only be accounted for by changes to Mars' climate. And although the induced changes do not match the level of activity of the Noachian period, the variations in insulation cause significant volatile activity: the atmosphere can become three to four times denser; water vapour levels increase by 100-fold; and polar ice moves *en masse* to lower latitudes guided by specific atmospheric and wind patterns.

The planet we see today seems superficially dormant, but in reality it is a place with a vast legacy and many more secrets to reveal about past and present activity. With such a tenuous atmosphere, and no capacity for solar heat storage, we could infer that Mars' climate does not change on timescales of days, months or years. Here, again, decades of observations of dust storms, and years of climatic monitoring by Viking and the modern reconnaissance orbiters, reveal significant climatic activity—and variations of activity—on a daily, seasonal, and annual basis. On a daily basis, the surface can vary from -80°C at night to above 15°C at noon at the equator. Further, so tenuous is the atmosphere that the temperature can drop by 15°C within a meter of the surface. Atmospheric pressure on Mars today is extremely low, typically just seven-tenths of 1% of Earth's, although carbon dioxide freezing out at the poles during winter can reduce the pressure by 30% in that hemisphere, causing planetary winds that draw carbon dioxide from one hemisphere to the other. Wind patterns on Mars are not dissimilar to those on Earth, where hot air at the tropical latitudes, for example, heats up and rises into the atmosphere, to be replaced by cold air from other latitudes. This convective cycling, coupled to the planet's rotation, induces wind patterns that circle the planet, not dissimilar to Earth's trade winds. We now suspect that local winds in the southern summer trigger local dust storms that link up to the planet's trade winds to produce dust storms that can, on occasion, encase the entire globe for months or years on end.

Despite what we have learned, we have a scant understanding of Mars' current climate. We cannot yet explain the true nature of the dust storms or why global dust storms arise in some years but not in others. Nor do we fully know the extent of climatic patterns on a seasonal or annual basis. Understanding such behavior is vital, however, for providing insight into longer timescale climate change. It allows us to characterize the planet more fully in order to: understand the changes that have occurred in the past; determine the possibility of life surviving to the current day; and prepare for future human missions. Hence one of the key priorities of all current reconnaissance missions is the extensive monitoring of a myriad of aspects to the current atmosphere and climate.

Interpreting Mars

Our recent explorations have been exhaustive and extensive. Although we are only starting out on our journey of exploration, we feel that a broad framework for the history of Mars is finally beginning to emerge. Mars probably possessed an ancient atmosphere prior to four billion years ago during the Phyllosian era that may have created a warm, moist climate and standing bodies of water. Through mass bombardment, most of that original atmosphere was lost to space. Later, during the Theiikian era (perhaps equivalent to the mid- to late-Noachian period), mass bombardment ceased and an active interior drove a planetary dynamo and magnetic field, plate tectonics, and widespread volcanism. A second atmosphere arose due to outgassing from Tharsis in particular, leading to extensive water and volatile activity as well as increased weathering, erosion, and deposition. Eventually, in the Siderikian era (perhaps coinciding with the Hesperian period), the planetary interior cooled, with the subsequent halting of the internal dynamo and plate tectonics. With a decline in atmospheric replenishment and loss of the planetary magnetic field, Mars' atmosphere started to decline, with a subsequent catastrophic change in climate that caused water to freeze and be lost to space or become locked within the crust. But the fundamental change to the atmosphere did not mean a complete cessation of climatic influences on the planet. Gigantic floods during the Hesperian period could flow across the surface of the planet, indicating at least brief returns to more active climatic conditions. Eventually the planet became a more quiescent world, but with quasi-periodic climate change due primarily to changes in the planet's obliquity causing a significant movement of volatiles as well as occasional tectonic and volcanic activity, which persist to the present day.

Though such a historical picture is slowly emerging, many fundamental unknowns remain about the timing, chronology, extent and nature of activity during the various eras. For example, despite all we have learned, we still cannot determine the prevailing climate during the Noachian period. Nor can we determine the timing of the various changes to the planet to within even hundreds of millions of years, let alone the broad character of the planetary chemistry prevailing at the height of activity. These and many other fundamental issues remain unresolved and there is an enormous amount of work to do before the full and intricate history of Mars is revealed.

None the less, what we have learned is significant. We now know, for example, that Mars *was* extremely active in its early history in ways broadly similar to Earth. We also know that that activity involved volatile materials similar to those on Earth. And, most significantly, we know that Mars was a planet influenced by water, which in itself constitutes a result with profound implications. It tells us that the types of chemistry needed for the emergence of life on Earth were also possible on Mars, and it confirms the possible existence of ancient prebiotic and life-relate chemistry. From a cosmic perspective, knowing that Mars possessed water tells us unequivocally that Earth is not unique in the Universe with regard to water. One of our great quest throughout the ages has been to discover if other planets could retain and be affected by water; and now we know the answer is "Yes." Of the eight major planets of our Solar System, two are watery worlds. The fact that Mars is now devoid of surface liquid water is irrelevant—in a billion years or so Earth may lose all its water. Every planet has its time; and with respect to the widespread presence of liquid water on Mars, that time was the Noachian period. Mars is a planet that was heavily influenced by water at a critical time in its history, relevant to the emergence of life as we know it. Such a realization tells us that liquid water residing on a planet is not as unique as was once held, and that other worlds across the Universe may also harbor water.

That Mars is understandable and explainable reveals it to be a natural world of less than extraordinary make-up. All that we have so far learned validates our suspicion that we can learn much from Mars about our young planet Earth and the emergence of life. And while essentially all of the significant work is still to be done, our efforts to date have revealed literally hundreds of possible habitats, both past and present, to visit. We are spoiled for choice, and our next steps are crystal clear.

17 Next Steps

Despite the astonishing progress to date, most of the major and fundamental questions regarding Mars remain essentially unanswered. We are in an excellent position, however, to take the next steps toward finding an answer. Currently, thousands of scientists across the world are pouring over the incredible amount of data being returned by the numerous space probes orbiting and on Mars. Unprecedented numbers of scientific papers and increasing frequent international conferences reveal the latest findings, and slowly but surely a comprehensive picture of the planet is taking shape. An entirely new generation of Mars scientists is currently emerging, armed with the hindsight and experience of those who came before, with brand new insights about the planet and with know-how on pursuing the next questions. The exploration of Mars is now one of the hottest topics in science. But the task ahead is truly enormous. It will take many more missions and many more years of sustained effort by the major science groups, universities, and national and international space agencies of the world before we can truly understand the planet's natural history and what it has to say about life.

New Motivations

While in the 1990s it was essential to identify fundamental questions about Mars and to devise an ingenious phased strategy to pursue them, several significant new factors now impact on how we should explore the Red Planet. Of course the phased strategy remains a central motivating factor: pursuing the global to the specific, past to present and geochemistry to biochemistry, in a phased manner, remains the cornerstone of our efforts. To date, orbital reconnaissance and geochemical lander missions have validated our initial conjectures on Mars' active past—the role of water, its broad connection with Earth, and that it may yet provide valuable insight into a planetary context for life—the results of our efforts demand that we continue with a phased strategy toward fully characterizing the planet.

K. Nolan, *Mars, A Cosmic Stepping Stone*,
DOI: 10.1007/978-0-387-49981-9_17,

The paradigm shift in our perception of the planet, brought about by current exploration (that of the complexity of Mars), must also be considered. Prior to Mars Global Surveyor, few could have envisaged the true complexity of Mars' landscape and history; and now we must factor this into our continuing program. We cannot yet declare that we have surveyed the planet adequately. The vast landscape that is the geological legacy of Mars remains largely elusive—at times even paradoxical—and we have much more to do. Literally hundreds of sites of potential biological interest have now been identified that demand Phase 3 biological landers and Phase 4 sample return missions; but the uncharted complexities of the planet at large also demand that ever improved orbital-, aerial-, and lander-based geological and geochemical reconnaissance must also continue.

Our explorations have also thrown up surprises regarding Mars' current behavior. The realization of the extent of (a) quasi-periodic climate change and the resulting discoveries of glaciation on the flanks of the great volcanoes, (b) geologically recent tectonic and water activity at Cerberus Fossae, (c) methane in the atmosphere, and (d) permafrost in the high latitude and polar regions are all relevant to our pursuit of the question of life on Mars. But they are also largely unexpected findings. Clearly Mars today is a dynamic world in ways that were not expected, and we must also make new efforts to understand such dynamism. The initial motivations behind our phased strategy were primarily about understanding Mars' past, but now we realize that there is a great deal to learn about its current state and behavior. Here again, new and innovative orbital and lander reconnaissance will be required.

Despite the new demands on the Mars program, our phased approach is perfectly suited to the task. None of the probes sent to the planet since Pathfinder has been ill-conceived, ill-timed, or scientifically inappropriate. All have worked and complemented one another very well, providing an ever-maturing understanding of the planet. And with time and opportunity to respond to current findings, the next generation of orbiters and landers will be appropriately designed, built, and equipped to accept the new challenges that Mars now presents—and there need be no conflict with our original aspirations. If the search for origins is important, then so is Mars; and this is not lost on the program designers, engineers, and planetary scientists currently involved with Mars exploration.

The success of the Mars program to date has drawn widespread commendation both from within and beyond the scientific community. Despite the hiccups and losses along the way, the overall program has worked extremely well, delivering spectacular results, with each probe outperforming its original expectations many times over. Also, the phased approach to the program has even provided a viable way forward for other space programs that are arguably stuck in a rut or suffering from a lack of direction. The current Mars program is the most focused and successful endeavor since the Apollo missions to the Moon and the Voyager reconnaissance of the outer Solar System. And where programs such as the Space Shuttle and the International Space Station are technically unprecedented, they have essentially remained exercises in the development of enabling technologies, but whose purpose has not yet been fully realized. In contrast, this has not been the case with the Mars program. From the outset it has been driven by well-defined and realistic goals, with the phased strategy progressing in a stepwise fashion toward their realization. All of these factors have led to a heightened political and social interest in Mars exploration that did not exist at the beginning of the program. An ownership of (and expectation for) Mars exploration is now emerging, not only by the science community but now also by politicians and the general public.

A renewed political interest in space exploration now exists within the USA, the EU, Japan, China, and India, with Mars as a high priority for many. The rewards are substantial—national achievement and standing, technical innovation driving science, education and the economy, as well as international acclaim and supremacy. And while the USA currently leads the way, the others are rapidly gaining ground. In this pursuit, Mars beckons. The success of the current phased strategy of robotic exploration has provided a template on how to safely and cost effectively pursue far-reaching space exploration. The possibility of exploring Mars, whether by robot or by human, is now real and many of the space-faring nations are interested. Already the USA and the other space-faring nations have politically declared a desire to send people to Mars in and around 2033; and whether this is feasible, such formally declared political will is unprecedented. Never before had any space-faring nation proposed sending people to Mars. Indeed the US government and NASA have previously been careful to draw the line at such a declaration, but this no longer applies.

As has already been discussed, major efforts are underway toward this end; with ESA's Aurora program now politically ratified by EU participating states and George Bush's Vision for Space Exploration (VSE) already impacting on budget allocation to future space exploration. The politically declared interest in Mars is real and is probably here to stay; and while it may be contentious and difficult to implement, never before has such a long-term and permanent interest in Mars been expressed. Whether we eventually send people to Mars or not, the intervening efforts, both robotic and human, will all be affected. Roadmaps, programs, missions, timetables, and budgets must all adapt—and are currently doing so—to accommodate the possibility of a human mission to Mars. And so the robotic phased strategy for Mars exploration must now function within a broader framework. There will be setbacks, as in the 2007 NASA budget announcements, that radically affects the time line for some upcoming robotic missions to Mars, but the nature of the phased program means that even where delays occur because of shifting priorities, we will always be able to continue precisely from where we left off, supported by the significant work already carried out.

Also emerging is a broader sociological and cultural dimension to Mars exploration. Initiated by the unprecedented interest in the Pathfinder mission there has not been such a healthy awareness of and interest in space exploration since the Voyagers. This was significantly augmented by the emergence of the World Wide Web, where, for the first time, we could all share in the spectacular findings on Mars in near-real time from our homes and offices. As in so many other facets to the modern world, the anarchy that is the World Wide Web has to some extent democratized the former privilege of space exploration, and an appetite for it has emerged. Such has been the success of Mars exploration—and of NASA's and JPL's desire and ability to disseminate their finding almost live to hundreds of millions of people—that there is now an emerging acceptance, if not expectation, for robotic exploration of Mars in particular. Today, such exploration is taken for granted, perhaps too much so, but it is more than tinged with an unspoken consensus that the price/performance of robotic exploration is now pitched correctly and should continue. This is something to which JPL in particular can readily respond. With their abounding successes to date, JPL can now respond almost as a matter of course in building ever more sophisticated Mars exploration robots. Such exploration has become accepted as a valid pursuit for the long term.

Current political and cultural interest now sets Mars in a broader context. The path to Mars is no longer purely scientific; it will increasingly be defined by political and sociological interests. Mars is rooted in modern life as never before, and with the successes to date and know-how at our disposal, pursuing such a broad agenda for Mars is not a problem. This is what sets us apart from all previous efforts to reach Mars. We know the full extent and scale of the challenges ahead, and can respond by implementing a phased strategy to overcome them. Our strategy is an *enabling strategy*, because it does not have a singular and troublesome end point of pending potential failure. We'll send people to Mars when it is safe to do so, is technically possible and is financially feasible, and in the meantime will build toward that goal and learn along the way. We have developed a sense for Mars exploration and it will always be a part of future society.

Objectives

Our objectives for Mars exploration are now many and varied. The path from robotic exploration to a human mission is long and tough but can, none the less, be capably pursued via the phased strategy now implemented through the Vision for Space Exploration and Aurora programs. For the next decade and a half, three particular objectives present themselves: to contine the robotic exploration of Mars with a view to achieving defining answers about its natural history; to conduct technology test bed missions with a view to a long-term human presence in space; and to initiate a new program of deep space human exploration to the Moon, Mars, and beyond.

Phase 1 and Phase 2 Robotic Explorations

Concerning the robotic exploration of Mars, we will continue with Phase 1 orbital reconnaissance and Phase 2 geochemical lander missions for some time to come. Here we wish to find defining answers about the exact nature of Mars' ancient climate, the extent of water activity and the types of chemistry occurring during the Noachian period in particular. We must also determine the evolution of the planet and precisely how and when it collapsed to its current state of relative dormancy.

Also of importance is the character of the planet today. Here we need to determine any prevailing seasonal and annular weather patterns and discover how global dust storms arise, as well as the nature of quasi-periodic climate change occurring over thousands of years. We must determine the internal structure of the planet and whether tectonic and volcanic activity persists, and if so their impact on the release of water and other volatiles. We must also determine the current story of water—whether underground ice and aquifers exist and, if so, their potential both as habitats for Martian life and as a resource for future human missions.

All of this can be achieved through various means. For example, many of the current reconnaissance orbiters can survive into the next decade, with their missions adjusted and even in some cases improved through software upgrades or changes in modes of operation. For example, Odyssey can theoretically continue until 2014 and Mars Express until 2010; and although Mars Global Surveyor has now failed, an innovative technique developed by Mike Malin and Ken Edgett in 2003 called "compensated pitch and roll targeted observation" or cPROTO improved MGS's MOC camera to 0.5-meter pixel resolution, demonstrating that, even in operation, significant space mission enhancement can be successfully made. Furthermore, even with the existing datasets, we are set to tackle many new and outstanding questions about Mars through sustained analysis in the coming years. As in the case of the Viking orbital dataset, painstaking analysis of the vast quantities of data already available today are likely to provide significant new finds.

New and upcoming orbital reconnaissance and geochemical lander missions will also provide valuable new data. The Mars Reconnaissance Orbiter is now the benchmark against which all future orbital missions must be measured. It routinely photographs the surface of Mars at a resolution of less than 1 meter—in many cases sufficient to provide definitive answers regarding the nature of sedimentary layering and other small-scale geological features.

Finally, NASA's Phoenix (2008) and Mars Science Laboratory (2009), Russia's Phobos-Grunt sample return mission to Phobos (2009), and ESA's ExoMars (2013) and proposed sample return mission to the surface of Mars (2016) will of course also provide extensive new insight into the geological, geochemical, and chemical nature of Mars, past and present.

Phase 3 and Phase 4 Robotic Explorations

We are also now ready to proceed to Phase 3 biological lander and Phase 4 sample return missions to Mars; and already the key missions toward this end (mentioned above) are in the development or the planning phase. Evidence from existing reconnaissance points to the possibility of prebiotic chemistry and even microbial life arising on Mars billions of years ago, as well as to the possibility of life surviving today in niche habitats or in a dormant state until climatic conditions become more favorable. Many ancient sites of past water and hydrothermal activity have been identified where prebiotic chemistry and microbial life may have arisen, and where fossilized evidence of that activity may persist. Meridiani Planum, Valles Marineris, Syrtis Major, Arabia Terra, Nili Fossae, and Marwth Vallis are but a few of the many sites now attracting interest.

Furthermore, orbital reconnaissance also points to the polar and polar-layered regions as well as the possible frozen sea south of Elysium, among many other locations, as potential habitats today where biomarkers and even microbial life might exist. To this end, the landers mentioned above—Phoenix, the Mars Science Laboratory, and ExoMars—represent a brand new phase set on exploring the biological potential on Mars. Each will be equipped with instruments capable of determining the surface chemistry, the presence and origin of organic materials as well as searches for fossil evidence of past life and biomarkers associated with current or dormant life. Unlike the Viking missions, these three landers will be sent to sites deemed likely to provide the best scientific return, with their instruments specifically tailored to their selected environment. Also, in the planning phase by ESA (and with strong interest by NASA though its 2007 budgetary commitments make such a mission less likely for the USA) is a sample return mission to the surface of Mars, to take place around 2016. Given the intriguing picture of Mars that is emerging, considering the great unknowns that still persist, the scientific community resolutely recognizes the need for an imminent sample return mission. Such a mission would return samples from Mars to Earth, significantly accelerating our ability to decipher the planet's long history and the quest for signs of life activity. The technical and logistical challenges presented by such a mission represent an excellent opportunity and necessary precursor step toward a human mission.

Technology Testbed Missions

Each and every mission launched into space represents an experiment in technical innovation. From the aerobraking capability of MGS to the airbag technology used to set Pathfinder, Spirit, and Opportunity onto the surface—and from the TES infrared mineralogical detector on board MGS to its derived mini-TES detectors on Spirit and Opportunity—each has not only played a vital part in the success of its own mission but has also contributed to enabling ever more sophisticated and innovative subsequent missions.

Overall, enormous strides forward have been taken since the 1990s, to a point where the robotic exploration of Mars is almost taken for granted. But if we are to fulfill the phased strategy of VSE and Aurora and send people to Mars within three decades, we must become equally apt at devising systems capable of delivering people safely to Mars, sustaining them on the surface and returning them to Earth perhaps three years after launch. We are so far from such a capability that a significant and relentless effort in technological innovation is required from this moment on, across a myriad of areas such as rocketry, habitation modules and vehicles, EDL systems, power and propulsion, guidance and navigation systems, communications, computing, life-support, and rover vehicles, among many others—all to be developed and capable of performing for many years without degradation.

Currently, several different approaches to achieving such capability are envisaged. First, each new mission to Mars is required to extend current science and so will, by definition, require technical innovation relevant to the future of the program. Second, part of the remit of each mission must test new technology necessary for future missions. Here, the technology being tested may not necessarily be central to the current mission, but will be vital to future missions. Third, specific testbed missions will be executed, whose role is purely to test new technology.

To this end, NASA is already actively engaging new technological innovations in all impending Mars missions. Furthermore, a new category of *scout missions*, of which Phoenix is the first (and which can be pitched for by other organizations outside JPL), will inspire technical innovation within institutions across the USA. The Mars Reconnaissance Orbiter, for example, will not only bring scientific reconnaissance to a new level of sophistication but will also test new modes of guidance, navigation, and low-powered communications. The Phoenix lander will rely upon a new thruster-based controlled landing system to touch down on the surface and will also use a new robotic arm and scoop mechanism for delivering soil samples to its various onboard experiments. The Mars Science Laboratory (MSL) will test an entire suite of new technologies: precision landing, long-range roving, self-guidance, and possibly radio-isotope nuclear power. Other NASA scout missions will also test a range of technologies from precision landing to orbital telecommunications.

ESA is following in like fashion, where its flagship missions ExoMars and the proposed Mars Sample Return Mission will by default require significant technical innovation and where its smaller arrow class missions will function only to test new technology. The two arrow class missions—an Earth re-entry capsule and an atmospheric friction-based orbital entry system—are already planned and will specifically test the systems needed by ExoMars and Sample Return Mission. ExoMars itself, as with the MSL, will test long-range roving, autonomous navigation, and guidance and drilling, while the Sample Return Mission will require a range of new technologies including the use of no less than five spacecraft over two missions—a Mars Ascent Vehicle (MAV) to launch the samples from the surface of Mars, Mars-orbital docking maneuvers, and return vehicles capable of traveling from Mars to Earth—all unprecedented and directly applicable to an eventual human mission.

Future Missions

As enthralling as current Mars exploration may be, the future years promise to take our engagement with the planet to even greater heights. With aspirations of acquiring defining answers about the planet and of laying the foundations for sustained human exploration, the missions pursued in the next decade will be of critical importance.

Phoenix

Phoenix is NASA's first scout mission to Mars and the first Phase 3 exobiological lander of the current era. As a scout mission, Phoenix is designed and operated by the University of Arizona's Lunar and Planetary

Figure 101: Artist's rendition of the Phoenix lander on the arctic plains of Mars just as it begins to dig a trench through the soil. The polar water-ice cap is shown in the far distance. [Credit: NASA/JPL/ Corby Waste]

Laboratory, having been selected in an external competition aimed at triggering innovation across the USA. Landing on Mars' frozen arctic northern lowlands, Phoenix is the first space probe to be sent to a location that was selected on the basis of the possible presence of water (Figure 101).

Launched in August 2007 and successfully landed on Mars on May 25, 2008, Phoenix is a fixed lander, using a robotic arm to dig into the icy soil and deliver samples to a suite of onboard science instruments and experiments. The primary aims of the mission are to study the history of water in the region and to examine the suitability of the soil for life. In particular, Phoenix will provide ground-truth information for Odyssey's GRS instrument which has already identified areas of up to 50% water within the soils of the northern arctic. Such ground-truth will allow for a more accurate interpretation of all GRS data, leading to a better understanding of the natural history of water across the entire region. By specifically analyzing the soil, Phoenix will also provide valuable insight into climate change from as recently as 100,000 years ago.

Although Phoenix will not specifically look for life, it will be able to survey the suitability of its landing site as a potential habitat and, in particular, determine ways in which it might act as a habitat. Using its onboard TEGA Furnace in tandem with a mass spectrometer, samples of soil will be heated to 1,000°C from which the presence of water, other volatile materials and life-giving elements such as hydrogen, carbon, nitrogen, and phosphorus can be identified. Furthermore, with a sensitivity of 1ppb (part per billion), any organic materials within a sample will be easily detected, perhaps revealing the presence of life or of biological by-products. Another onboard instrument called MECA will conduct wet-chemistry experiments, determining both the acidity and salinity of the soil and revealing potential sources of chemical metabolic energy. By dissolving soil samples in water, MECA will also be able to detect the presence of a range of chloride, bromide, and sulfate salts, as well as the presence of oxygen and carbon dioxide. Finally, with an optical microscope accurate to 4 microns and an atomic-force microscope accurate to an incredible 10 nanometers, any hydrous or clay materials in the soil will be directly detectable, revealing the history of water activity at the landing site. With stereo-imaging and meteorology detectors, Phoenix will provide significant insights into the history and current activity of water at its landing site and, for the entire northern arctic region, reveal vital details on quasi-periodic climate change, determine the suitability of the site as a possible habitat, and detect any organic materials present. Indeed, within four weeks of having landed, Phoenix provided unequivocal evidence of water-ice residing at the surface, just below a thin layer of dust. The discovery, of historic importance, validates our search for water and life-related activity on Mars and heralds a new era of Mars exploration, in which Phoenix's findings will play a major role.

Mars Science Laboratory

The enigmatic Mars Science Laboratory (MSL) will be launched to Mars in 2009, arriving on the surface in 2010. Weighing in at four times that of Spirit or Opportunity, the MSL exobiological rover is designed to travel more than 10 kilometers from its landing site during its 668-day (almost a Martian year) primary mission (Figure 102). MSL represents another significant step forward in Mars exploration and an unprecedented opportunity to search for life on that planet.

On arrival, MSL will use a guided precision system to land closer to its intended target than any previous rover. The target chosen will be one deemed to be of greatest biological potential. Although MSL will not bring metabolic and growth experiments similar to those on board Viking, it will carry the most sophisticated science package ever sent to another planet, capable of identifying a great number of life signs, past and present. MSL's science goals include:

Figure 102: Artist's impression of NASA/JPL Mars Science Laboratory (MSL). MSL will be larger than the MER rovers, be able to travel over 10 kilometers from its landing site, and look for evidence of biology on Mars. [Credit: NASA/JPL]

- Characterize the geology of an entire region on Mars to the microscopic level.
- Analyze the mineralogical, chemical, and isotopic processes shaping rocks in the region.
- Access the long-term evolution of the planet.
- Investigate the cycling of water and carbon dioxide.
- Investigate past processes in the region relevant to past prebiotic chemistry and life.
- Identify and analyze fixtures possibly harboring evidence of life, including the possible identification of any of thousands of different organic molecules including microfossils.
- Determine the toxicity and radiation hazards for humans on Mars.

To accomplish this, MSL will carry a science package including a number of exquisite imagers: a panoramic stereo camera, a contact microscope, X-ray and neutron detectors of higher specification than those on the MERs, as well as new and revolutionary experiments such as ChemCam (a laser-fired chemical analyzer), SAM (a gas chromatography mass spectrometer (GCMS)), CheMin (an X-ray fluorescence spectrometer), and DAN (a neutron detector) capable of identifying the presence and structure of subsurface water to one-tenth of 1%. Also on board are two instruments—a

Radiation Assessment Detector (RAD) that will measure cosmic radiation levels on the ground, and a Spanish-built Rover Environmental Monitoring Station (REMS), capable of monitoring ultraviolet levels, air temperature, barometric pressure, humidity, and wind speeds and directions. Using the chemical detectors in combination with REMS and RAD, the Mars Science Laboratory will be able to determine for the first time the toxic and radiation hazards likely to impact upon a human mission.

MSL will perform three fundamental types of analysis. First, it will scoop soil and rock samples and bring them on board for analysis by SAM, ChemCam, and CheMin. Second, it will carry out contact investigations using its onboard imagers, microscopes, and spectrometers. Finally, it will conduct remote-sensing analysis (without physical contact) using its imagers, radiation, and meteorological instruments.

It will thoroughly investigate its landing region for evidence of past and present water and life-related activity, determine past "water activity" levels, and search for thousands of organic molecular types that may constitute biomarkers of past or present life. It will also look for rocks and clays perhaps retaining evidence of prebiotic chemistry or microfossils, as well as perform morphological, mineralogical, chemical, and organic chemistry experiments to identify the precise origin and nature of any such evidence. The results from MSL will impact significantly upon current ideas about Mars, providing significant new insights into the nature of the planet and, in particular, heralding a new era in astrobiology field studies.

ExoMars

Europe will launch it first Aurora flagship mission in 2011. Called ExoMars, this exobiological rover will set down on Mars in 2013 using an air-bag and parachute system, at which time MSL may well be still operational (Figure 103). Despite being ESA's first rover, the ExoMar science package—appropriately titled *Pasteur*, because it will look for evidence of life—already has a long pedigree. Many of Pasteur's instruments are derived directly from Beagle 2 and the French Net-Lander network of Mars landers. Despite the fact that none of the instruments has been used on Mars, they have none the less had an extraordinarily long development and test period, ensuring their optimal operation.

ExoMars will not only complement but also greatly extend the biological analyses of MSL. Its science goals include: the characterization of the site geology, mineralogical, and chemical analyses of rocks, soil, sand, and dust; the detection and characterization of any organics present; and the identification of chemical and biological biomarkers. Similar to MSL, ExoMars will have a number of sophisticated cameras, microscopes, and an

Figure 103: The ExoMars rover will be ESA's field biologist on Mars. It aims to characterize the biological environment on Mars in preparation for robotic missions and then human exploration. [Credit: ESA]

APXS for surveying the site geology as well as mineralogical and morphological studies of surface materials. A number of other features unique to ExoMars, however, will make it the most sophisticated biological lander ever sent to Mars. With two onboard drills, ExoMars can extract samples as deep as 2 meters below the surface, thereby accessing soil that has not been affected by harmful radiation or the oxidizing surface. Samples will be delivered to the onboard Pasteur mini-laboratory for analysis by a number of instruments and experiments, and by this capability Pasteur will excel like no other lander in history. Included within Pasteur are a GCMS capable of detecting organics to part per billion levels, a separate GCMS with a unique *biomarker microchip* for detecting amino acids and aromatic hydrocarbons, and yet another and separate biomarker microchip to allow for the identification of a range of chemical and biological biomarkers from past, dormant, or present life. Through ExoMars, we are confident that if evidence of past or present life exists at the landing site, ExoMars will find it.

Aerobots

Although they have not yet been scheduled, aerial robots or *Aerobots* will be sent to Mars at some time in the next decade. ESA has already put out a call for designs of such craft from students across the ESA nations, with the top 25 being considered for further development. Extensive work in both light aircraft and balloons for Mars has also been conducted by The Planetary Society and NASA.

The advantage of Aerobots, whether aircraft or balloons, is that they can carry out aerial reconnaissance work over large regions of Mars potentially to hundreds of times the resolution of an orbiter. Such a unique perspective could deliver spectacular views of the landscape and provide extremely high resolution analyses of more of the Martian surface than is possible from orbit. Aerobots will be critical to the identification of resources and in providing detailed landscape surveys for pending or already occurring robotic rover and human missions.

With an atmosphere less than 1% as dense as Earth's, aerial craft are extremely difficult to design for Mars. Given current materials, a light aircraft would have to be dropped into the atmosphere from, say, a descending lander, where it might glide for an hour or so. Even in that time, however, the glider (or gliders) could provide significant image and mineralogical data of a vast swath of Mars' surface in the vicinity of the lander, thus hugely enhancing its mission (Figure 104).

Using balloons, there are two viable options. First, an Ultra Duration Balloon (ULDB) could be dropped from an entering lander, where it would quickly inflate with helium. A second type of balloon is a Solar Montgolfier Balloon (named after the French brothers Joseph and Jacques-Ètienne Montgolfier who pioneered ballooning), which is open to the Martian atmosphere and where buoyancy is achieved by solar-powered heating of the air within the balloon. With comprehensive payloads on board, balloons might be constructed with a buoyancy that varies between day and night. With the aid of a long tether, the balloons may never descend fully to the surface and remain afloat for long periods of time. Instruments attached to

Figure 104: Artist's impression of a light aircraft gliding through the atmosphere of Mars. [Credit: NASA/John Frassanito and Associates]

the base of the tether might even be able to extract ground samples at the current location for analyses.

Mars Sample Return

Despite the ferocity with which we are pursuing robotic exploration on Mars, it may take decades to attain a full picture of the nature of Mars, past and present. In this respect, any means at our disposal that maximizes our ability to understand its nature must be seriously considered, and engaged where possible. As sophisticated as robotic explorers may be, they are arguably too limited to fulfill our aspirations for Mars. First, they are a mode of exploration in which those involved are one step removed from the actual process of exploration. Also, for the foreseeable future, robots will remain more cumbersome and less flexible than humans in field operations. In the case of a stationary lander, for example, there is nothing that can be done if a point of interest is simply beyond its reach. Rovers also have limits to their maneuverability in terms of the roughness of the terrain and the steepness of slopes that can be engaged. Even a change in direction can be a major ordeal, taking hours, if not days, to fully complete. There are also severe limitations to the sophistication of the instruments and experiments on board robotic rovers in particular, mainly due to weight constraints. Even some of the most sophisticated equipment to be sent to the surface of Mars in the coming decade will not be the equal of its Earth-based laboratory counterpart.

Furthermore, where the capabilities of onboard equipment are sufficient for the task, the nature of the beast is that answers are rarely 100% conclusive. They usually lead to more questions and a requirement for further investigation, and in this respect robots are of little use. In the case of Opportunity, for example, where it has already delivered spectacular results regarding past geochemical and water activity in the region, it is utterly incapable of conducting the types of exobiological investigations intended for MSL and ExoMars. In reality, unless it is selected as a target for a future mission, it will be decades, or longer, before we return to Meridiani Planum to conduct such investigations. Also, imagine the situation if Phoenix, MSL or ExoMars were to detect the presence of life activity. Will a single rover at that location fully satisfy the flood of new questions that will follow about the origin, nature, evolution and full extent of such life across that entire region and, indeed, the planet at large? Even in less dramatic circumstances, we are already finding a proliferation of fascinating environmental scenarios that prompt new questions, many of which we are powerless to follow up.

Using robots alone, every new question about Mars requires a new mission, and eventually this becomes too cumbersome and ineffective. Our use of robots for initial surveys is entirely appropriate and to date they have been (apart from the study of the SNC meteorites) our only means of conducting analyses on Martian materials. But already we are finding that Mars is so complex and our questions are so sophisticated that other means of examining Martian materials is becoming an imperative. For these reasons we envisage enhancing how science is done on Mars by two other methods: returning samples from Mars to Earth and via human missions to Mars.

The Mars sample return method would probably involve two simultaneous robotic missions in which the first collects rock, soil, sand, dust, and air samples and the second returns them to Earth. Having available samples from a known location on Mars whose regional context is also known would be hugely significant. We could apply the full weight of the best analytical techniques available, which are perhaps decades ahead of anything we can currently send to Mars. The returned samples could also be shared among many of the leading institutions and experts across the world, providing quicker answers and better insight. Newly arising questions could be immediately followed up, even if they require different or new analytical techniques. Also, as technological advancement provides better analysis tools and techniques than we have today, new insights into Mars could be gleaned from such samples.

Implementation

Currently, a sample return mission is planned for about 2016. The European Space Agency has already declared that their second Flagship mission will be a sample return mission. While NASA has also expressed a desire to conduct a sample return mission at the earliest possible moment—even as early as 2013—recent budget reallocations (to allow completion of the James E. Webb Space Telescope, for example), and the return of people to the Moon by 2020 have affected some missions of the Mars robotic program, including a Mars sample return mission. Despite this setback, NASA remains committed in the long run to a sample return mission, but with no specified date. Such has been the outrage expressed by the space science community at the shelving of such a crucial mission (especially in view of ongoing findings that increase the possibility that Mars may have supported life) that it is difficult to see how NASA's program for Mars can remain sustainable (and competitive). The most likely scenario, therefore, is for a single international sample return mission to take place toward the end of the next decade.

How would such a mission occur? There are several possible implementations, but among the most likely is a two-mission strategy. Here, the first mission travels to the surface of Mars, where a lander and/or robot gathers samples of many types of rock, drilled underground soil, surface soil (Figure 105), sand, and dust, as well as air samples collected at various times of the day, and, if the lander is there for a long enough period, at various times during the Martian year. The samples are then placed into a sealed capsule and transferred to a Mars Ascent Vehicle (MAV) which blasts off the surface into Mars orbit (Figure 106). The MAV may remain in orbit or it may put the capsule into orbit for future collection. A second space probe subsequently travels to Mars, collects the samples in orbit, and returns to Earth. Alternatively, the second vehicle might dock with the MAV to collect the samples. However such a mission is conducted, if successful it will be a milestone in space exploration. Not only will the returned samples be of historic importance, but the mission itself would have successfully demonstrated all stages of the journey for a human mission, providing a significant technical boost to the possibility of such a mission taking place.

Containment

Sample containment is one of the most contentious issues regarding a sample return mission. This arises because of the nature of our search for evidence of life and, in particular, *microbial* life. Hence, if we are serious about the possibility of life on Mars, we need to be even more serious in

Figure 105: Artist's impression of a deep drill lander searching for evidence of life, past or present on Mars. It is likely that such a mission will take place in the next decade. [Credit: NASA/JPL]

evaluating and dealing with the risk to Earth's biosphere that is posed by the return of such samples.

There are those who argue that even if microbial life is found to exist on Mars and is present in returned samples, its impact would be essentially zero even if released to the general environment. Because the life-forms are from

Figure 106: Artist's impression of the launch of a Mars Ascent Vehicle (MAV), bringing samples into Mars orbit where they will be collected and returned to Earth. It is likely that such a mission will take place in the latter half of the next decade. [Credit: NASA/JPL]

Mars, they argue, they will bare no significant commonality to life on Earth. Toxicity aside, there could be no meaningful interaction between Martian and Earthly biology constituting a virulent or pathogenic threat. They argue further that there could be no meaningful competition between Martian biology within Earth ecosystems—and with the natural advantages available, Earth life-forms in their natural environments would win any competition for resources. Finally, they argue, even in the unlikely event of meaningful biological interaction taking place, it must already have occurred because we know that about 500 kilograms of Martian material fall to Earth every year, having been blasted into space by comet and asteroid impacts.

Those opposed to such arguments point out that we are still often unclear on the details of how terrestrial microorganisms adapt and evolve. And we already know of many microorganisms that have evolved in very short periods of time; for example, *Deinococcus radiodurans*, as mentioned on page 27, developed the capacity within only a matter of years to withstand otherwise lethal doses of nuclear radiation. And so we cannot say with certainty what the pathogenic or environmental competitive capabilities of Martian microbes might be. Furthermore, even if there was scientific certainty regarding the safety of such microbial life, would it be reasonable to ask the world community to accept such a recommendation of safety? Should the World Health Organization and national and international agricultural agencies accept such complacency? Would it not be reckless and ethically dubious if Martian samples were not treated with the greatest caution possible, to ensure maximum protection for our planet?

The chance of life having arisen on Mars is slim. The chance of life surviving through to the present is also slim and the chances of microbes living within returned samples extracted from close to the Martian surface are even more so. None the less, the worst case scenario—where living microbes released to the open environment reek havoc with some part of Earth's biosphere—cannot be completely ignored and therefore constitutes a significant risk and a requirement for extreme measures in containment.

In dealing with containment, leading agencies are developing policies and systems that are appropriate to the task. It might be a requirement, for example, that before landing on Earth, the samples will be sterilized, or even burnt. Although they will then be mostly destroyed, we may simply have to live with what we can determine about Martian biology from such sterilized or burnt remains of the sample—still of significant scientific value—until subsequent technological advances and return missions build sufficient confidence in returning pristine samples to Earth. Or perhaps they might be returned to a lunar outpost or an orbiting space station. There is no doubt that we are capable of engineering a containment system that can ensure the safe return of Martian samples to Earth. But the question is not simply one of engineering, it is also about public acceptability and world responsibility. Irrespective of the debate, the prospect of returning Martian samples to Earth will demand extreme measures in containment, and this is already being addressed by the appropriate world bodies, as we will see in detail in Chapter 19.

18 Because It's There

We explore when we uncover somewhere new to go to that also challenges us. So it was for the exploration of The New World, the Arctic, the Antarctic, and the Moon. For the next phase of human exploration, Mars is the challenge.

As coined by mountaineer George Leigh Mallory, we often say that we explore a new domain simply "because it's there." Perhaps this is because exploration is so instinctive to us. It is certainly innately human, connecting our sense of curiosity that leads to increased awareness of the Universe around us with a desire to challenge ourselves to go there, to learn, and to become something we currently are not. Through such a process we extend and redefine who we are.

Even if we are unable or are not inclined to partake in great feats of exploration, we all explore in our own individual ways—through personal goals and ambitions or through a desire to better our family, community, or society, for example. The act of exploring a new cave or of climbing a hitherto undiscovered mountain route is not displaced from the desire to master a new language, pursue a new direction in music, or gain a better insight into the history of civilization. A sense of exploration manifests itself in many ways, from which none of us needs to be excluded.

As spectators to great feats of exploration, however, few are left untouched. Who among us is not affected by the vision and guile of Christopher Columbus, the endurance and bravery of Ernest Shackleton, or the boldness and mettle of Neil Armstrong, Buzz Aldrin, and Michael Collins? Even though such exceptional people engage in daring adventures that most of us will never attempt, we never the less give them our full support. The world waited with bated breath as Neil Armstrong took his first step onto the Sea of Tranquility. Through him, we also took a collective leap forward onto a new world. Through the Apollo program, the history of humanity to that point was, in part, vindicated.

And so it has always been. Columbus discovered the Americas, but not in isolation. The world explores through special individuals and programs, and

K. Nolan, *Mars, A Cosmic Stepping Stone*,
DOI: 10.1007/978-0-387-49981-9_18,

Columbus emerged from a tide of consciousness that felt a need to extend itself. An ethos of exploration is deep rooted in the unfolding human story. It is a symptom of the capabilities of a people at a given time, through which new worlds are handed down to new generations.

The Value of Space Exploration

The value of space exploration is often called into question, cited as being too expensive and, where human expeditions are concerned, too risky. It has often been argued that the money would be better spent on other worldly causes. These are valid concerns: space exploration can indeed be too expensive and too risky. But does it mean that we should not pursue any space exploration, and human space expeditions in particular? How do we value such ventures, and what is deemed too risky? These are pivotal questions in the perpetual debate surrounding our pursuit of space.

Space exploration is just exploration none the less, and the desire for people to go into space is no less underpinned by the instincts that motivated every other expedition through history. It is as valid as any other form of exploration and if we recognize the central role of exploration in the past, it is impossible to deny its currency. Hence, the value of space exploration should not be questioned in isolation. It does not cost more than many other questionable ventures and there are many greater risks in life that go unchallenged. If we are to debate the merits of space exploration, it must be in context; and if we are to question the cost of space exploration, should we not also question the costs of bureaucracies and world conflicts? The question perhaps should be: What are the values, costs, and risks of any given venture? Here, the answer will often be subjective, but there can be little argument that space exploration in principle is as valid as any other human pursuit.

There are tangible and valuable reasons for space exploration. There are, of course, the significant practical benefits, from satellite communications and global positioning technology to the increasingly important planetary monitoring required in such environmentally precarious times. Space exploration is also one of the few pursuits that literally *drive* revolutions in technical and scientific innovation. Perhaps most significantly of all—as a species who do not yet know where we came from—we have found that many parts of the answer to that question actually reside in space. Our understanding of the origin and bulk nature of our planet, for example, lies mostly in its cosmological context. And now we suspect that part of the answer for our biological context can be found on planets, such as Mars. We may be wrong, but increasingly we think it may be the case, and indeed is worth checking out.

Very often, however, such benefits are not clear. A population in a poor region of the world, struggling to survive, cannot see space exploration as a priority. Many in wealthier countries similarly see it as a waste of effort and resources as long as there are serious human problems to solve. But what should we do? Should we shut down all avenues of pursuit not central to human survival until all suffering and survival issues have been resolved? We know the answer to that. There cannot be validity in attacking space exploration over other non-survival, or even survival-based human pursuits. It would be akin to asking for a cessation of art or music until the world finds time to appreciate them. Such a mindset will not work—we cannot wait until all the world's problems are solved before enlightenment becomes a valid pursuit. What gives one person the right to assume that he should spend his life in comparative comfort in a world that is otherwise filled with people struggling for the basic necessities? No general decree can, or should, curtail human potential, whether through space exploration or otherwise. That potential is real and tangible; it cannot be denied or dampened.

There are, however, genuine contentions regarding the perceived value of space exploration; and here the space community must carry its share of responsibility. Too often does the space community concentrate only on important and sophisticated science, deeming all other considerations to be of lesser importance. Too often the space community concentrate exclusively on the science, giving scant regard to communicating the value of their work to the wider community. None of us is expert in all fields, and due effort must be made to enlighten all concerned, whether actively engaged or not, and especially when they are footing the bill. Space efforts are not about the community or just about science. They are also of sociological importance.

Similarly, the space community has often deemed that, by its very nature, space exploration is somehow *special* and hence deserves special attention. But this is not always the case. Space exploration is not special by default; and most people do not see it as special in any way. It is just another facet to human activity, yet it is of immense value; and if it is to be granted resources, the community to whom they are entrusted must earn and justify their requirements. Space exploration is hugely expensive, and the community must be in a position to justify the required resources and put them to legitimate and effective use. And while space exploration is inherently risky, society can be tolerant of those risks if they are identified and discussed, and if every conceivable effort is made to reduce them. This will, however, remain contentious. None of us is immune to the chilling tragedy of losing people to space, especially when we realize that those tragedies could have been avoided by improved effort or by simply waiting for less risky opportunities.

The Human Exploration of Mars

While an ethos of exploration may be innately human, every new venture into the unknown brings with it a great number of contentious issues. Not all expeditions are wise or timely; nor should exploration necessarily be pursued on the basis of ethos alone. Even the most daring individual mountaineer will go to extraordinary lengths to ensure the success of the expedition and the wellbeing of all concerned. And so, while Mars beckons, the question of deep-space human exploration must be thoroughly scrutinized and examined, not only because such expeditions appear to be increasingly likely, but also because they involve society in general and are so immensely risky, technically challenging, and expensive.

A human mission to Mars will arguably be the most challenging feat of exploration in history. Indeed such a mission is beyond us at present. We are far from ready and thc demands are truly colossal—to hurl a group of people across interplanetary space toward Mars, to set them down onto the surface for up to two years, to launch them back into space, and to return them safely to Earth some three years after they first set out. Even if a human mission to Mars were agreed today, it can only happen when a strategy has been devised, when we are technically ready, when all risk and safety issues have been addressed, and when it is affordable. The bottom line is that if all issues and concerns are not robustly addressed and resolved, we cannot send people to Mars irrespective of our desires.

None the less, there are genuine reasons to think that a human mission will happen in the not too distant future. First, there are already significant scientific, political, and sociological motivations to do so—with an already declared political intent on the part of the USA and Europe. Second, and significantly, we have by now identified most of the major challenges. We have a true sense of the scale of the task *and* are beginning to get the measure of it. We have identified a workable strategy, the major risks and safety issues, and have even devised possible mission implementations, of which NASA's *Design Reference Missions* are of particular note. Furthermore, through the phased Vision for Space Exploration and Aurora programs, many of the risk, safety, and technical readiness issues will be tackled in an affordable and capable manner, well in advance of any eventual human mission. Such is the scale of the task that only after a lengthy period of preparatory work through VSE and Aurora will we be in a position to know if we are then ready to initiate a specific program to send people to Mars, perhaps a decade or two later.

We now find ourselves in an unprecedented point in history regarding a human mission to Mars. There is a politically declared interest in Mars by both the USA and Europe, with ESA and NASA committing much of their resources and future missions toward that goal. There are significant scientific reasons for such missions, with many of the key issues identified and, through VSE and Aurora, a phased robotic and human space program set on resolving many of those key issues over approximately the next decade. We are now laying the foundations that will enable a decision to be made in the near future regarding a human mission to Mars.

Political Motivation

In this era of globalization, the USA, Europe, Russia, Japan, China, and India have each identified space as a significant playing field for competition. While direct political rivalry, especially between the USA and China, may become significant, the not unrelated economic and technological benefits are also seen as increasingly important. In particular, the steady progress of China, both as a major political and economic power and as a space-faring nation, has in part prompted the USA (and probably Europe) to refocus their long-term aims for space exploration, lest they be overtaken. There can be little doubt that the rise of China is a significant motivating factor behind the Vision for Space Exploration and Aurora programs. Arguably, the first nation to declare an intention to send people to Mars—and resolutely mean it—will gain significant stature on the world stage for the ensuing decades.

There are, however, other political reasons. In the USA, for example, the Challenger and Columbia tragedies, along with the various high-profile robotic mission losses have prompted all concerned to seriously examine the viability of the Space Shuttle program and, indeed, the entire traditional approach to space exploration. From such assessments it was realized that a new and far-reaching approach is needed to serve the country's long-term political, technological, and scientific space exploration goals through the twenty-first century. Most significantly, it was realized that for a long-term strategy to be viable, it must have clear focus and direction—and involve unprecedented human exploration of the Moon, Mars, and beyond. Hence, the Vision for Space Exploration is a statement of intent by the USA to pursue space exploration for political as well as scientific gain, but it is also a statement of a desire for safe, sustainable, and worthwhile space exploration, with Mars as a significant goal.

Similarly, as the European Union (and ESA) has grown, so has a genuine ambition to contribute on the world stage with regard to science, technology, and innovation. Europe, in transitioning from a relatively stable group of individual nations to a singular global power, recognizes the need for landmark projects, with space exploration seen as both ideologically and economically significant. Cosmic Vision and Aurora both represent the currently accepted way forward, with Aurora representing an unprecedented political declaration for long-term exploration of the Solar System, placing a European person on Mars as pivotal.

Mars is the planet we think we can reach, strategically and technically, in the coming decades. It now represents the next big challenge to those who are confident that, if successful, it will deliver huge rewards. Emerging from a period of relative quiescence regarding human space exploration, global powers seem to be developing an appetite to challenge Mars.

A Human Science Expedition

The scientific case for going to Mars is overwhelming. The question of how best to pursue that science—by robotic explorer or by human expedition—is now being hotly debated. As already indicated, the key issues are cost and risk. At present we could send a fleet of robots to Mars and to many other locations of interest in the Solar System for the cost of just one human expedition to Mars, and to many such a cost is unacceptable.

There is no right or wrong answer to this debate, but there exists a valid case for pursuing cost-effective robotic reconnaissance throughout the Solar System, including Mars. We still know so little about our solar neighborhood, and robotic exploration will remain *the* most effective means of conducting valuable scientific reconnaissance for decades to come. But there are also investigations that are best pursued by direct human involvement. Indeed, in general, it is better if people can be directly involved because of our superior mobility, dexterity, cognitive ability, and so on. We wouldn't dream of sending robots to explore the remote Amazon rainforest or the Antarctic, for example; not just because we can now travel to such remote locations, but also because of the nature of the work—field tests, in-situ analyses, and real-time adaptation of strategy and techniques. There is also the *human factor*: the interaction with things that bewilder, excite, or inspire scientific creativity. All of these would all be severely curtailed if the the work was only performed remotely.

Where possible, we prefer to do science hands-on; and when necessary, we insist upon it. Scientific exploration is not an emotionless process through which simple and static answers are revealed. Rather, it is a very "human" process that demands the sum total of our curiosity, imagination, ingenuity, and inventiveness. Usually, answers only emerge after many rounds of painstaking physical and mental effort. In this context, human science expeditions to Mars take on validity. When we consider the nature of our quest—to understand an entire world over its four-billion-year history and potentially to the molecular complexity of life—it becomes clear that it will require *both* robotic and human expeditions to have any chance of success. And when we consider the implications—uncovering clues to our origins and broadest natural context—human expeditions seem inevitable. This quest is about us. It is about using our analytical capacity to find out who we are, from an arena of investigation that will inspire new ways of looking at and understanding nature. We do not know the outcomes, or the answers we will derive, but we do know from experience that the journey itself will be vital and that currently unimaginable evidence will be uncovered along the way. We are only now beginning to devise ways to send people to Mars, and such is the weight and scope of the investigation that this will become highly desirable at some point in the future, just as it has become desirable to do science hands-on at the four corners of our own world. The path toward human expeditions has already been largely defined by the nature and scope of the quest. It has also been laid down by the nature of Mars itself—a place of far-reaching importance that is also hospitable enough for us to visit and conduct valuable scientific exploration, when we are ready.

For the moment, however, we will not send people to Mars. There is so much still to do that can be done by robots, whether from orbital reconnaissance or with rover-based analysis at the molecular level. Sending people to Mars at this stage would be at the expense of the many robotic expeditions that are needed. For the present, and for the coming decades, robotic exploration is paramount. But at some point in the future the balance will shift. When we have sufficient knowledge of Mars, we will want to explore it in a way that is appropriate to human expeditions. Whatever answers robots provide—whether tantalizing evidence that is difficult to access, or spectacular evidence of past and present life—the answers gleaned will be neither final nor sufficient. We will then want, or probably need, to investigate further, upturning every stone, digging at every site and exploring every valley, crater, and mountain range in search of conclusive answers. When that time arrives, it will be essential to begin human expeditions. Indeed, we have already in part reached this point. There have no doubt been many occasions when the Jet Propulsion Laboratory has considered many highly desirable investigations for both Spirit and Opportunity, yet rejected their pursuit because the tasks cannot be done by robots. How many vital clues have both rovers passed by because JPL did not spot them, or because they are simply too inaccessible. At some stage the cost, however reasonable, of sending yet another robot will seem like poor value for money and where sending people, however expensive, will be more acceptable owing to the incalculable potential gains. When that time comes, the scientific case for human exploration of Mars will become a reality.

There are other motivations, especially political, that will affect a decision to send people to Mars, but the nature of Mars will become sufficiently clear over the next decade or two to confirm the scientific case for a human expedition. In preparation for such a mission, both NASA and ESA are, through VSE and Aurora, extensively preparing themselves.

Sociological Considerations

The decision to send people to Mars will be one of historic proportions. Not since the sailing ships that sought out the ends of the Earth will people have engaged in such far-reaching exploration.

Sending people to Mars will be a statement of intent: of humans set on a long-term exploration, and on shaping, as best they can, their own future. Whether engaged by just one superpower or by the global community, it will be a world project, a *future-enabling* endeavor with incalculable rewards. Political gains aside, arriving at a deep understanding of the origin, nature, and cosmic context for life will have a major long-term impact. Humanity will see itself in a new and unprecedented context—and however that context evolves, it will be different to how we see ourselves today. It may not immediately impact on our daily lives, but profound discoveries about our nature can only but have a profound impact on our long-term view of ourselves.

The act of setting people down onto another world will also be of significance. There will be no going back. Humanity will have visited a different planet and that will affect how we see ourselves. From launch to their eventual return, those who go to Mars will be subject to incomprehensible stresses, dangers, elation, and awe such as no one alive has ever encountered. Their survival will not be guaranteed, however well prepared, and the help they may require at times might not be forthcoming from Earth. It will be a journey whose outcome cannot be assured before they set off. It will be exploration in its purest form, and we will all take part. The excitement and trepidation, among many other emotions felt by the crew along the way, will surely be echoed back on Earth. We cannot yet predict how a successful landing and safe return, or a disastrous outcome, will impact on society in general. The risks will be enormous and the stakes will be even higher.

The prospect of a mission will also no doubt spark intense debate about who we are as people and what our priorities are. As the mission becomes a reality and unfolds, all the sociological issues regarding our history, our future, our origins and nature will raise their heads. There will also be issues of environmental planetary protection and questions of the ownership of celestial territory.

From the moment the small band of people are launched toward Mars, it is likely that, however contentious, the vast majority of the planet will be firmly and passionately behind them. We will become a single global community with a shared concern for their wellbeing, protection, and safety. They will become a symbol of unity, hope, and goodwill, which is rare today. The mission could become a catalyst that precipitates the best qualities in people and society, providing some measure of belief that we do have a future worth working toward, that we are part of an unfolding human story, and that we can contribute in a positive way for the benefit of future generations.

Planetary Protection 19

Since the dawn of the space age, contamination, biological or otherwise, carried from Earth to outer space and brought back to Earth from outer space has been an issue of concern. Contamination of other celestial objects originating from Earth is called *forward contamination*, while that from other objects upon Earth is called *back contamination.*

With the launch of Sputnik 1 in 1957 the International Council of Scientific Unions, now the International Council for Science, established in 1958 its Committee on Space Research (COSPAR). The purpose of COSPAR was, and continues to be, to promote an international level for scientific research in space, with emphasis on the exchange of results, information, and opinions; and to provide a forum for the discussion of problems that may affect scientific space research. In its first years, COSPAR played a significant role as a bridge between East and West regarding cooperation in space. The most significant outcome of its efforts was the establishment in 1967 of The International Space Treaty, which, apart from creating a basis for international fairness and parity regarding space science and exploration, developed the first Planetary Protection policy to which all space-faring parties signed up, stating in part that "Parties to the Treaty shall pursue studies of outer space, including the Moon and other celestial bodies, and conduct exploration of them so as to avoid their harmful contamination and also adverse changes in the environment of the Earth resulting from the introduction of extraterrestrial matter, and where necessary, shall adopt appropriate measures for this purpose."

At the time, the only human missions into space were into low Earth orbit, except of course the first landings on the Moon between 1969 and 1972. Although COSPAR-based back contamination countermeasures were generally followed, it was quickly realized that the surface of the Moon did not pose a toxic or biological contamination treat. Furthermore, with no human missions beyond low Earth orbit in the intervening decades, the lion's share of concerns about contamination have concentrated on forward contamination of the other worlds of the Solar System visited by our many

K. Nolan, *Mars, A Cosmic Stepping Stone*,
DOI: 10.1007/978-0-387-49981-9_19, © Praxis Publishing, Ltd. 2008

robotic explorers. Indeed our approach here has been quite specific, requiring stringent limitations on forward contamination of all celestial objects, but especially those providing insight into questions of origins and of the possibility of life elsewhere, to ensure that scientific exploration or the advancement of knowledge is not impeded. Although this approach seems (and is to an extent) laudable, it is worth noting that the emphasis has been on safeguarding our ability to conduct scientific research, not on safeguarding the purity of any prospective environment we visit. The requirement is for contamination (microbial or otherwise) carried from Earth on board our spacecraft to remain below the detection limits of our instruments when looking for clues for life-related activity on other worlds; but there is no absolute requirement for the protection of potential indigenous life on those worlds. It has been realized, none the less, that it would be impossible to completely sterilize our space hardware; and while we may not specifically cite the protection of potential microbial life on other worlds of the Solar System, the COSPAR treaty and resulting recommended countermeasures are stringent.

Overall, there are five categories to consider with regard to contamination. Category 1 deals with objects such as our Sun and the planet Mercury, of which no containment countermeasures are required. Category 2 considers missions whose trajectory approaches no celestial objects and again no particular countermeasures are needed. Category 3 includes fly-by craft and orbiters to targets of biological interest and where a risk of contamination therefore exists. Countermeasure procedures here include a requirement of the biological-burden (microbial life likely to reside on the visiting craft) to be determined, documented, and declared, as well as some clean assembly and trajectory measures to minimize the risk. This is the first category relevant to the exploration of Mars. Category 4 deals with lander missions to targets of chemical evolution and/or origin of life interest. In the special case of Mars, this category has been divided into three subcategories, with anticontamination measures compared against the Viking landers; such was the stringency of countermeasures applied during their assembly:

- Category 4(a): Missions to sites not carrying instruments to search for extant life must carry a biological burden not greater than the Viking lander presterilization levels (i.e. the probe is clean but not sterilized).
- Category 4(b): Landers looking for extant life must be clean-assembled and completely sterilized to at least the post-sterilization levels of biological-burden of Viking—or higher if the investigation requires it.
- Category 4(c): Lander missions to special regions require post-sterilization to Viking lander missions for all aspects of the landing

assembly—the descent capsules, parachutes, airbags, etc.—and not just the probe carrying out the scientific investigation.

Finally, Category 5 deals with all Earth return missions, which are called *restricted Earth return missions*. Here the outbound part of the hardware must comply with Category 4(b) regulations to avoid *false-positive* identification of life and for any acquired sample to remain completely sealed. Furthermore, the sequencing of mission vehicles must break the chain of contact between those in contact with the target and those returning to Earth, while the mission must be continuously monitored, with immediate analysis upon return to Earth to determine the potential biohazard.

Redeveloping Planetary Protection Policy

Forward Contamination

Over the decades, COSPAR guidelines, ESA's and NASA's own internal procedures, and recommendations from such organizations as the Space Science Board (part of the US Nations Research Council) have all provided for a generally robust and widely accepted level of measures to ensure minimal forward and back contamination. But there have been, as we have seen throughout this book, significant recent developments in our understanding of the robustness and diversity of microbial life, our understanding of Mars, and the imminent prospect of deep-space human missions there, for example, that all prompt serious re-examination on how planetary protection is perceived, managed, and pursued.

With recently renewed recognition of a possibility of past and present life on Mars, issues of both forward and back contamination take on a new relevance. It is our hope—based on scientific grounds—to find evidence of prebiotic chemistry, extinct, and even extant life on Mars, and in this light issues of contamination need to be pursued in a more sophisticated manner. For example, given our new awareness of the diversity of microbial life on Earth and the sophistication of current instruments, we now realize that Viking post-sterilization levels of contamination likely to be on board a Mars lander are not sufficient to guarantee that a false-positive result will not arise. So sensitive are our instruments, so tentative may be the evidence, and so adaptable may be Earthly microbes, even on Mars, that upcoming missions will require greatly enhanced clean-assembly, sterilization, and auditing of the bioburden upon our exobiological landers. Here we will have to research the nature, survival, and growth of bioburdens upon spacecraft in conditions matching interplanetary space, as well as the surface of Mars, from which new and sophisticated techniques of assembly and sterilization must be developed.

Ethical issues also arise with the prospect of detecting life-related activity on Mars. For example, should a positive identification of life be made on Mars by robots, the question of whether to send people there would take on a new significance. Currently, there is no clear policy on such an eventuality, although it is unlikely that people would be sent to Mars until any detected life was well characterized and the potential biohazards determined and managed. Indeed, both Aurora and VSE aim to determine the existence of life on Mars and will only subsequently make a decision to send people there. Even so, sufficient countermeasures are not yet in place and must be developed before any human mission can take place.

Also, it is currently unclear how we value alien life. Indeed no legal framework and/or code of behavior yet exists regarding the protection of Martian microbial life, should it be found. This may at first seem unimportant, but should such a discovery be made it will become immensely critical not only with regard to treatment of the life itself but also regarding any international agreement on how to deal with the discovery. There are two broad approaches to the ethical dimension to Martian life. One school of thought proposes that Mars should then be left alone and that, however basic or sparse, any existing Martian ecosystem has a right to exist, develop, and evolve of its own accord. Furthermore, where benefits can derive from studying Martian life—such as understanding our origins, the nature of life there, and a general context for all biology with the potential medical and technological spin-offs—our interests will best be served by pursuing our investigations with the utmost care and by ensuring that there is no fundamental impact upon the Martian biosphere. A second school of though argues that, being part of nature and the evolutionary process, all actions by us, however invasive upon Martian life, are also part of nature and are therefore beyond ethical questioning. Hence, a course of action that provided maximum benefit to humans is reasonable.

However we pursue Mars, even our purest intentions will be, and have already been, invasive. As discussed above, our science interests to date have only acted to safeguard the Earth, not prospective Martian life, and it is already certain that the space probes that have landed on Mars have carried microbial life from Earth that may survive in the long term. And although the rate at which we are exploring Mars may seem low, there is no telling how we might impact upon the planet over a prolonged period, despite current countermeasures. We are already placing Mars in a long-term position that is unclear regarding the survival and spread of earthly microbial life there, irrespective of the existence of indigenous life. When we consider the full gauntlet of potential interest in Mars—human missions and outposts and resource utilization in the coming decades—ethical issues may not be considered. If our record on Earth is a benchmark, then unless there is a significant change in how we value life and the natural environment, Mars will fair no better.

But here there is an opportunity. If so much of what is done badly with Earth's environment is simply bad habits, poor planning, and historical legacy, then we can, by default, explore and pursue Mars in an improved manner from the outset, better safeguarding its environment while also protecting our long-term interests. Indeed, when we consider the possible benefits even for the medium term, such a strategy is not only wise, but probably necessary. With those benefits mostly being of a scientific and sociological *information* nature, and with potential biological and medical benefits (should evidence of even sparse life be found), it is surely incumbent upon all concerned, irrespective of ethical stance, to vastly improve how we deal with forward contamination on Mars. Indeed, the scientific community can take the lead in this respect. Even if society at large cannot agree on an ethical stance, or if economic and political forces eventually come to the fore, the head start that the scientific community now enjoys can provide an opportunity for a statement to be made—and for the foundations to be laid—regarding the best way to pursue Mars for the maximum benefit of the Martian environment and for our hopes and aspirations.

Back Contamination

With the prospect of a Mars sample return mission within a decade, and human missions within three decades, the policy surrounding back contamination also needs to be redeveloped to a far greater degree. Two issues arise in this area: (1) the biohazard risk associated with robotic sample return to Earth, whether or not we know it to contain life; and (2) the risk to humans who travel to Mars and the associated risk to Earth upon their return. In both cases, it is incumbent upon the space science community to recognize the scale of these risks and to take appropriate action, irrespective of the short-term impact upon the quality of the science. As with other areas of earthly environmental science, the risk is not simply a function of the likelihood of microbial life within a given sample, but a product of that likelihood and the worst case scenario's impact upon Earth's environment. While the likelihood of life on Mars and in any returned samples is extremely low, a worst case scenario—where, for example, microbial life was accidentally released into Earth's biosphere and impacted catastrophically on human life or agricultural yields—constitutes a truly unacceptable risk with potentially horrendous consequences. In the absence of valid testing or of any meaningful information, our communal responsibility demands that we regard the risk as extremely high, and for extreme back contamination measures to be implemented. In any case, national and international governing bodies, environmental protection agencies, agricultural boards, and the general public would all find it utterly unacceptable to allow any potential biohazard anywhere near the planet without extraordinary countermeasures put in place.

Mars Sample Return

Any sample return mission from Mars will represent a significant biohazard; and while COSPAR's Category 5 containment recommendations are sufficient, their implemention represents a significant challenge (but is now being hotly pursued by COSPAR and the major space agencies of the world).

While we can develop a mission strategy that can break the chain of contact near Mars, it is in the secure handling of the samples *en route* back to Earth, their safe arrival on Earth, and their subsequent secure handling for scientific analysis across many locations on Earth that pose the major challenges. We will have to develop proven failsafe containment systems that completely guarantee safe delivery of samples from Mars to contained environments on Earth for subsequent analysis. Other measures will also be required, such as combustion or sterilization of the samples before their arrival back on Earth. Currently, several groups are working to develop verifiable containment systems, to be ready before our first sample return mission some time in the next decade.

People Visiting Mars

Protecting people who visit Mars, and subsequently protecting Earth from their return, will present an enormous challenge. As with current robotic exploration, where forward contamination cannot be fully prevented, absolute guarantees of protection of people who land on Mars will never be possible. The decision to send people there will therefore constitute a risk, where the best we can hope to achieve is to minimize that risk through initial surveys of the planet as a whole, the surface environment, the atmosphere, and finally any chosen site, followed by appropriately designed anti-contamination measures. If life is detected prior to sending people to Mars, however, a far-reaching analysis of that life will need to be made to access its potential threat, perhaps delaying a human mission by years, if not decades, until the threat is fully appraised and can be adequately dealt with. However, if we detect only the presence of past, fossilized life or no life at all, we will ultimately have to make a judgment based on incomplete information as to the potential biohazard to the crew. Similar to the situation regarding forward contamination of Mars by robots, we will probably reach a point where, even with incomplete information, we will deem the risk worth taking for the sake of exploration and scientific advancement. We will in any case be able to take many precautions. We are rapidly becoming technologically and logistically sophisticated enough to reduce any threat to potentially molecular levels, meaning that we can detect even molecular level threats and can safeguard against them. Even where contamination was to occur, despite the ethical dilemmas arising, the lengthy nature of the stay on Mars would probably provide sufficient time to gain insight into the nature of the contamination and threat to people and other terrestrial biology. Furthermore, any remaining uncertainty regarding the nature of the threat could then be dealt with within the vicinity of Earth, where the crew and/or other materials to be accessed could be performed on lunar bases, in Earth orbit, or within adequately contained environments on Earth.

Future Actions

It is now clear that many of the issues of forward and back contamination, and planetary protection in general, remain unresolved and need urgent action. The existing laws, legislation, policies, and procedures are inadequate to deal with our aspirations of sample return and human missions to Mars. Indeed these and the many ethical issues regarding our view of, and approach to, Mars also require urgent widespread and public debate, and here there may be a role for such outreach organizations as The Planetary Society as well as the leading astronomical organizations of the world.

In deference to the outstanding issues however, all the leading space agencies, science organizations, and even the United Nations (through COSPAR) are now frantically engaging the issues. NASA, for example, recently requested the US Space Science Board to conduct a review of planetary protection policy, followed by recommendations. ESA is also actively engaging a number of workshops on how best to proceed from here. Issues to be addressed include greatly reducing forward contamination on Mars, developing verifiable failsafe containment systems for a sample return mission, as well as initiating a far-reaching debate regarding the ethical issues surrounding life on Mars and back contamination. Subsequently, national and international treaties, policies, and laws will have to be updated if our newly developed procedures and methodologies at safeguarding both Earth and Mars are to be effective in this time of increased interconnection between both worlds.

Following NASA's request, the Space Science Board conducted an extensive review, publishing their findings in August 2005, with the following recommendations:

- For the development of superior sterilization techniques—to the molecular level—and reflecting our increased understanding of the robustness and diversity of microbial life on Earth. These include steam, gamma radiation, and hydrogen peroxide plasma sterilizing technologies.
- For formal recognition of the increased importance of Mars, including making all of Mars a *special region.*
- For the development of verifiable containment systems for a sample return mission.
- For research to be conducted into the nature of the biological burdens that are carried to Mars on board our space probes and for research into how those particular microorganisms behave in Martian environments.
- For a wide debate, involving the public, regarding the ethical issues of alien life in general, regarding making all of Mars a special region and regarding back contamination.
- For NASA to fund research into new sterilizing technologies and the development of failsafe containment systems.
- For NASA to hold workshops with COSPAR, ESA, among others regarding all planetary protection issues.

The Space Science Board also recognized an urgency to these issues, given our rapidly changing and improving understanding of Mars and our already operating long-term robotic program, imminent sample return, and possible human missions. To this end it also recommended some interim steps that need to be taken immediately if a sample return mission in particular is to take place within the next decade:

- For current planetary protection policy to be updated to reflect recent findings regarding microbial life on Earth and about the nature of Mars.
- In particular, for major efforts to distinguish special regions, or to consider all of Mars a special region.

- Special measures to be implemented to protect already known special regions.
- NASA to initiate an adequately funded transition to more resilient planetary protection measures.
- NASA to immediately initiate new workshops and research into a long-term planetary protection policy to match its ambitions on Mars.

Furthermore, the following time line was proposed by the Space Science Board:

- A new planetary protection road map to be immediately developed.
- Testing of new sterilization techniques to be completed by 2008.
- Analysis of space probe biological burden, and its likely impact on Mars to be determined by 2012.
- Fail safe containment systems to be developed prior to a sample return mission.
- All interim policies and systems to be in place by 2016.

Combined with the practical necessity to protect Earth as best as possible, we have much to think about and decide upon regarding how we value life on Mars, and indeed all life. These issues force our hand to examine life in a broader context, to think of ourselves in a broader context, and to think about our long-term wellbeing. The decisions we come to will in part define who we are. In this light we have an unique opportunity to take a valued stance on an issue new to our time, but it is also one on which we will be judged by future generations. A direct ethical connection has been established between our humanity and the possible microbial make-up of Martian soil.

20 A Human Mission to Mars

Both NASA and ESA have set their sights on sending people to Mars. Indeed all launch opportunities and mission timings until 2048 are already computed and known, with 2033 standing out as an optimum opportunity. And although that's a long way off, even if both agencies were to combine their efforts and work flat out from this moment on, they would be extremely fortunate to reach mission readiness by that date. Sociological, political, and scientific commitment aside, there are truly enormous technical and human health and safety obstacles to overcome before we will be ready to send people to Mars. It will be so difficult compared to sending people to the Moon, for example, that no meaningful comparison between the two can be made. It will be among the most ambitious undertakings, ever, by humanity.

So challenging will such an expedition be that we will have to harness—and obey—the laws of celestial mechanics in ways traditionally seen as fitting only for robotic craft. Yet we will have to do so in the most exquisite way, with no margin for error. Furthermore, the particular routes taken to and from Mars will critically impact on all other aspects of the mission—propulsion systems, the required amount of fuel and quantities of human consumables, whether to stay on Mars for 30 days or 500 days (the only two currently feasible options), and even the velocity with which the astronauts enter Earth's atmosphere on their return home. Every facet to the mission will be interconnected in what will be the most extensively orchestrated space operation ever undertaken. The individuals on board will have to live under conditions never before endured: confinement in a completely sealed environment for upward of three years, months *en route* to Mars under the detrimental influence of near-zero gravity, followed immediately by having to perform strenuous survival and exploration activities on the surface, where health, psychological, and mental endurance will be pushed beyond currently known limits. The sum total of effort in reaching the summit of Everest, the North and South Poles, the depths of the oceans, and even the Moon pales by comparison. In every conceivable way, a successful human mission to Mars will be an achievement of historic proportions.

K. Nolan, *Mars, A Cosmic Stepping Stone*,
DOI: 10.1007/978-0-387-49981-9_20, © Praxis Publishing, Ltd. 2008

Although we have pondered such a mission throughout the space age, it is only recently that we have been in a position to seriously and realistically engage the idea—and then at a stretch. The mission is now possible owing to the extensive legacy of work done in Earth orbit and on the Moon over the past 50 years, together with the experience of prolonged stays in space on board Skylab, Mir, and the International Space Station (ISS), human missions to the Moon, and extensive orbital, lander, and rover reconnaissance on and around Mars. The wealth of information and insight accumulated from all those efforts has brought us to the point where we can begin to grasp the magnitude of a human mission to Mars, the obstacles and issues to be encountered, and what we must be do to succeed.

Furthermore, since the 1990s, several in-depth studies of possible mission scenarios, strategies, and implementations have been conducted. Coupled with current and projected advances in technology over the next few decades, as well as an unprecedented interest in Mars across the civic divides, we can now consider the possibility of implementing a human mission more earnestly than ever before. And it is within such a context that both the Vision for Space Exploration and the Aurora programs have found legitimacy in laying the foundations for such a mission. For the first time, the various factors required to make a human mission to Mars possible are beginning to align.

Since the 1990s, many of the defining characteristics and requirements for traveling to Mars have been known to us. A number of separate in-depth design reference missions (DRMs) such as NASA's *Design Reference Mission 3*, Robert Zubrin's *Mars Direct*, The Mars Society's *Mars Society Mission* and ESA's *Study for a Human Mission to Mars*, have all extensively analyzed the major issues and offered several (but not dissimilar) viable options for a human mission to Mars. Indeed, Robert Zubrin has inspired many of the features now common to all considered mission strategies, being among the first of recent times to bring forward the notion of a realistic human mission to Mars in his book *The Case for Mars*. Critically, all studies have identified the already mentioned inextricable interdependencies among the various aspects of the mission. Also, proposed routes to Mars within all DRMs follow those first suggested by the German mathematician W. Hohmann in 1925. Given the rocket technology that is likely to be available over the next 50 years (chemical rockets and possibly nuclear thermal rockets), there are essentially two available routes. The first is called a *conjunction class* route, referring to the fact that Earth and Mars are on opposite sides of the Sun during launch. This mission would require the least amount of fuel because the Sun's gravity would provide some assistance *en route*. In this scenario, the crew would take six months to travel to Mars, stay on the surface for 18 months (about 550 days) and take six months to return home, giving a total mission time of about 900 days. In a variation of this type of journey, called *a long stay, rapid transit* journey, the astronauts would travel to Mars along the same route, but more quickly, requiring more fuel or new and more efficient rockets. While the total length of the mission would remain the same (governed by the motion and positions of Earth and Mars), the quicker journey would enable the astronauts to be less exposed to radiation *en route* and spend more time on the planet's surface.

Another option, called an *opposition class* mission, would launch when Mars is on the same side of the Sun as the Earth. With a total mission time of between 400 and 650 days, the crew would again travel to and from Mars in about six months, but with their stay on the surface restricted to 30 days or so. In a variation on the opposition class journey—called a *Venus fly-by* mission—the crew would still travel to Mars while it is in opposition, but would fly near the planet Venus and use it as a gravity assist, either *en route* to Mars or on the return journey. While this provides a significant saving on fuel, the crew can spend even less than 30 days on the surface, and would also be exposed to increased radiation when traveling so close to the Sun.

Initial Considerations

The most favored of all missions is for a conjunction class trajectory, requiring a stay on the surface of about 550 days. This method demands minimum energy (and fuel) but provides maximum science and exploration returns. For such a lengthy mission, however, an enormous amount of infrastructure will be needed to support up to six people while traveling to and from Mars and during their time on the surface. Apart from the spacecraft that will carry the crew to and from Mars, the mission will require a Mars lander and Mars Ascent Vehicle (MAV), surface regolith moving vehicles, surface habitation modules, surface power generators, in situ resource utilization equipment, robotic rovers, pressurized drivable rovers, scientific equipment, computers, communications and navigation equipment, water, food, and many other consumables. Not only do all design reference missions generally agree on such an inventory, but they all indicate that the overall mass of those support materials will be in the region of 80 tonnes, comprising approximately 40 tonnes of surface infrastructure and equipment and 40 tonnes of consumables.

Knowing this figure provides important constraints for devising a mission architecture. We must remember that all materials ending up on

Mars originate on Earth and require an enormous effort to get there. Everything sent to Mars—whether people, vehicles, equipment, or consumables—must follow the same journey. Consider a cargo mission to Mars, for example. All materials, including the mission vehicle, its fuel, and the cargo itself must first be launched from Earth's surface into low Earth orbit (LEO), where some orbital assembly may be required. Once ready to head toward Mars, a second launch maneuver called a Trans Mars Injection (TMI) allows the entire vehicle to break Earth orbit and establish its route to Mars. During this maneuver an enormous amount of fuel will be used. If using chemical propellant, upward of two-thirds of the entire mass of the spacecraft must be fuel for the Trans Mars Injection maneuver alone. On arrival at Mars, the spacecraft will settle into orbit, either by using yet more fuel to slow it down, or by performing aerobraking maneuvers similar to those performed by the Mars Global Surveyor, only more intensively. Finally, once in orbit the cargo will be released from the Mars orbit and will land on the surface via a controlled touch down, again requiring fuel.

In all of this, virtually the entire mass that originally launched from Earth will comprise fuel and the spacecraft. Indeed, we know that for every 1,000 kilograms launched from Earth, just 66 kilograms will make it into LEO and only 12 kilograms can be set down onto the surface of Mars. In other words, approximately one-hundredth of whatever leaves Earth can be placed on Mars as usable cargo; or one-fifth of the material that leaves low Earth orbit reaches Mars, known as the *five-to-one* rule. This figure tells us, that to send 80 tonnes of equipment and consumables to Mars, we will need to place five times that mass or 400 tonnes into LEO, of which 320 tonnes will be fuel and the spacecraft itself, with the remaining 80 tonnes comprising the cargo.

To put this into perspective, the Saturn V rocket that took people to the Moon was capable of placing about 120 tonnes into LEO. If we wished to send people to Mars by a single launch from Earth, we would require a rocket perhaps four to five times more powerful than the Saturn V. Even today, however, we cannot envisage building such a rocket; and so it is already clear that a human mission to Mars via a once-off single rocket launch from the surface of Earth to Mars will not be possible. Rather, we will need to lift all that travels to Mars into low Earth orbit in smaller units and carry out some assembly in a manner not dissimilar to the construction of the International Space Station. Indeed, it is now envisaged that the approximate 40-tonne surface infrastructural cargo and 40-tonne consumables cargo will be sent to Mars separately, requiring two separate Mars cargo spacecraft, each of mass around 200 tonnes, and to be constructed in low Earth orbit; with the human crew traveling on a separate, third mission, which will also carry a similar mass payload.

If the objective is to set about 40 tonnes of material down on Mars, each of the Mars spacecraft constructed in low Earth orbit will need to have a mass of 200 tonnes (applying the five-to-one rule from LEO to Mars). Even 200 tonnes is too massive to place into low Earth orbit all at once however; and so for each mission to Mars (whether human or cargo) we envisage the use of two rockets as powerful as Saturn V, each launching about 100 tonnes into LEO; where a rendezvous and assembly maneuver of the two segments will form a single Mars spacecraft of 200 tonnes mass ready to go to Mars (Figure 107). Indeed, the USA is already planning for such an eventuality, with two new rockets currently under development within its Vision for Space Exploration's *Constellation Program*. Given the general title of *Shuttle Derived Vehicles* (SDVs)—because they derive much of their capability from Space Shuttle technology—the larger of the two, called the *Cargo Launch Vehicle* (CLV) is capable of launching up to 125 tonnes to LEO, supporting both goals of returning people to the Moon as well as the construction of Mars spaceships in LEO. The rockets are due to be ready by 2012.

Hardware, Vehicles, and Habitation

The hardware involved in a human mission to Mars will be of staggering proportions. The assembly of each Mars spacecraft in low Earth orbit will in itself require a new generation of orbital-assembly techniques even greater in scale than for the ISS. With regard to the mission itself, a great number of different vehicles and large-scale hardware will be needed.

For example, if we envisage a human expedition to Mars comprising three missions in total—two initial cargo missions followed by one human mission—all three will require a 200-tonne spacecraft (discussed above) for traveling from Earth orbit to Mars orbit. This spacecraft, called the Mars Transfer Vehicle (MTV), will comprise two main sections but can be configured appropriately for each individual mission. For example, while transferring cargo, an MTV will comprise a propulsion section (either chemical or nuclear thermal rockets) and a cargo section, but when transferring people to and from Mars, it will comprise a propulsion section and a Transfer Habitation Module (THM).

The hardware to be deployed on and around Mars ahead of the crew will also be substantial. For example, a Mars lander will need to be placed into orbit in advance by a cargo mission, to ensure that when the crew arrives, they can dock with it and travel to the surface. Furthermore, along with the

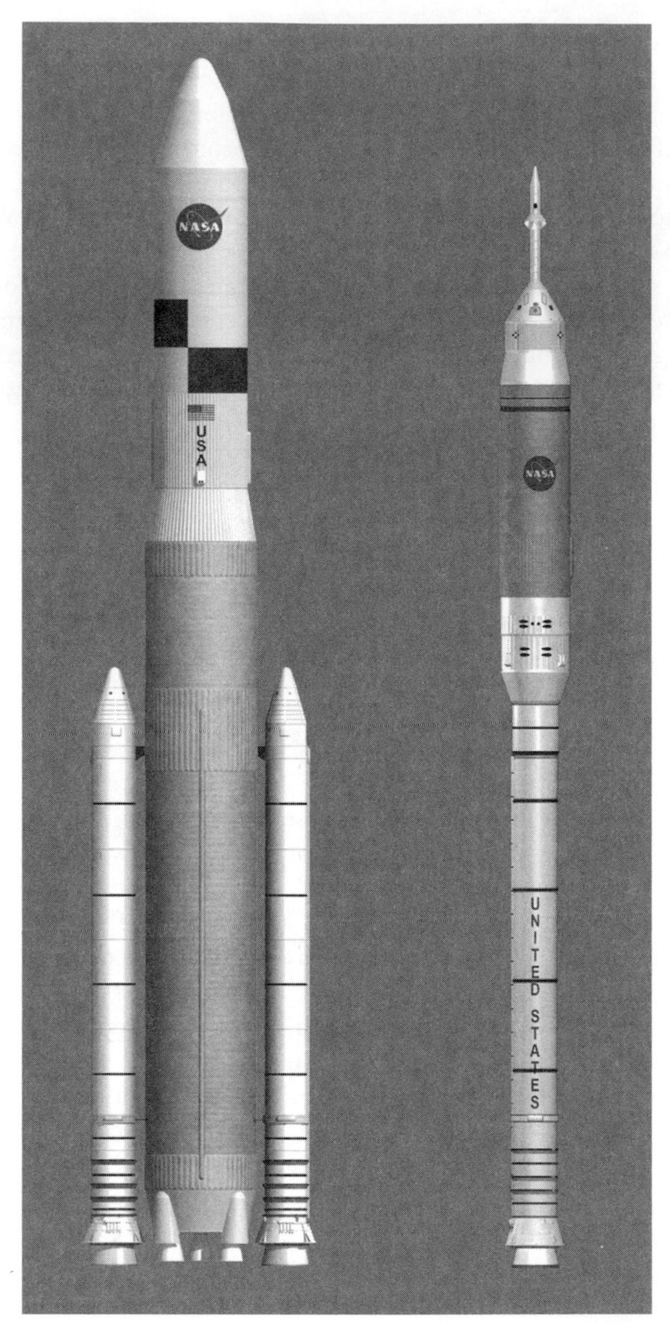

Figure 107: NASA's two new SDV rockets, the Cargo Launch Vehicle—Ares V (left); and the Crew Launch Vehicle—Ares I (right). Ares V can lift up to 125 tonnes into low Earth orbit, adequate to build the spacecraft required for a human mission to Mars. [Credit: NASA/John Frassanito and Associates]

surface habitation unit, support infrastructure and consumables and a Mars Ascent Vehicle (MAV) must also be placed on the surface in advance to eventually launch the crew from the surface when the expedition is over. Also proposed by NASA's DRM3 (but not by Mars Direct or the Mars Society Mission)—and to be deployed in advance into Mars orbit by a third cargo mission—is a separate Earth Return Vehicle (ERV) that will return the crew from Mars orbit to Earth orbit.

Among the most important issues to resolve is that of propulsion. Incredibly, two-thirds of the required propellant to get to Mars will be used in an initial Trans Mars Injection phase. Here, the Mars spacecraft must blast out of orbit, escaping Earth's gravity altogether to establish a viable velocity and trajectory to reach Mars. Currently, the only way to do this is via traditional chemical rockets that use a mix of hydrogen and oxygen, which chemically combine to create rocket thrust. And while chemical rockets are feasible, more efficient rockets are highly desirable for Mars missions, especially for the human part of the expedition. NASA, in particular, proposes the use of a new and altogether different type of rocket, called a nuclear thermal rocket (NTR). An NTR is not based on nuclear fusion or fission, but rather on radioactive decay. In this system, nuclear-based radioactivity heats gas to an extremely high temperature, producing an extraordinary rocket thrust. An NTR can produce the same thrust as a traditional rocket, but requires only one-third the amount of fuel. Hence NTRs, considered feasible given one or two decades of development, would provide a viable option either for more fuel-efficient conjunction class minimum energy trajectories when sending cargo, or for more powerful and hence quicker conjunction class rapid transit journeys when sending people.

While NASA currently indicates its intention to use nuclear thermal rockets, many questions remain about their viability. First, with no indication of serious development work before 2020, the use of NTRs would arguably push a human mission out to at least 2040. Furthermore, with the political and environmental sensitivities of carrying radioactive materials into LEO, it is generally regarded that an NTR would only be allowed to ignite at a higher altitude—perhaps 1,000 kilometers or more. Hence, the added requirement of raising an NTR-powered vehicle from LEO to a higher altitude by using additional chemical fuel and/or some sort of tug vehicle greatly negates the fuel-saving dimension to using NTRs. Neither the Mars Direct nor MSM propose using NTRs, and have built their designs around the use of chemical propellant.

Although the substantial lack of cloud cover near the equator on Mars means that solar power is an excellent option for small robotic explorers such as Spirit and Opportunity; it would be inconceivable to expect solar power alone to provide the power requirements of an 18-month expedition supporting six people. Hence a need has been unanimously identified across all design reference missions for at least one and perhaps two small nuclear reactors, each capable of delivering about 150 kilowatts, to be placed on the surface of Mars in advance of the human mission. These small yet powerful units would provide significant power for all the needs of the mission, and with a life span of perhaps 15 years or more, may even serve several complete expeditions to the surface over a decade or two, without the need to deploy further power requirements for each subsequent expedition. Of course, the deployment of nuclear power on the surface of Mars will be an issue of great international contention, demanding significant development to ensure safe launch, transport, and operation. When finally on the surface, it is envisaged that the reactors would reside at a distance of about 1 kilometer from the habitation site, with robust power cables running across the landscape to the habitation site.

Another requirement, of course, will be for fuel to power the Mars Ascent Vehicle that lifts the crew into Mars orbit for their return home. All design reference missions have identified that approximately 30 tonnes of chemical propellant will be needed to achieve orbit. Again applying the five-to-one rule, delivering this fuel from Earth would require approximately 150 tonnes to be placed into LEO—essentially another full cargo mission just to deliver the MAV fuel. However, given the recent identification of water within the uppermost layers of soil across much of Mars, and the abundance of carbon dioxide within the atmosphere, a chemical process called the Sabatier Process may be exploited to extract hydrogen and oxygen from the water, and carbon from atmospheric carbon dioxide, yielding methane and oxygen that can be used as rocket propellant. Hence, it is now generally regarded that when the MAV is sent to Mars dry, its fuel will be manufactured on the planet. Although at first this may sound far fetched, the process can readily yield the quantities of propellant needed if the proposed nuclear reactors are deployed onto the surface. Indeed, so straightforward is this process when compared to many of the more complex technical challenges facing a human mission, that it is the preferred option across all current design reference missions. Here, an in-situ resource utilization (ISRU) unit brought by one of the initial cargo missions would produce the necessary fuel for the duration of the human mission, but would be operating well in advance of the departure time, ensuring that the MAV was fully fueled and capable of achieving orbit.

Despite the identification of plausible mission strategies and of the required hardware, the challenges ahead are significant. The political sensitivities of realistic nuclear thermal rockets and portable nuclear reactors alone will be significant hurdles to overcome. But the development of habitats, a range of vehicles, ISRU units and the myriad of surface support infrastructure—all required to operate fault free for years on end—represent equally daunting challenges that are well beyond our current capability. None the less, both ESA and NASA are now preparing to tackle each of these issues seriously, and if they succeed they will have carried the entire space program into a new era that makes us ready for widespread space exploration well beyond low Earth orbit and to many other destinations. With new heights in global cooperation, a long-term program for Mars, and the spectacular technical innovations at our disposal, many other opportunities in space exploration will, by then, also be possible.

Design Reference Mission 3

Through the 1990s, both Robert Zubrin and NASA's Johnson Space Center (JSC) devised design reference missions for Mars. Although Robert Zubrin's Mars Direct offers as viable a proposal for sending humans to Mars, we will examine JSC's Design Reference Mission (DRM) version 3, for no reason other than it has driven much of NASA's current thinking on the issue. However, no particular DRM is at this stage more valid than the others. Critically, they have all influenced the specification of the pending SDVs (Shuttle Derived Vehicles), but beyond that, DRMs at this stage simply serve as guidelines for implementing a human mission to Mars, with a view to triggering further ideas and innovations that may eventually enable an actual mission to take place.

JSC's DMR 3 outlines a human mission to Mars whose objective is to carry out scientific investigations on the surface and to examine the feasibility of a long-term outpost on Mars. It envisages four missions in total: three cargo missions and one human mission. The timing of these missions will be crucial. Every 26 months Earth and Mars return to approximately the same relative position with respect to each other. Hence, launch windows or *opportunities* from Earth to Mars arise only every 26 months. For DRM 3, all three cargo missions will be launched within just weeks of one another during one particular launch opportunity, called Opportunity 1. DRM 3 proposes that all cargo missions will use nuclear thermal rockets and follow a conjunction class minimum energy trajectory. The first cargo mission, called ERV-1, will deliver a chemical propulsion Earth Return Vehicle (ERV) into Mars orbit, where it will remain for approximately four years, waiting to return its human occupants to Earth. Mission two, called Cargo-1, will deliver a host of equipment to Mars—the dry MAV, surface solar power, robotic rovers, unpressurized and pressurized human rovers, plants and feedstocks, science laboratory equipment, and ISRU equipment. Mission three, called Hab-1, will deliver the habitation unit, nuclear power unit(s), and consumables.

With all support infrastructures in place on and about Mars, the human mission will launch 26 months later during Opportunity 2. Again using NTRs, this mission also follows a conjunction class trajectory, but may be a rapid transit journey to reduce the length of time the crew remains in interplanetary space under the detrimental effects of radiation exposure and near-zero gravity. Upon launching toward Mars, the crew will travel from Earth into LEO on board a Crew Exploration Vehicle (the same type of CEV that is now being developed to return people to the Moon in 2018), where they will dock with their Mars Transfer Vehicle (MTV). Following the Trans Mars Injection (TMI) maneuver, the MTV settles into its path to Mars, called the Earth–Mars Cruise phase, taking approximately 180 days to get there. On arrival at the planet, the crew will execute a maneuver to slow the vessel into a Mars capture trajectory that will eventually bring them into orbit about Mars. Subsequently the crew will transfer into the Mars lander (either previously deployed in orbit or brought at the same time) and execute a controlled and precise Entry, Descent and Landing (EDL) sequence that sets them down onto the surface, in close proximity to their already deployed support infrastructure.

When they reach the surface the crew will immediately have to make ready and/or assemble their habitat, as well as drive across the surface to the nuclear power unit and lay the power cabling back to the habitat. A myriad of other set-up operations—for example, setting up communications and scientific equipment, configuring the ISRU unit to initiate the manufacture of MAV propellant, and establishing a small greenhouse for the growth of plants—will keep the crew occupied for weeks until the site begins to stabilize as a secure habitat for the remainder of their 18-month stay. Eventually, the crew will develop a stable routine and find time for scientific expeditions and laboratory analysis.

Upon completing their 500- or 600-day stay, the crew will begin their journey back toward Earth by launching from the surface in their by now fully fueled MAV, where they will rendezvous with the ERV (or their original MTV if a separate ERV is not used). They will then execute a Trans Earth Injection maneuver and settle into a Mars–Earth Cruise trajectory to take them home. A significant proportion of the return journey will be spent preparing the crew for their return to Earth, where the significantly stronger gravity will pose problems for a crew who will by then have spent over two years in lower gravity. After 180 days, the crew will approach Earth. Several days before arriving, they will transfer to a small Earth Return Vehicle (probably a CEV) that will slow down toward Earth and enter the atmosphere at over 11,000 kilometers per hour. The crew will enter Earth's lower atmosphere and be directed to a land target under the guidance of a paragliding parachute in a not-dissimilar maneuver to the Apollo astronauts arriving back from the Moon. The Mars astronauts, no doubt in a weakened state in Earth's gravity, will have completed one of the most historic exploration expeditions in human history.

The Human Dimension

Despite the enormous technical challenges to be overcome just to get people to and from Mars, even more challenging will be guaranteeing the survival, physical health, and mental wellbeing of the crew. Indeed the challenges here are so great that it is not clear that they can be overcome, given current design reference missions, technological capability, know-how, and countermeasures. If a decision is made not to send people to Mars, it is less likely to be on the grounds of technical inability or political sensitivity than on a fundamental inadequacy in our understanding of human health and our physiological and psychological requirements for a three-year deep-space mission. The most vulnerable "components" of the entire mission will obviously be the crew, and until launch date the majority of our efforts must be channeled into developing adequate survival and sustaining systems, as well as extensive countermeasures to the many hazards that will be encountered during the expedition. Several critical factors—of which we currently know very little simply because nobody has ever had to endure such a feat—could have such an adverse effect on the long-term wellbeing of the crew that the mission itself could be jeopardized. First is complete confinement for a three-year period and its resulting detrimental effects on the physical condition of the crew, their mental and psychological condition, and even on how illness is brought on and coped with. Second is the significant solar and cosmic radiation hazard to be endured during all stages of the expedition—*en route* to Mars, on the surface and on the return journey. Third is the effects of microgravity (near-zero gravity) on the human physiology, especially concerning muscle and bone mass.

And so before any commitment is made to a human mission to Mars, all the above issues need to be thoroughly addressed and understood, with completely effective countermeasures in place to ensure the mission's success and the wellbeing of all individuals on board. When we examine the components of the various threats, however, it becomes soberingly apparent just how enormous will be the efforts to overcome them.

Confinement and Isolation

Confinement of the astronauts for a three-year period will impact hugely on their physical and psychological health. This will constitute one of the greatest endurances ever undertaken. Spending months in a submarine or enduring an Antarctic winter are insignificant by comparison. Indeed, such relentless confinement, ever-increasing and absolute isolation from Earth, and knowing that rescue is not possible should the mission go badly wrong, all represent major unknowns with respect to how the crew will cope. Coupled to this will be separation from loved ones, lack of privacy, forced perpetual contact with other crew members, and monotony—all of which can adversely affect the psychological wellbeing and mental capability of each crew member.

Even though each will be in peak physical and mental condition when setting off, the assumption must be that even the most resolute crew member is vulnerable and will require adequate countermeasures. To this end, there is a need for far-reaching research into the human condition; identifying all possible effects, symptoms, and impacts on individual and group operations, followed by identifying and formulating effective countermeasures. Much of this research can be achieved by long periods of isolation by volunteers at remote sites on Earth, such as those currently run by The Mars Society and by NASA near Haughton Crater in Greenland. Also of enormous value will be the prolonged stays on the International Space Station, as well as lunar outposts where crews could stay for long periods of time and endure many of the hardships likely to face a Mars expedition, but would still be close enough to Earth should any serious issues require to be quickly addressed. Indeed, this is one of the primary motivations behind the decision by both the Vision for Space Exploration and Aurora programs to return to the Moon as a stepping stone to Mars.

Countermeasures might include careful habitat design to provide optimum ergonomics, comfort and privacy, careful crew screening and selection, careful planning of work schedules and practices, as well as the provision of free time, rest time, and personal activities. Technological assistance will be pivotal, providing, for example, adequate communications back to loved ones as well as access to hobbies and entertainment. But these represent just the tip of the iceberg in terms of seriously sustaining the wellbeing of the crew—a challenge that will no doubt occupy mission planners for many years to come and will demand every new technical innovation available.

Radiation

One of the most hazardous aspects of a human mission to Mars will be exposure to high-energy electromagnetic radiation such as x-rays and gamma-rays and from high-energy particle radiation that emanate from the Sun and the galactic background. The crew will be bombarded by such radiation while *en route* to and from Mars and on the surface (where its tenuous atmosphere offers little protection). Of course, the trajectories and length of time to and from Mars will critically govern how close to the Sun the crew must travel and for how long they will remain in interplanetary space—all affecting radiation dosages. While we currently have insufficient data to know the precise radiation dosages to be encountered and the countermeasures to apply, we are already confident that those dosages will be lethal without adequate protection. The measurement of radiation causing a biological effect is the Sievert (Sv), where a radiation dosage corresponding to 8 Sieverts is lethal, 3.5 Sieverts is termed a career dosage, and about 1.5 Sieverts is an unacceptable dosage on blood-forming organs. Given these limits, current radiation shielding technology would not be adequate to ensure the wellbeing of the crew—all would exceed a career dosage and probably suffer long-term health effects.

Hence much research is needed to analyze the full nature of the entire range of radiation types and expected dosages to be encountered along various trajectories and while on Mars, as well as at different times throughout the Sun's 11-year radiation cycle. Once the character of the radiation is known, we can then more accurately determine what the risks are and what countermeasures must be put in place. Once again research on board the ISS and on lunar outposts will be pivotal, as will robotic missions to Mars which can carry detection equipment to determine the radiation encountered *en route*, as well as on Mars or while in orbit. Countermeasures will include: synchronizing human missions to periods of minima of the Sun's 11-year activity cycle; effective shielding on board all craft and perhaps reinforced shielding in certain regions of each vessel; genetic, age, and gender selection of radiation-resistant individuals; radio-protective chemical supplements; and nutrition management. Currently, however, all such precautions implemented using our current technology would *not* adequately protect a human mission to Mars, and solving this issue will be among the greatest challenges to be overcome if a human mission to Mars is ever to occur.

Microgravity

Another critical factor is the effect of microgravity on the crew. It is already known that prolonged periods in space, where the force of gravity is minute and results in weightlessness, has serious detrimental effects such as loss of muscle and bone mass, reduced cardio-muscle capacity, reduced motor skills, and an increased risk to general diseases. These, in turn, can lead to a multitude of short- or long-term detrimental effects. When we consider that a Mars mission will demand peak fitness at all times, the full negative impact of prolonged periods in microgravity constitute a significant risk. Among the most critical issues is the 180-day journey from Earth to Mars where the crew will be weightless, followed immediately with a demand for extensive and strenuous yet critically important physical effort on the surface to establish a secure habitat. Even Mars' gravity, at only one-third that of Earth, will pose a significant challenge unless specific countermeasures *en route* are taken to maintain muscle and bone mass as well as to maintain the peak fitness of the crew. Apart from the risks to the mission, any person traveling to Mars and back will have spent a total of 900 days away from the influence of Earth's gravity, with potentially disastrous long-term health consequences. Here again, any mission to Mars must provide sufficient countermeasures to sustain the long-term health of its crew.

Before effective countermeasures are determined, extensive research must be conducted, both on board the ISS and on the Moon—and over prolonged periods equivalent to those to be encountered by a crew to travel to Mars. We must determine the precise effect of prolonged periods in microgravity on bone and muscle mass, on the endocrine and nervous systems, and their associated effects on metabolic activity, nutritional intake, digestion, and all manner of injuries and illnesses. Already, lengthy stays on Skylab, MIR, and the ISS provide valuable information and some pointers toward countermeasures, which include nutritional supplements to maintain positive energy balance, strict fitness regimes, Lower Body Negative Pressure (LBNP) sessions as well as fluid and salt intake before returning to high gravity and antigravity suits during the first few days of their return to Earth's gravity (Figure 108). It is not clear if these measures are sufficient for a Mars mission, however, and to this end both the Mars Direct and Mars Society propose the necessity of having a spacecraft that is capable of generating an artificial gravity through a rotating section of the craft that would produce a centripetal force equivalent to Earth's gravity. The addition of such a feature would add yet further complexity to the mission, however. As with other human health and safety attributes to the mission, identifying the full implications of such a prolonged stay in low and microgravity, and providing adequate countermeasures, is not a luxury, it is critical to the mission.

Figure 108: An artist's concept of the International Space Station, which will play a pivotal role in researching how to adequately protect humans during a mission to Mars. [Credit: NASA]

The Logistics of Living and Exploring

In addition to all the critical, technical, and human faculties related to a successful mission, there are a myriad of other issues to be considered and adequately addressed. For example, currently there is no way of determining the full biological dimension to a human mission to Mars. Complete immunization of the crew against illness will be impossible, and it has already been projected that there may be approximately 17 bouts of illness per year. The full nature of those illnesses, from infection to physiological ailments—as well as their behavior and contiguity within a completely sealed environment over a three-year period—are not known and represent a significant risk. There will, however, be other biological considerations, and it is generally recognized, for healthy eating and as a psychological aid, that plants should be grown on Mars within a sealed greenhouse environment. Hence a range of issues will also arise associated with the ability to grow plants successfully; the crew's dependency upon that form of "agriculture" and the effect should something go wrong. Finally, the use of microorganisms for cleaning and waste recycling, even if not necessary, is highly desirable on such a lengthy mission; and again managing such systems over a three-year period represents yet another unprecedented challenge. Overall, the management of all biological systems, from the health of the crew to growing plants and the use of microorganisms in cleaning and recycling, represent yet more challenges that will require extensive precursor research and development.

Another issue will be the management of exploration on the surface of Mars. Here, a range of technologies—such as light and flexible Mars spacesuits with sophisticated wearable computers and navigation systems; support robotic rovers, as well as pressured drivable rovers; sophisticated laboratories with computer networks; networked communications back to Earth; and a range of geology, geochemical, and life science analytical equipment—will be available on the surface. But many issues, such as the chain of command, the optimum duration and distance to be traveled on given field trips, the extent of communication with Earth, and the most effective way to use people in the field and in the laboratory, must all be determined prior to the missions and adapted according to experience. Already, simulated Mars missions and field trips at Haughton Crater in Greenland by NASA and The Mars Society are providing essential insight and highlighting many well-intentioned but ill-conceived ideas that actually cripple effective exploration; for example, involving too many people in the decision-making process can seriously hamper field expeditions. In all of this, optimized methodologies and codes of practice must be determined to ensure that realistic scientific exploration can take place.

Of course, by the time humans land on Mars, significant resources will await them on and around the planet. Furthermore, such a far-reaching expedition will only occur within a framework of a long-term and extensive program that had already been in operation for a decade or two back on Earth. Those first Mars astronauts will be the final step in a gigantic effort; and although they will be physically on their own on Mars, in every other respect they will have at their disposal some of the most extensive support systems ever devised.

The Moon: A Stepping Stone to Mars

We now find ourselves in a position, for the first time in history, to be able to appreciate the extent, nature, and weight of issues to be resolved in order to send people to Mars. Hence, for the first time we are also in a position to start addressing the right problems in the right order and in a prioritized way; and this can only bode well for the chances of a real mission taking place at some time in the foreseeable future. Understanding the issues is *the* necessary first step toward sending people to Mars. Today, however, we cannot say with sufficient certainty that it is possible to send people to Mars given projected advancements in science and technology. There are too many fundamental issues such as propulsion, radiation, and microgravity countermeasures, not to mention the mission logistics, political sensitivity, and cost to be investigated before we can declare that such a mission is achievable within several decades and, hence, announce a human program for Mars. None the less, we are cautiously optimistic that they can all be overcome. And so VSE and Aurora, for the foreseeable future, will perform such acts as the construction of new rockets, lunar missions and outposts, robotic testbeds and a sample return mission to Mars, as well as conduct far-reaching research into propulsion, power, and human wellbeing in space before directly engaging in a human mission to Mars. A defining goal of VSE and Aurora is to send people to Mars, but a human mission will only become a real and practical option after critical interim programs have been successfully completed.

Some claim that if we wish to go to Mars, we need to engage in a program that can achieve this within about 10 years; and that by going to the Moon first, or by taking decades to do so, only hampers the prospect of a human mission to Mars ever taking place. In light of the fundamental outstanding issues outlined above, however, it would be a daring and arguably reckless nation that decided to initiate a specific human Mars program with the present technology. The technical and human outstanding issues are, currently, simply too great. It would be irresponsible to send people to Mars in such precarious circumstances. We are not ready to make that decision, nor do we need to be. Sending people to Mars will be contentious, costly, and risky even when we think we are ready. Mars is potentially a devastatingly hostile environment, with no room for recourse should a serious accident or mishap occur. The harshest desert on Earth is literally a paradise compared to even the most clement places on Mars. We cannot send people to Mars if there is even a hint of doubt regarding mission success and crew safety. We cannot send people to Mars simply because of the ambitions of the space community or particular interest groups. At stake are not only the lives of a gallant crew, but also the course of history and the future of our space activities. It may have been in the past, and it may still be today, that if space programs are spread out over long period of time then they lose their focus; but some projects just take a long time. Cathedrals of the past took hundreds of years to build, no doubt initiated by those who knew they would never see the finished product but sensed that their compatriots and descendents would complete the project. We must surely adopt this approach when we consider sending people to Mars. It is irrelevant whether we send people to Mars in 2040 or 2140; the critical importance, however, is that we recognize when the time is right and have the foresight and sense of human development and history to undertake the project. But while today we cannot be sure of a commitment to a human mission to Mars by 2030 or 2040, we can be sure that everything being undertaken today by VSE and Aurora—including serious and in-earnest internal shifts in emphasis by both NASA and ESA—is with a view to sending people to Mars at the earliest available opportunity. We can also be confident that as long as a space program exists, it would be futile not to make Mars the prime target. It genuinely has more to offer than any other celestial object. If a long-term human space program persists, a human mission to Mars is inevitable.

In recognition of the magnitude of the effort, as well as the publicly declared intention by VSE and Aurora to send people to Mars, both NASA and ESA have already responded with a number of far-reaching programs. NASA, in particular, has made the decision that a return to the Moon would be a crucial first step toward Mars. Although contentious, this decision was taken in light of the significant hurdles to be overcome, and in this respect the Moon represents a unique opportunity. By returning people to the Moon and establishing a permanently occupied lunar outpost, many of the outstanding issue regarding a human mission to Mars could be directly and robustly tackled, and the proximity of the Moon would provide a safety net should unexpected and serious mishaps occur. First, returning people to the Moon will require the development of the Shuttle Derived Rockets also needed for LEO assembly of Mars spacecraft. Lunar missions and outposts will also need the range of hardware and technologies required for Mars—habitats, power systems, regolith moving vehicles, in-situ resource utilization, and microbial-based cleaning and recycling systems—and they will all be field tested for use on the planet. Long stays on the lunar outpost will also provide vital information regarding a range of human wellbeing issues, from the logistics of living and working for extended periods on another world to issues of confinement, radiation, and low gravity. Furthermore, in preparation for an extensive lunar program that will feed into a Mars program, NASA and ESA have both extensively reorganized. They have adopted *systems-of-systems* and *spiral-transformation* approaches to technological development and program management where entirely new generations of space technologies will be applied across a myriad of missions from small robotic rovers to human-rated deep-space expeditions, and well-defined technical and capability roadmaps will point the way forward—all vital for an eventual human Mars mission.

NASA published an extensive *Exploration Systems Architecture Study* in late 2005 and announced its intention to send people to the Moon in 2020 as

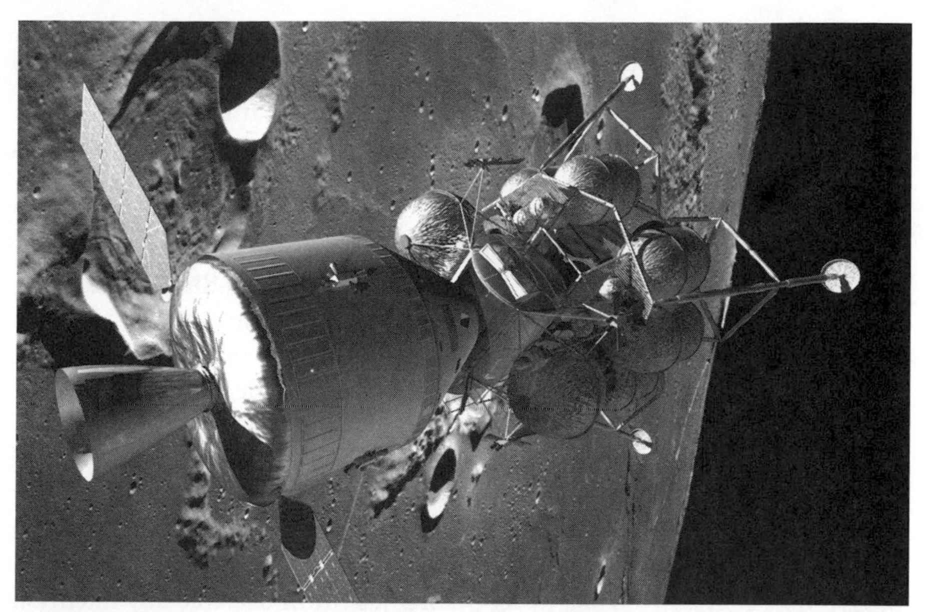

Figure 109: NASA's Vision for Space Exploration is set on sending people eventually to Mars. The first step is to return to the Moon by 2020. Here the Lunar Crew Vehicle is docked with the lunar lander in Earth orbit, ready to travel to the Moon. [Credit: NASA/John Frassanito and Associates]

Figure 110: Astronauts are scheduled to return to the lunar surface by 2020. [Credit: NASA/ John Frassanito and Associates]

well as develop, for readiness, the Shuttle Derived Rockets and a Crew Exploration Vehicle by about 2012. This will be followed immediately by the developments of habitat, ISRU, and regolith moving technology for a lunar outpost (Figures 109 and 110). Similarly, the Aurora program has received enthusiastic endorsement on a ministerial level across most of ESA's participating nations, allowing for the development of a Mars Sample Return Mission to begin, with a view to launch in or around 2016.

Even though both agencies are engaging in their programs toward Mars as rapidly as possible, given budgetary constraints, their declared aspiration of setting people down on Mars sometime shortly after 2030 is already being viewed as too ambitious. The proposed development work to take place between now and 2020 arguably leaves too many outstanding issues to be resolved post 2020 for a human mission to occur approximately a decade later, especially concerning the development of nuclear thermal rockets and the problem of radiation and other human wellbeing countermeasures. Some are already skeptical of the value of NASA's commitment to the Moon, and that it may derail any commitment to Mars (or perhaps expose a lack of appetite for sending humans to Mars in the first place). In NASA's defence, it now publicly and optimistically declares, post 2020, a dual strand program of sustaining lunar missions and outposts while also ramping up a human program for Mars. Given that never before has NASA declared any intention of sending people to Mars, this in itself represents a fundamental shift in declared intentions. And, as has already been discussed, the lunar program is not an end in itself but instead is seen as a stepping stone to Mars.

The Vision for Space Exploration and Aurora programs may not survive changes in political administrations, and emphasis may sway within space agency budgets; but there can be no denying that the overriding interest in Solar System exploration now has Mars firmly at the focus. If all present efforts are not to be jettisoned, it may well come to pass within three or four decades that the first human expedition will set out for Mars.

A Sense of It

A most inspiring description of space exploration was presented by Neil Armstrong during a live audience interview in Ireland in 2004. He beautifully described the decision to leave Earth orbit and head toward the Moon. Upon firing their rockets, what he, Buzz Aldrin, and Michael Collins witnessed when looking out the window of the Command Module was simply the slowly spinning blue globe of Earth falling away from them while they seemed to stay motionless in the vastness of space. All three knew there and then that despite the support infrastructure on Earth, they were essentially on their own. He recalled none the less, Michael Collins, the command module pilot, reassuringly telling him (and Buzz) to sit back and enjoy the ride; that he would get them to the Moon safe and well. On being asked "Which was more important to him, landing on the Moon or splashing back down on Earth?," Neil Armstrong replied without hesitation, "Splashing down on Earth, because John F. Kennedy had, in his 1962 Rice University speech, proposed sending a rocket to the Moon and then returning it safely to Earth; and so splashing down on Earth had fulfilled that effort."

It was clear to that audience in 2004, as no doubt it was clear to the NASA selection panel, that not everyone is born to explore. Despite the hazards and unprecedented nature of the expedition, that crew acted in a calm, professional, and selfless way, always mindful of the entire program and indeed humankind that were with them all the way. But what Neil Armstrong also provided was a unique insight that only he and a few others can provide—a true sense of the wonderment of exploration. His experiences could never have been imagined by anyone who has not also undertaken the monumental steps he took; and it is only through explorers like Neil Armstrong that the rest of us can have any sense of the new realities that lie beyond our everyday experience.

One wonders what thoughts will go through the minds of the first Mars astronauts. As they leave Earth orbit, they too will see its glistening blue globe fall gently away from beneath them. But unlike the view to the lunar explorers, it will not remain as a globe for long. It will continue to fade, becoming smaller and smaller until eventually it is but a brilliant speck of light in the sky. It will never disappear completely but it will none the less seem incomprehensibly far away. As they are hurled across interplanetary space many time faster than a bullet, they will be truly alone and dependent on millions of technical components for their survival and wellbeing. One serious technical mishap or one serious mistake might spell disaster for the mission. That first interplanetary journey to Mars will be of such historic proportions, irrespective of any Martian excursions that follow. It will not be for the faint-hearted.

All being well, that gallant crew will approach Mars some six months later and be subservient to the laws of celestial mechanics that have maintained the planets orbiting of the Sun for billions of years. What had appeared as a bright red speck of light will, day by day, become ever more brilliant, until one day one of the crew will glance out of the window and discern a distant red globe. Just as the blue globe of Earth had fallen from beneath them six months earlier, the striking orange-red globe of Mars will reveal itself in the distance, then begin to fall toward them. These people will truly know the distance between the planets, and as they gaze upon the ever-enlarging red globe they will be in awe of seeing with their own eyes a world only vaguely seen through telescope eyepieces and on computer screens throughout the ages. The full glory of another entirely new world will have been witnessed by people for the very first time in history. Yet for all its glory, the crew will also be relieved. Having spent six months or more in the weightlessness of space, it will surely be a relief to set down onto the Martian surface, step onto solid ground, experience gravity, and look up into a sunlit sky. They will have arrived at their new home for upward of the next two years—a not insubstantial fraction of their entire lives. Despite the feeling of familiarity that the surface of Mars may bring, the crew will have ventured onto what is also a very alien landscape. At less than half its intensity on Earth, even the noon Sun may seem curiously dim and with an orange sky acting as a permanent reminder that this is not Earth. And there will be spectacular sights. With only a tenuous atmosphere, the sky will be spectacularly clear, and with Mars' two moons, Phobos and Deimos, racing across the sky in only a matter of hours on a nightly basis, as well as a host of artificial satellites—reconnaissance orbiters, telecommunications satellites, and even their Earth Return Vehicle, silently circling the planet for some 18 months, awaiting their arrival. Seeing these human-made objects will surely be a comforting reminder of home.

The crew will also have to work hard and quickly to establish their base. A multitude of tasks—configuring their habitat, routing power from their near-by nuclear power unit, setting up a greenhouse, and unloading vehicles, consumables, scientific equipment, and the ISRU unit will occupy them for weeks before any sense of normality sets in (Figure 111). Survival will be the first port of call, with any serious technical problems to be quickly identified and resolved. As already indicated, even when the astronauts settle in, they will have to follow strict protocols and work practices. In particular, optimum planning and implementation will be needed for effective scientific exploration and investigations, both in the field and when handling Martian samples within the laboratory.

It is incredible to think that life on Mars may become ordinary and routine. If this transpires, it will be a measure of the phenomenal success of the mission to date, having overcome the enormous hurdles to get there and establish a living and operating base where groundbreaking scientific research is taking place (Figure 112). If the crew lands at a site of potential prebiotic and life-activity interest, they will be able to conduct fundamentally important and defining research in a manner never before possible. As complex as the locale may be, they will be able to bring the sum total of human analytical science to bare and slowly build up a comprehensive picture of the natural history of the site with far-reaching implications regarding the nature of Mars in its entirety. Spectroscopes, powerful microscopes, mass-spectrometers, and a host of biological experiments can be conducted on a range of samples from many sites and under various

Figure 111: Artist's impression of a base on Mars. NASA and ESA hope to send humans to Mars some time after 2030. [Credit: ESA]

Figure 112: Two astronauts on Mars prepare robotic rovers during a scientific field expedition. [Credit: NASA]

configurations, with results gleaned leading to repeated cycles of ever more specific testing. There is no doubt whatsoever that the presence of a human science team on Mars, interlinked to the full scientific community back on Earth, cannot but acquire spectacular and defining results about the natural history of Mars and of life there. Even one human expedition to Mars will be of historic scientific importance. And even in a scenario where no evidence of life whatsoever is detected at a site chosen for its life potential, such an extensive characterization of the planet will provide critical insight into the types of chemistry occurring there and equally valuable constraints on the requirements for the origin of life. The science outcomes from such an expedition can only be positive as they radically alter our understanding of the origin of life on Earth and a general planetary context for the emergence of life, to say nothing of what we will then specifically learn about Mars itself.

Eventually, after perhaps 500 or 600 days, the astronauts will have to pack their equipment and head for home. Having been well settled in, they will have to uproot their lives once again, preparing for a launch on board their fully fueled MAV while also setting their base camp into a stand-by mode, maintained by nuclear power until the next crew arrives some years later. What welcome surprise will they leave for the next generation of compatriots? Only they will surely know what will constitute an appropriate surprise!

Upon launching from the surface of Mars the crew will, in a matter of minutes, attain Mars orbit where they will dock with the Earth Return Vehicle that had patiently awaited the arrival of its travelers for approximately four years. Blasting out of orbit, the Trans Earth Injection maneuver will set the crew on a trajectory toward Earth. They will once again look down upon the red globe of Mars—their home for nearly two years—and bid farewell. Even if young and fit enough to return, the crew members, having been exposed to their career dosage of radiation on this voyage, can never set foot on Mars again. As the red globe falls away from beneath, they will all surely be overwhelmed by emotions of sadness mixed with pride for a job well done, and tinged with a sense of excitement in anticipation of returning home to Earth.

Earth, for all its trials and tribulations, will be waiting. And after another 180 days or so, the crew will once again see the sparkling blue globe of Earth slowly emerge in front of them. What must they feel? What perspective will that voyage bring upon that crew regarding their home world? Nobody can yet answer that.

But the crew will know that billions of people will be looking in their direction, wishing them well and willing them home. With all the skill, expertise, and fortitude that humanity can muster, the crew will touch down on Earth two and a half years after departure. Including years of intensive training, this may be the end of a decade of effort for each. But they will arrive home on a world experiencing a global upturn in spirit, action, and a sense of history. Surely the gift they bring back to Earth will be of a renewed sense of purpose and capability that has not been experience in the modern era.

21 Our Life in the Universe

The quest for knowledge, and acknowledging the validity of that quest, is a powerful purpose for humanity. To know who we are and from whence we came is at the root of that purpose. Our generation, as with every previous and future generation, has a role in this unfolding human story. We do not yet have anything like a complete picture of our origins and nature, but we now know they are rooted in their cosmological context, and that Mars in particular has something important to reveal on that subject. In realizing this, we have come far. The path to Mars has been long and winding, but now we are on the right track, hopefully on the verge of uncovering its ancient legacy and discovering what it has to tell us about life. If one day we manage to kneel in the dirt of Mars, it will be on behalf of those throughout time who pondered the nature of that red point of light in the sky, in honor of the many Mars explorers over the centuries since Brahe and Kepler; and in reverence of all people—past, present, and future—whose entitlement it is to know the truth of their origin and nature.

Mars exploration is about us, not just about Mars. In this respect it is inherently valuable to each and every one of us. And it is especially valuable because it represents such a significant, all-encompassing search regarding our own origins, life on Mars itself, a generalized picture of biology, and a cosmic context for life.

Finding microbial life on Mars is not mandatory for success. By having a reason and opportunity to ask the difficult questions, and a means to seek answers, we can make progress in understanding our nature; and Mars is capable of providing all of these. It is the quest itself that slowly drags us ever further from our incomplete view of our nature and Earth's context in the grand scheme. We have being asking questions for many generations and now, through the exploration of Mars, we have a means of finding some of the answers.

We are already well on our way. From tentative evidence of life within Martian meteorites spring-boarding the discovery of hitherto unknown microorganisms on Earth, to having to dig deep—physically, analytically,

K. Nolan, *Mars, A Cosmic Stepping Stone*,
DOI: 10.1007/978-0-387-49981-9_21, © Praxis Publishing, Ltd. 2008

and philosophically—on Mars itself to pursue difficult yet far-reaching questions about a planetary context for origins, all such efforts have already reaped rewards. We also know that we will learn so much more by fully deciphering Mars' true legacy from the planetary scale to the molecular level over four billion years—whatever it turns out to be. Our hope, of course, is for the greatest prize of all—that of uncovering evidence of prebiotic chemistry, and fossilized evidence of ancient life, or even of current microbial life. Such findings will rank among the most profound discoveries of all time, providing direct insight into our origins, a cosmological context for all life, and a generalized understanding of the intricate workings of biology. In all of these ways, Mars is truly a stepping stone that may help humanity to uncover its own ancient heritage and see itself in the broadest natural context.

After many false starts, we are now on the right path toward understanding Mars. It is a planet sharing a complex birth and early history, perhaps similar to its sister planets, Earth and Venus. Patient and painstaking analysis of the Mariner and Viking data led to our first clear picture about Mars and prompted a more sophisticated and mature return. And while most of the work has still to be done, what we have learned since the arrival of Pathfinder has not only validated our instincts but also revolutionized our understanding and perception of Mars as a planet of significant past and present complexity, where perhaps—just perhaps—life may have arisen and may even persist today.

While we do not yet have those answers, one result of cosmic significance has now been confirmed: Mars is a watery world. We now know that for millions of years in its early history, water played a significant role on a planetary scale as surface liquid and ice, within subsurface aquifers, and probably within hydrothermal settings. Furthermore, the story of water on Mars is not some strange coincidence. On considering the enormous range of planetary scenarios across even our own Solar System, the possibility of water on Mars within a planetary setting that is quite similar to Earth warrants serious attention. Mars may have evolved within the early climatic conditions that brought about the type of water-based chemistry that is conducive to life as we know it. Certainly the evidence of wide-scale internal planetary activity and global tectonic and volcanic activity all point to such a possibility. The debate concerning the presence of water on Mars is now over, and we must turn our attention to determining the precise conditions under which it existed and the chemical activity that supported it.

The consequences of this result are still only dawning on us. And although we must be cautious of running ahead of ourselves regarding the connection between the presence of water and the possibly of life on Mars, let us resolutely state here that it is now absolutely true that at least two of the eight major planets of the Solar System are significantly influenced by water. As Mars, too, is a planet affected by water, that removes the uniqueness of Earth in this respect, and with water being so central to the emergence of life as we know it, this is a conclusion with cosmological implications.

The fact that water activity on Mars was confined to its early history does not detract from or diminish its importance. Water activity may have persisted for upwards of a billion years *and* during the epoch during which life was emerging on its sister Earth. In a billion years or so from now our Sun may become so hot as to vaporize water on our world, rendering it dry and barren. Will that occurrence have any bearing on Earth's original ability to give rise to life so centrally defined by water? Every planet has its time, and the story of water on any world—including Earth—is one that is finite in duration. In this light Mars, too, is a watery world, whose time has past but whose early history was in part shaped by its planetary supplies of water. In an ironic twist of fate, it may come to pass that, as Earth becomes too warm to retain its water in a far distant future, Mars may heat up to a point where its remaining planetary water supplies are unleashed from their global permafrost, perhaps retuning some measure of dynamism and clemency to the planet during the autumn of our Sun's life. While Earth loses its grasp on water, Mars may become the planet most influenced by flowing water in our Solar System. The story of water on Mars is perhaps not over.

Indeed, as we have explored Mars in ever-increasing detail, we have also come to realize that not only was Mars very active in its early history, but that today it is also not as dormant as once thought. As vital as our quest to understand Mars' past may be, we now realize that there is much to discover about what is happening there today—seasonal and annual weather, planetary dust storms, climate change over durations of thousands of years, possible subsurface water activity, polar layered deposition and erosion, geologically recent (albeit sporadic) surface water, and tectonic and volcanic activity. And with ever more sophisticated instruments, we continue to detect new and more subtle details about present activity—glaciers and glacial deposition on the flanks of volcanoes, possible water gullies on the flanks of craters, wide variations in surface environments, and weather that can give rise to ground frost and water vapor clouds. Even today, the closer we look at Mars, the more hitherto unconsidered activity we find. Mars is still active in a great number of ways, many of which are still unknown to us and have to be discovered—especially in relation to the volatile behavior of water and carbon dioxide and, perhaps, in ways relevant to life. Whatever the full extent of Mars' current activity, we can now also draw this conclusion: Mars is not like our Moon or the other dormant worlds of the Solar System. It is a planet that, despite the catastrophic collapse of high activity in its early history, has remained active in many ways that involve planetary dynamics and volatile behavior not unlike those occurring on Earth.

Although we have made progress in the exploration of Mars, we still have most of the major work to do. We are far from acquiring definitive answers about its nature and the possibility of life there, but we know that whatever the answers, they will be significant. The question is no longer whether Mars was or is an active planet, rather it is about the full nature and evolution of that activity, from which we will learn a great deal about our own legacy. It is staggering to think of what we have learned since the arrival of Pathfinder in 1997. While Mariner 9 and the Vikings provided exceptional overviews of the planet, the subsequent prolonged surveying and detailed reconnaissance over recent years have begun to unveil the true legacy of Mars and the subtlety of change taking place there today. These efforts are a clear indication that we cannot draw conclusions about an entire world simply by looking at it on a planetary scale, or through brief snapshots. Only when we gain knowledge of many specific locations over a long period of time can we know it well; and it is through such efforts that Mars is being revealed to us. Mars has become a real and tangible place and each and every location we visit has an ancient and intricate story to tell. If we wish to know the story of Mars, we must visit many more sites and study them in equally exquisite detail. And now we are ready to do just that. The efforts of previous generations, our current phased strategy for Mars exploration, and the emergence of the Vision for Space Exploration and Aurora programs can fulfill those aspirations.

And through such efforts we must recognize our role. In every important sense what we are doing is no different to the hunter–gathers who pondered the nature of the planets and their place in the greater scheme of things, or to Schiaparelli on trying to determine the true nature of the surface of Mars. They were pursuing questions of origins and their broader natural context in the best way they knew how, and so are we. Mars is a litmus test of our mettle regarding our desire to discover more about who we are, and whether we are prepared to fulfill our place in history by building upon the efforts of those who came before and by laying the foundations for those who will come after and perhaps want to push forward even further. Scientists, philosophers, and priests of old would have given anything to know what we know today; and arguably we have a responsibility to the issues we are now uncovering. We, too, must ask the best possible questions and do all at our disposal to pursue new answers. And of late we have done this. The Mariner and Viking generation of scientists were initially blamed for being so daring, but from their efforts we were able to build the first coherent picture of Mars. We have been impressively patient; learning from their efforts, asking new and relevant questions, and pursuing them in a prioritized and cost-effective way through our robotic phased strategy, and soon by VSE and Aurora. A significant amount of laborious groundwork has been required, and we have done it. We are not only asking tough questions and pursuing their answers, one by one, but we are also building on each question with ever more pertinent ones and truly innovative exploration programs.

Slowly we are becoming a permanent space-faring civilization because we find more and more that is relevant in space, with Mars of special value. How we have pursued Mars exploration could not have been done much better; and given the resources at our disposal, our efforts have been spectacularly successful, to date. In the coming decades, however, we will be faced with new challenges and responsibilities. Our science agenda, if pursued through ever-increasing numbers of robotic landers, sample return missions, or even by sending people to the planet, constitute a new level of intrusion upon Mars and new interactions between that planet and Earth. And such future exploration will not just be about science; if a mission as far reaching as sending people to Mars is to take place, then political, economic, and sociological motivations will also increasingly bare upon it. Even Mars will ultimately yield to the full gauntlet of human engagement, whether positive or negative.

In this sense, everything we do from here on brings with it new responsibilities. As Mars is a fantastic and unique, although one-off, opportunity to extend ourselves and learn about our origins and nature, it is surely incumbent upon all of us to pursue that opportunity as responsibly as we possibly can. If we pursue sloppy or reckless agendas, we may damage that opportunity forever. Every good habit practiced now in exploring Mars will build toward an enthralling future with the planet; ever poorly conceived, careless, or dubious agenda will contribute ever increasingly toward a detrimental future for Mars and our opportunities there. This may all seem far-fetched and alarmist, but we must do no more than look at our behavior on this planet to see how badly we can go wrong. Even our current exploration, on sending landers with a known biological burden, may be damaging. We have no current way of knowing. The affect of a handful of landers over a few decades may not be noticeable, but dozens or even hundreds of missions over many decades may have a significant impact.

In this respect, the manner in which the space science community pursues the future exploration of Mars takes on a new significance. Of paramount importance in the coming years will be the development of robust planetary protection policies that safeguard both Mars and Earth, and we have time and opportunity to get this right. Coupled with developments in cleaning and sterilizing technologies, robust and internationally agreed protection policies can adequately protect both planets. The space science community must also continue—and be allowed to continue—to ask new questions and pursue the necessary science, but it must also continue to be cost-effective and strive to bring the relevance of this endeavor to the attention of all. With regard to a human mission to Mars, this should only take place when all effective robotic science has been completed and the mission has been shown to be cost-effective and extremely safe. The world cannot afford a premature human mission to Mars, nor will it easily recover if the mission fails; but if it is timely and successful, a human mission to Mars may catalyze a more optimistic future on a global scale. The stakes are high and the potential rewards are enormous; but there should be no hurry to send people to Mars until the time is right. Maverick entrepreneurs and fanatical space-interest groups may be desperate to get to Mars at all costs, but there can be no justification for promoting human missions to any planet on the basis of such self-interest. As astronauts such as Neil Armstrong have shown so clearly, the first explorers to Mars will be strong, self-motivated and ambitious individuals, but they will also be acutely aware of the broader social and historical context within which they pursue their goals. They know that exploration is not about them—it's about all of us.

This is what is so enthralling and exciting about Mars exploration: it is about us as humans trying to discover who we are. We often fumble in the darkness, but every so often we find a path forward, and when we do we recognize the opportunity that is suddenly presented to us. And so it is with the exploration of Mars. No longer can the scientific answers we find be relegated to some obscure scientific journal. Every answer gleaned from Mars will have direct relevance to every person who ever lived. We are a global community who now know that we have a global and indeed cosmological context and are now trying to fully understand it.

We realize that we do not know everything about our nature and we also realize that nearly all the significant and fundamental questions have yet to be answered. But we also realize that we have a role to play in trying to answer them, and that is what Mars exploration is all about. Mars is providing us with an opportunity to add to our unfolding human story, and it is a reminder that there is indeed a story to unfold as Humanity has yet to discover its place in the Cosmos.

The answers about life on Mars, whatever they are, will be of immense importance. When we have surveyed Mars to completion, we will know to a much greater extent our own origins and nature. Even if our efforts prove that there is no life on Mars, the insight provided will still reveal substantially more about who we are and how we originated. But the discovery of prebiotic chemistry, fossilized evidence of life, or current life, will provide such far-reaching insight into the specific nature of life that we will never view it in the same way again. In any case, the answers will be of value, because we value the truth in nature. Knowing the truth of our natural origin—whether it suggests we are alone or are one among many—will be a profound addition to the human condition. It will profoundly affect how we see ourselves in the grand scheme of the Universe. We will know, better than anyone who came before, the full scope of the cosmic rules, bounds, and context within which we exist. We may know that there was a beginning, a starting point; and may know more of the precise details of that beginning. Our identity will be complete, as with an adopted child who has come to know its biological parents.

And while such insight might not impact on society in an instant, it will surely have a long-term and deep-rooted sociological impact. We will be a people who, for the first time in history (and as a result of our efforts through history), truly know who we are. Such insight may not solve the world's problems or remove our trepidation about the future, but at least we will engage such problems and trepidations from a better and more informed vantage point, perhaps providing better perspectives and hitherto unforeseen solutions.

If we discover that Mars gave rise to life, whether long since gone or still surviving today, it will become a powerful beacon of hope for the future. As we gaze upon Mars at night, we will know that life exists there—and probably beyond. We will also know that a truly enigmatic future for humanity, within a cosmos where life resides elsewhere, awaits us if we wish to explore it. And perhaps any life persisting on Mars will bring about a new respect and reverence for life. We will know that even humble microorganisms on another entirely separate world had journeyed with us, breaching the enormous gulf to form another living part of the Cosmos.

The gap from our industrious minds to possible microscopic life under the surface of Mars is diminishing. This is where we are. If that life exists, we will find it in the not too distant future. As you gaze upon that red point of light in the sky at night, you are looking at a place with an ancient story to tell of immense significance to us—a place where life may exist right now.

And so the coming years will be so exciting. The quest to find answers will be a predominantly positive one, pursued in an ever-increasingly international context and for highly significant reasons. We will marvel at the technical might of our missions and at pushing ourselves to truly great heights, in the process learning ever more comprehensively about who we are and the composition of that great dark expanse with which our true and ancient connection is becoming ever clearer and ever more accessible.

Nor need you wait. You are privy to Mars findings today and every day for decades to come. The moment is now. There are currently missions orbiting Mars, on Mars, *en route* to Mars, in preparation, and on the drawing board. Keep your eyes and ears open—there is much for you to be aware of. A genuinely new era of human space exploration is being undertaken on your behalf. It is an era of immense possibilities yet to be realized, but one by one they are becoming clearer, forming a gargantuan line of stepping stones all the way to Mars.

Suggested Reading

Mars, By Perceval Lowell.
http://www.wanderer.org/references/lowell/Mars/

On Mars, Exploration of the Red Planet 1958–1978, by Edward Clinton Ezell and Linda Neuman Ezell.
http://history.nasa.gov/SP-4212/on-mars.html

The Cosmic Code, by Carl Sagan. Cambridge University Press, ISBN 13: 9780521783033 | ISBN-10: 0521783038

Cosmos, by Carl Sagan. Ballantine Books, ISBN: 978-0-345-33135-9 (0-345-33135-4)

Destination Mars, by Martin Caidin, Jay Barbree, and Susan Wright. Penguin Studio Books, ISBN: 0670860204

The Case for Mars, by Robert Zubrin. Touchstone, ISBN: 0-684-82757-3

The Exploration of Mars, by Piers Bizony. Aurum Press, ISBN: 1 85410 584 1

Planet Mars: Story of Another World, by François Forget, François Costard, and Philippe Lognonné. Springer-Praxis, ISBN: 978-0-387-48925-4

Expedition Mars, by Martin J.L. Turner, Springer-Praxis, ISBN: 1-85233-735-4

Suggested Websites

Two Weeks on Mars
http://mthamilton.ucolick.org/public/TwoWeeksOnMars/writings/

Lowell Observatory
http://www.lowell.edu/

The Jet Propulsion Laboratory Mars Page
http://mars.jpl.nasa.gov

ESA's Mars Page
http://www.esa.int/SPECIALS/Mars_Express/index.html

The Mars Society
http://www.marssociety.org/

The Planetary Society
http://www.planetary.org

Astrobiology Web
http://www.astrobiology.com

NASA Origins
http://origins.jpl.nasa.gov

Vision for Space Exploration
http://www.nasa.gov/mission_pages/exploration/main/index.html

ESA Aurora
http://www.esa.int/esaMI/Aurora/

Mars Global Surveyor
http://mars.jpl.nasa.gov/mgs

Mars Exploration Rover Mission
http://marsrovers.nasa.gov/home

ESA – Mars Express
http://www.esa.int/SPECIALS/Mars_Express

Mars Reconnaissance Orbiter
http://mars.jpl.nasa.gov/mro

Phoenix Mars Mission
http://phoenix.lpl.arizona.edu

Index

(named surface features on Mars are included in main alphabetical listing)